Computational Modelling and Advanced Simulations

Computational Methods in Applied Sciences

Volume 24

Series Editor

E. Oñate
International Center for Numerical Methods in Engineering (CIMNE)
Technical University of Catalonia (UPC)
Edificio C-1, Campus Norte UPC
Gran Capitán, s/n
08034 Barcelona, Spain
onate@cimne.upc.edu
www.cimne.com

For further volumes:
http://www.springer.com/series/6899

Computational Modelling and Advanced Simulations

Edited by

Justín Murín
Slovak University of Technology in Bratislava, Slovakia

Vladimír Kompiš
Armed Forces Academy of General Milan Rastislav Štefánik, Slovakia

and

Vladimír Kutiš
Slovak University of Technology in Bratislava, Slovakia

Springer

Editors
Prof. Justín Murín
Slovak University of Technology
in Bratislava
Department of Mechanics
Ilkovičova 3
812 19 Bratislava
Slovakia
justin.murin@stuba.sk

Vladimír Kompiš
Armed Forces Academy of General
Milan Rastislav Štefánik
Demänová 393
031 19 Liptovský Mikuláš
Slovakia
vladimir.kompis@aos.sk

Vladimír Kutiš
Slovak University of Technology
in Bratislava
Department of Mechanics
Ilkovičova 3
812 19 Bratislava
Slovakia
kutis@elf.stuba.sk

ISSN 1871-3033
ISBN 978-94-007-3476-0 ISBN 978-94-007-0317-9 (eBook)
DOI 10.1007/978-94-007-0317-9
Springer Dordrecht Heidelberg London New York

Printed on acid-free paper

Springer is part of Springer Science+Business Media (www.springer.com)

Introduction

This book contains selected, extended papers presented in the thematic ECCOMAS conference on Computational Modelling and Advanced Simulations (CMAS2009) held in Bratislava, Slovakia, in June 30–July 3, 2009.

Modelling and simulation of engineering problems play a very important role in the classic and new composite material sciences, and in design and computational prototyping of modern and advanced technologic parts and systems. According to this, the existing numerical methods have been improved and new numerical methods have been established for modelling and simulation of more and more complex and complicated engineering problems. The present book should contribute to the effort to make the modelling and simulation more effective and accurate.

All the extended papers were reviewed by members of the Conference CMAS2009 Scientific Board, and these papers are arranged into following 18 separated chapters:

In the first chapter a boundary element method is developed for the nonlinear dynamic analysis of beam-columns of arbitrary doubly symmetric simply or multiply connected constant cross-section, partially supported on tensionless Winkler foundation, undergoing moderate large deflections under general boundary conditions, taking into account the effects of shear deformation and rotary inertia. Numerical examples are worked out to illustrate the efficiency, wherever possible the accuracy and the range of applications of the developed method.

In the second chapter a new plate theory based on the direct approach is introduced and applied to plates composed of functionally graded materials (FGM). The material properties are changing in the thickness direction. Solving some problems of the global structural analysis it will be demonstrated that in some cases the results significantly differ from the results based on the Kirchhoff-type theory.

In the third chapter the indirect Trefftz collocation method is applied to solve the inverse Cauchy problem of linear piezoelectricity. The static analysis of coupled fields in 2D piezoelectric solids is considered.

In the forth chapter new phenomenological material model for solid foam materials is presented. This model describes the uniaxial compressive stress-strain curves of foam materials. Parameters of this model specifically control the shape of compressive stress-strain curve and they are function of foam density.

The fifth chapter is focused mainly on polypropylene based composites filled by mineral fillers, but some results can be generalized for other particle composites. The stiffness and toughness of the composite is modelled using a three-phase continuum consisting of the polymer matrix, mineral particles and an interphase between them. It is shown that the effect of the interphase on the macroscopic characteristics of the composite is decisive.

Chapter 6 deals with computational models for heat transfer in composites reinforced by short fibres having very large aspect ratio. Method of Continuous Source Functions (MCSF) were employed and used for simulation of temperature fields in such materials. Strong interaction of fibres and large gradient in temperature fields are good approximated using 1D MCSF along fibre axes.

In the seventh chapter a basic theoretical model of a distributed fibre optic system for detection and location of ammonia leaks is proposed and employed in computer simulations of principal system characteristics.

In the eighth chapter a fourth-order differential equation of the functionally graded material (FGM) beam deflection with longitudinal variation of the effective material properties has been derived where the second order beam theory has been applied for establishing the equilibrium- and kinematics beam equations. Not only the shear forces deformation effect and the effect of consistent mass distribution and mass moment of inertia but also the effect of large axial force has been taken into account.

Numerical experiments will be done concerning the calculation of the eigenfrequencies and corresponded eigenmodes of chosen one-layer beams and multilayered FGM sandwich beams. Effect of the axial forces on the free vibration has been studied and evaluated. The solution results will be compared with those obtained by using a very fine mesh of 2D plane elements of the FEM software ANSYS.

The ninth chapter deals with shear stresses analysis of fully saturated Nevada sand layer by means of Wavelet Transform. The layer modelled as two-phase Biot's porous medium, is subjected to the impulse wave. Results, obtained from program called Swandyne, were analyzed using MathCad.

In the tenth chapter the prediction of the laminate strength is carried out by evaluating the stress state within each layer of the laminate based on the classical lamination theory. Finite Element Method (FEM) is used as a tool to predict the laminate strength.

Chapter 11 studies the plane strain problem of micropolar elastostatics assuming that the governing equations are given in terms of stress functions of order one. The conditions of single valuedness are clarified and the fundamental solution is constructed for the dual basic equations. The integral equations are then established by the direct method. Numerical examples illustrate the applicability of these integral equations.

In the Chapter 12 two alternative methods are proposed to solve problem of equations nonlinearity. Both methods require solving linear biharmonic boundary value problem. For this purpose the Method of Fundamental solutions is implemented. Some numerical examples are presented to confirm that the proposed methods are good tools to solve nonlinear problems.

The thirteenth chapter presents advanced application of computational code HELIOS 1.10 and several application problems of this computer code caused by mathematical solutions of reactor theory physical methods in the source program.

The fourteenth chapter shows that a variationally-based boundary element formulation, obtained as a generalization of the Hellinger-Reissner potential to take the time effect into account, may be applied to finite elements. In such a generalized formulation, the only difference between a boundary element and a finite element is the use of singular or non-singular fundamental solutions as domain interpolation functions.

In the fifteenth chapter the development of a new quadrilateral membrane finite element with drilling degrees of freedom is discussed. A variational principle employing an independent rotation field around the normal of a plane continuum element is derived. This potential is based on the Cosserat continuum theory where skew symmetric stress and strain tensors are introduced in connection with the rotation of a point.

In the sixteenth chapter the results of investigation in the area of designing of two-dimensional composite materials subjected to service loading are presented. To solve the problem of optimal design of these structures, the series hybrid optimization algorithm composed of a sequence of evolution and gradient-oriented procedures is proposed.

The seventeenth chapter presents the computation of effective elastic properties and the analysis of stress intensity factors for representative volume elements (RVE) with randomly distributed microcracks. The RVEs are subjected to static and dynamic loadings. The microcracks randomly distributed, parallel or randomly oriented, having the same length, are considered. The structures with microcracks are modelled by using the boundary element method (BEM). The time dependent problems are solved using the Laplace transform method.

Finally, in the eighteenth chapter the heat transfer processes proceeding in domain of living tissue are discussed. The typical model of bioheat transfer bases, as a rule, on the well known Pennes equation. An approach basing on the dual-phase-lag equation (DPLE) is considered. This equation is supplemented by the adequate boundary and initial conditions. To solve the problem, the general boundary element method is adapted. The examples of computations for 2D problem are presented in the final part of the chapter.

We would like to thank the members of scientific board of the CMAS2009 conference for careful and consistent review of the selected extended papers.

Justín Murín
Vladimír Kutiš
Vladimír Kompiš

Contents

Contributors

C.A. Aguilar Marón Civil Engineering Department, Pontifical Catholic University of Rio de Janeiro, Rio de Janeiro, Brazil, carlosampe@aluno.puc-rio.br

H. Altenbach Lehrstuhl für Technische Mechanik, Zentrum für Ingenieurwissenschaften, Martin-Luther-Universität Halle-Wittenberg, D-06099, Halle (Saale), Germany, holm.altenbach@iw.uni-halle.de

Mehdi Aminbaghai Institute for Mechanics of Materials and Structures, Vienna University of Technology, Vienna, Austria, mehdi.aminbaghai@tuwien.ac.at

J. Aubrecht Department of Solid State Engineering, Faculty of Nuclear Science and Physical Engineering, Czech Technical University in Prague, Prague, Czech Republic

A. Borowiec Division of Soil – Structure Interaction, Faculty of Civil Engineering, Cracow University of Technology, ul. Warszawska 24, 31-155, Kraków, Poland, anabo@pk.edu.pl

P. Darilek VUJE, Inc., Okruzna 5, Trnava, Slovakia, petr.darilek@vuje.sk

K. Dems Department of Technical Mechanics and Informatics, Faculty of Material Technologies and Textile Design, Technical University of Lodz, Lodz, Poland, krzysztof.dems@p.lodz.pl

J. Dudra Bay Zoltan Foundation for Applied Research, Institute of Logistics and Production Engineering, Miskolctapolca, Hungary, dudra.judit@bay-logi.hu

N.A. Dumont Civil Engineering Department, Pontifical Catholic University of Rio de Janeiro, Rio de Janeiro, Brazil, dumont@puc-rio.br

G. Dziatkiewicz Department of Strength of Materials and Computational Mechanics, Silesian University of Technology, Konarskiego 18A, 44-100, Gliwice, Poland, grzegorz.dziatkiewicz@polsl.pl

V.A. Eremeyev South Scientific Center of RASci South Federal University, Rostov on Don, Russia, victor.eremeyev@iw.uni-halle.de

P. Fedelinski Department of Strength of Materials and Computational Mechanics, Silesian University of Technology, Konarskiego 18A, 44-100, Gliwice, Poland, Piotr.Fedelinski@polsl.pl

Peter A. Fotiu Department of Applied and Numerical Mechanics, University of Applied Sciences Wiener Neustadt, Wiener Neustadt, Austria, pf@fhwn.ac.at

V. Goga Department of Mechanics, Faculty of Electrical Engineering and Information Technology, Slovak University of Technology, Bratislava, Slovakia, vladimir.goga@stuba.sk

P. Hutař Institute of Physics of Materials, Academy of Sciences, Žižkova 22, 616 62 Brno, Czech Republic, hutar@ipm.cz

L. Kalvoda Department of Solid State Engineering, Faculty of Nuclear Science and Physical Engineering, Czech Technical University in Prague, Prague, Czech Republic, ladislav.kalvoda@fjfi.cvut.cz

A.E. Kampitsis School of Civil Engineering, National Technical University of Athens, Athens, Greece

R. Klepáček Department of Solid State Engineering, Faculty of Nuclear Science and Physical Engineering, Czech Technical University in Prague, Prague, Czech Republic

Z. Knésl Institute of Physics of Materials, Academy of Sciences, Brno, Czech Republic, knesl@ipm.cz

Vladimír Kompiš Armed Forces Academy of General Milan Rastislav Štefánik, Demänová 393, 031 19 Liptovský Mikuláš, Slovakia, vladimir.kompis@aos.sk

E. Kormaníková Department of Structural Mechanics, Faculty of Civil Engineering, Technical University of Košice, Košice, Slovakia, eva.kormanikova@tuke.sk

Stephan Kugler Department of Applied and Numerical Mechanics, University of Applied Sciences Wiener Neustadt, Wiener Neustadt, Austria, kugler@fhwn.ac.at

Vladimír Kutiš Department of Mechanics, Slovak University of Technology in Bratislava, Ilkovičova 3, 812 19 Bratislava, Slovakia, kutis@elf.stuba.sk

E. Majchrzak Department for Strength of Materials and Computational Mechanics, Silesian University of Technology, Konarskiego 18a, 44-100 Gliwice, Poland, ewa.majchrzak@polsl.pl

Z. Majer Faculty of Mechanical Engineering, Brno University of Technology, Brno, Czech Republic

Z. Murčinková Faculty of Manufacturing Technologies, Technical University in Košice, Prešov, Slovakia, zuzana.murcinkova@tuke.sk

Justín Murín Department of Mechanics, Faculty of Electrical Engineering and Information Technology, Slovak University of Technology, Bratislava, Slovakia, justin.murin@stuba.sk

L. Náhlík Institute of Physics of Materials, Academy of Sciences, Brno, Czech Republic; Faculty of Mechanical Engineering, Brno University of Technology, Brno, Czech Republic

V. Necas Faculty of Electrical Engineering and Information Technology, Department of Nuclear Physics and Technology, Slovak University of Technology in Bratislava, Ilkovicova 3, Bratislava, Slovakia, vladimir.necas@stuba.sk

M. Očkay Academy of Armed Forces of General Milan Rastislav Štefánik, Liptovský Mikuláš, Slovakia, milos.ockay@aos.sk

D. Riecky Department of Applied Mechanics, Faculty of Mechanical Engineering, University of Žilina, Žilina, Slovakia, daniel.riecky@fstroj.uniza.sk

E.J. Sapountzakis School of Civil Engineering, National Technical University of Athens, Athens, Greece, cvsapoun@central.ntua.gr

O. Sýkora Department of Solid State Engineering, Faculty of Nuclear Science and Physical Engineering, Czech Technical University in Prague, Prague, Czech Republic

Gy. Szeidl Department of Mechanics, University of Miskolc, 3515 Miskolc, Hungary, gyorgy.szeidl@uni-miskolc.hu

A. Uscilowska Institute of Applied Mechanics, Poznan University of Technology, ul. Piotrowo 3, 60-965 Poznan, Poland, anita.uscilowska@put.poznan.pl

J. Wiśniewski Department of Technical Mechanics and Informatics, Faculty of Material Technologies and Textile Design, Technical University of Lodz, Zeromskiego 116, 90-924 Lodz, Poland, jacek.wisniewski@p.lodz.pl

R. Zajac VUJE, Inc., Okruzna 5, Trnava, Slovakia; Faculty of Electrical Engineering and Information Technology, Department of Nuclear Physics and Technology, Slovak University of Technology in Bratislava, Ilkovicova 3, Bratislava, Slovakia, radoslav.zajac@vuje.sk

M. Žmindák Department of Applied Mechanics, Faculty of Mechanical Engineering, University of Žilina, Žilina, Slovakia, mifaa.zmindak@fstroj.uniza.sk

Chapter 1
Nonlinear Dynamic Analysis of Partially Supported Beam-Columns on Nonlinear Elastic Foundation Including Shear Deformation Effect

E.J. Sapountzakis and A.E. Kampitsis

Abstract In this paper, a boundary element method is developed for the nonlinear dynamic analysis of beam-columns of arbitrary doubly symmetric simply or multiply connected constant cross section, partially supported on tensionless Winkler foundation, undergoing moderate large deflections under general boundary conditions, taking into account the effects of shear deformation and rotary inertia. The beam-column is subjected to the combined action of arbitrarily distributed or concentrated transverse loading and bending moments in both directions as well as to axial loading. To account for shear deformations, the concept of shear deformation coefficients is used. Five boundary value problems are formulated with respect to the transverse displacements, to the axial displacement and to two stress functions and solved using the Analog Equation Method, a BEM based method. Application of the boundary element technique yields a nonlinear coupled system of equations of motion. The solution of this system is accomplished iteratively by employing the average acceleration method in combination with the modified Newton Raphson method. The evaluation of the shear deformation coefficients is accomplished from the aforementioned stress functions using only boundary integration. The proposed model takes into account the coupling effects of bending and shear deformations along the member as well as the shear forces along the span induced by the applied axial loading. Numerical examples are worked out to illustrate the efficiency, wherever possible the accuracy and the range of applications of the developed method.

Keywords Nonlinear dynamic analysis · Large deflections · Timoshenko beam · Tensionless foundation

---

E.J. Sapountzakis (✉)
School of Civil Engineering, National Technical University of Athens, Athens, Greece
e-mail: cvsapoun@central.ntua.gr

J. Murín et al. (eds.), *Computational Modelling and Advanced Simulations*,
Computational Methods in Applied Sciences 24, DOI 10.1007/978-94-007-0317-9_1,

1.1 Introduction

The study of nonlinear effects on the dynamic analysis of structural elements is essential in civil engineering applications, wherein weight saving is of paramount importance. This non-linearity results from retaining the square of the slope in the strain-displacement relations (intermediate non-linear theory), avoiding in this way the inaccuracies arising from a linearized second – order analysis. Thus, the aforementioned study takes into account the influence of the action of axial, lateral forces and end moments on the deformed shape of the structural element.

Besides, the admission of tensile stresses across the interface separating the beam from the foundation is not realistic. When there is no bonding between beam and subgrade, regions of no contact develop beneath the beam. These regions are unknown and the change of the transverse displacement sign provides the condition for the determination of the contact region. Moreover, due to the intensive use of materials having relatively high transverse shear modulus and the need for beam members with high natural frequencies the error incurred from the ignorance of the effect of shear deformation may be substantial, particularly in the case of heavy lateral loading. The Timoshenko beam theory, which includes shear deformation and rotary inertia effects has an extended range of applications as it allows treatment of deep beam (depth is large relative to length), short and thin-webbed beams and beams where higher modes are excited.

When the beam-column deflections of the structure are small, a wide range of linear analysis tools, such as modal analysis, can be used, and some analytical results are possible. During the past few years, the linear dynamic analysis of beams on elastic foundation has received a good amount of attention in the literature with pioneer the work of Hetenyi [1] who studied the elementary Bernoulli-Euler beams on elastic Winkler foundation. Rades [2] presented the steady-state response of a finite rigid beam resting on a foundation defined by one inertial and three elastic parameters in the assumption of a permanent and smooth contact between beam and foundation considering only uncoupled modes. Wang and Stephens [3] studied the natural vibrations of a Timoshenko beam on a Pasternak-type foundation showing the effects of rotary inertia, shear deformation and foundation constants of the beam employing general analytic solutions for simple cases of boundary conditions. De Rosa [4] and El-Mously [5] derived explicit formulae for the fundamental natural frequencies of finite Timoshenko-beams mounted on finite Pasternak foundation. Explicit expressions for the natural frequencies and the associated amplitude ratios of a double beam system and the analytical solution of its critical buckling were also derived by Zhang et al. [6]. Moreover, semi-analytical closed form solutions [7] and numerical methods such as the method of power series expansion of displacement components employing Hamilton's principle [8], the Galerkin method [9], the differential quadrature element method [10], a combination between the state space and the differential quadrature methods [11], double Fourier transforms [12] and the finite element technique [13–15] have also been used for the vibration and buckling analysis of beam-columns on one- or two– parameter linear or nonlinear (tensionless) elastic foundations taking into account or ignoring shear deformation effect.

As the deflections become larger, the induced geometric nonlinearities result in effects that are not observed in linear systems. Contrary to the good amount of attention in the literature concerning the linear dynamic analysis of beam-columns supported on elastic foundation, very little work has been done on the corresponding nonlinear problem, such as the nonlinear free vibration analysis of multispan beams on elastic supports presented by Lewandowski [16] employing the dynamic finite element method, neglecting the horizontally and rotary inertia forces and considering the beams as distributed mass systems.

In this paper, a boundary element method is developed for the nonlinear dynamic analysis of beam-columns of arbitrary doubly symmetric simply or multiply connected constant cross section, partially supported on tensionless Winkler foundation, undergoing moderate large deflections under general boundary conditions, taking into account the effects of shear deformation and rotary inertia. The beam-column is subjected to the combined action of arbitrarily distributed or concentrated transverse loading and bending moments in both directions as well as to axial loading. To account for shear deformations, the concept of shear deformation coefficients is used. Five boundary value problems are formulated with respect to the transverse displacements, to the axial displacement and to two stress functions and solved using the Analog Equation Method [17], a BEM based method. Application of the boundary element technique yields a nonlinear coupled system of equations of motion. The solution of this system is accomplished iteratively by employing the average acceleration method in combination with the modified Newton Raphson method [18, 19]. The evaluation of the shear deformation coefficients is accomplished from the aforementioned stress functions using only boundary integration. The proposed model takes into account the coupling effects of bending and shear deformations along the member as well as the shear forces along the span induced by the applied axial loading. The essential features and novel aspects of the present formulation compared with previous ones are summarized as follows.

1. Shear deformation effect and rotary inertia are taken into account on the nonlinear dynamic analysis of beam-columns subjected to arbitrary loading (distributed or concentrated transverse loading and bending moments in both directions, as well as axial loading).
2. The homogeneous linear half-space is approximated by a tensionless Winkler foundation.
3. The beam-column is supported by the most general nonlinear boundary conditions including elastic support or restrain, while its cross section is an arbitrary doubly symmetric one.
4. The proposed model takes into account the coupling effects of bending and shear deformations along the member as well as shear forces along the span induced by the applied axial loading.
5. The shear deformation coefficients are evaluated using an energy approach, instead of Timoshenko's [20] and [21] definitions, for which several authors [22, 23] have pointed out that one obtains unsatisfactory results or definitions given by other researchers [24, 25], for which these factors take negative values.

6. The effect of the material's Poisson ratio ν is taken into account.
7. The proposed method employs a BEM approach (requiring boundary discretization) resulting in line or parabolic elements instead of area elements of the FEM solutions (requiring the whole cross section to be discretized into triangular or quadrilateral area elements), while a small number of line elements are required to achieve high accuracy.

Numerical examples are worked out to illustrate the efficiency, wherever possible the accuracy and the range of applications of the developed method.

1.2 Statement of the Problem

Let us consider a prismatic beam-column of length l (Fig. 1.1), of constant arbitrary doubly symmetric cross-section of area A. The homogeneous isotropic and linearly elastic material of the beam-column cross-section, with modulus of elasticity E, shear modulus G and Poisson's ratio ν occupies the two dimensional multiply connected region Ω of the y, z plane and is bounded by the Γ_j $(j = 1, 2, ..., K)$ boundary curves, which are piecewise smooth, i.e. they may have a finite number of corners. In Fig. 1.1b Cyz is the principal bending coordinate system through the cross section's centroid. The beam-column is partially supported on a tensionless homogeneous elastic soil with k_x, k_y and k_z the moduli of subgrade reaction for the x, y, z directions,

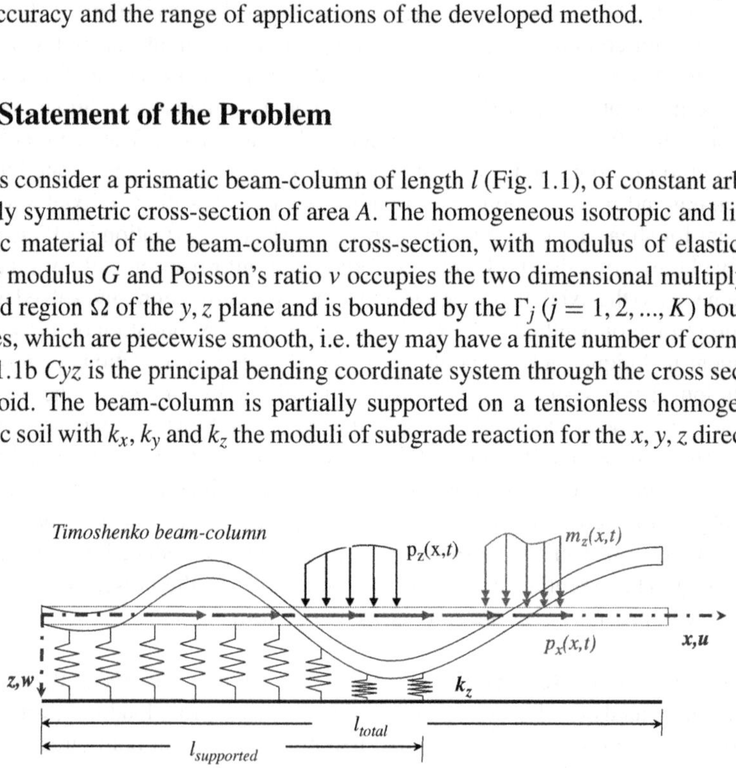

(a)

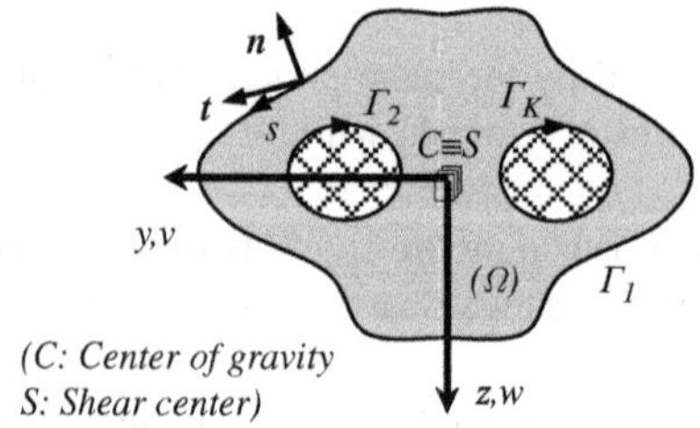

(b)

Fig 1.1 x-z plane of a prismatic beam-column in axial – flexural loading (**a**) with an arbitrary doubly symmetric cross-section occupying the two dimensional region Ω (**b**)

respectively (Winkler spring stiffness). Taking into account the unbonded contact between beam and subgrade, the interaction pressure at the interface is compressive and can be represented for the horizontal and vertical directions by the following relations

$$p_{sx} = U_u(x)\, k_x u \tag{1.1a}$$

$$p_{sy} = U_v(x)\, k_y v \tag{1.1b}$$

$$p_{sz} = U_w(x)\, k_z w \tag{1.1c}$$

where $U_i(x)$ is the unit step function defined as

$$U_i(x) = \begin{cases} 0 & \mathit{if}\ i < 0 \\ 1 & \mathit{if}\ i \geq 0 \end{cases} \quad i = u, v, w \tag{1.2}$$

The beam is subjected to the combined action of the arbitrarily distributed or concentrated time dependent axial loading $p_x = p_x(x,t)$, transverse loading $p_y = p_y(x,t)$, $p_z = p_z(x,t)$ acting in the y and z directions, respectively and bending moments $m_y = m_y(x,t)$, $m_z = m_z(x,t)$ along y and z axes, respectively (Fig. 1.1a).

Under the action of the aforementioned loading, the displacement field of the beam taking into account shear deformation effect is given as

$$\bar{u}(x,y,z,t) = u(x,t) - y\theta_z(x,t) + z\theta_y(x,t) \tag{1.3a}$$

$$\bar{v}(x,t) = v(x,t) \tag{1.3b}$$

$$\bar{w}(x,t) = w(x,t) \tag{1.3c}$$

where $\bar{u}$, $\bar{v}$, $\bar{w}$ are the axial and transverse beam displacement components with respect to the Cyz system of axes; $u(x,t)$, $v(x,t)$, $w(x,t)$ are the corresponding components of the centroid C and $\theta_y(x,t)$, $\theta_z(x,t)$ are the angles of rotation due to bending of the cross-section with respect to its centroid.

Employing the strain-displacement relations of the three-dimensional elasticity for moderate displacements [26, 27], the following strain components can be easily obtained

$$\varepsilon_{xx} = \frac{\partial \bar{u}}{\partial x} + \frac{1}{2}\left[\left(\frac{\partial \bar{v}}{\partial x}\right)^2 + \left(\frac{\partial \bar{w}}{\partial x}\right)^2\right] \tag{1.4a}$$

$$\gamma_{xz} = \frac{\partial \bar{w}}{\partial x} + \frac{\partial \bar{u}}{\partial z} + \left(\frac{\partial \bar{v}}{\partial x}\frac{\partial \bar{v}}{\partial z} + \frac{\partial \bar{w}}{\partial x}\frac{\partial \bar{w}}{\partial z}\right) \tag{1.4b}$$

$$\gamma_{xy} = \frac{\partial \bar{v}}{\partial x} + \frac{\partial \bar{u}}{\partial y} + \left(\frac{\partial \bar{v}}{\partial x}\frac{\partial \bar{v}}{\partial y} + \frac{\partial \bar{w}}{\partial x}\frac{\partial \bar{w}}{\partial y}\right) \tag{1.4c}$$

$$\varepsilon_{yy} = \varepsilon_{zz} = \gamma_{yz} = 0 \tag{1.4d}$$

where it has been assumed that for moderate displacements $(\partial\bar{u}/\partial x)^2 << \partial\bar{u}/\partial x$, $(\partial\bar{u}/\partial x)(\partial\bar{u}/\partial z) << (\partial\bar{u}/\partial x) + (\partial\bar{u}/\partial z)$, $(\partial\bar{u}/\partial x)(\partial\bar{u}/\partial y) << (\partial\bar{u}/\partial x) + (\partial\bar{u}/\partial y)$.

Substituting the displacement components (1.3) to the strain-displacement relations (1.4), the strain components can be written as

$$\varepsilon_{xx}(x,y,z,t) = u' + z\theta_y' - y\theta_z' + \frac{1}{2}\left(v'^2 + w'^2\right) \tag{1.5a}$$

$$\gamma_{xy} = v' - \theta_z \tag{1.5b}$$

$$\gamma_{xz} = w' + \theta_y \tag{1.5c}$$

where γ_{xy}, γ_{xz} are the additional angles of rotation of the cross-section due to shear deformation. Considering strains to be small, employing the second Piola – Kirchhoff stress tensor and assuming an isotropic and homogeneous material, the stress components are defined in terms of the strain ones as

$$\begin{Bmatrix} S_{xx} \\ S_{xy} \\ S_{xz} \end{Bmatrix} = \begin{bmatrix} E & 0 & 0 \\ 0 & G & 0 \\ 0 & 0 & G \end{bmatrix} \begin{Bmatrix} \varepsilon_{xx} \\ \gamma_{xy} \\ \gamma_{xz} \end{Bmatrix} \tag{1.6}$$

or employing (1.5) as

$$S_{xx} = E\left[u' + z\theta_y' - y\theta_z' + \frac{1}{2}\left(v'^2 + w'^2\right)\right] \tag{1.7a}$$

$$S_{xy} = G \cdot \left(v' - \theta_z\right) \tag{1.7b}$$

$$S_{xz} = G \cdot \left(w' + \theta_y\right) \tag{1.7c}$$

On the basis of Hamilton's principle, the variations of the Lagrangian equation defined as

$$\delta \int_{t_1}^{t_2} (U - K - W_{\text{ext}})dt = 0 \tag{1.8}$$

and expressed as a function of the stress resultants acting on the cross section of the beam in the deformed state provide the governing equations and the boundary conditions of the beam subjected to nonlinear vibrations. In (1.8) $\delta(\cdot)$ denotes variation of quantities, while U, K, W_{ext} are the strain energy, the kinetic energy and the external load work, respectively given as

$$\delta U = \int_V \left(S_{xx}\delta\varepsilon_{xx} + S_{xy}\delta\gamma_{xy} + S_{xz}\delta\gamma_{xz}\right)dV \tag{1.9a}$$

$$\delta K = \frac{1}{2}\int_V \rho\left(\delta\dot{\bar{u}}^2 + \delta\dot{\bar{v}}^2 + \delta\dot{\bar{w}}^2\right)dV \tag{1.9b}$$

$$\delta W_{\text{ext}} = \int_L \begin{pmatrix} p_x\delta u - U_u k_x u\delta u + p_y\delta v - U_v k_y v\delta v + \\ + p_z\delta w - U_w k_z w\delta w + m_y\delta\theta_y + m_z\delta\theta_z \end{pmatrix} dx \tag{1.9c}$$

Moreover, the stress resultants of the beam are given as

$$N = \int_{\Omega} S_{xx} d\Omega \tag{1.10a}$$

$$\mathrm{M}_y = \int_{\Omega} S_{xx} z d\Omega \tag{1.10b}$$

$$\mathrm{M}_z = -\int_{\Omega} S_{xx} y d\Omega \tag{1.10c}$$

$$Q_y = \int_{A_y} S_{xy} d\Omega \tag{1.10d}$$

$$Q_z = \int_{A_z} S_{xz} d\Omega \tag{1.10e}$$

Substituting the expressions of the stress components (1.7) into (1.10), the stress resultants are obtained as

$$N = EA\left[u' + \frac{1}{2}\left(v'^2 + w'^2\right)\right] \tag{1.11a}$$

$$M_y = EI_y \theta'_y \tag{1.11b}$$

$$M_z = EI_z \theta'_z \tag{1.11c}$$

$$Q_y = GA_y \gamma_{xy} \tag{1.11d}$$

$$Q_z = GA_z \gamma_{xz} \tag{1.11e}$$

where A is the cross section area, I_y, I_z the moments of inertia with respect to the principle bending axes given as

$$A = \int_{\Omega} d\Omega \tag{1.12}$$

$$I_y = \int_{\Omega} z^2 d\Omega \tag{1.13a}$$

$$I_z = \int_{\Omega} y^2 d\Omega \tag{1.13b}$$

and GA_y, GA_z are its shear rigidities of the Timoshenko's beam theory, where

$$A_z = \kappa_z A = \frac{1}{a_z} A \tag{1.14a}$$

$$A_y = \kappa_y A = \frac{1}{a_y} A \tag{1.14b}$$

are the shear areas with respect to y, z axes, respectively with κ_y, κ_z the shear correction factors and a_y, a_z the shear deformation coefficients. Substituting the stress components given in (1.7) and the strain resultants given in (1.5) to the strain energy variation δE_{int} (1.9a) and employing (1.8) the equilibrium equations of the beam are derived as

$$-EA\left(u''+w'w''+v'v''\right)+\rho A\ddot{u}+U_u k_x u=p_x \tag{1.15a}$$

$$-\left(Nv'\right)'+\rho A\ddot{v}-GA_y\left(v''-\theta_z'\right)+U_v k_y v=p_y \tag{1.15b}$$

$$-EI_z\theta_z''+\rho I_z\ddot{\theta}_z-GA_y\left(v'-\theta_z\right)=m_z \tag{1.15c}$$

$$-\left(Nw'\right)'+\rho A\ddot{w}-GA_z\left(w''+\theta_y'\right)+U_w k_z w=p_z \tag{1.15d}$$

$$-EI_y\theta_y''+\rho I_y\ddot{\theta}_y+GA_z\left(w'+\theta_y\right)=m_y \tag{1.15e}$$

Combining (1.15b, c) and (1.15d, e), the following differential equations with respect to u, v, w are derived

$$-EA\left(u''+w'w''+v'v''\right)+\rho A\ddot{u}+U_u k_x u=p_x \tag{1.16a}$$

$$\begin{aligned} &EI_z v''''-\rho I_z\left(\frac{Ea_y}{G}+1\right)\frac{\partial^2\ddot{v}}{\partial x^2}+\rho A\ddot{v}+\frac{EI_z}{GA_y}\left(Nv'\right)'''-\left(Nv'\right)'\\ &-\frac{\rho I_z}{GA_y}\left(\frac{\partial^2\left(Nv'\right)'}{\partial t^2}-\rho A\ddddot{v}\right)+\left(k_y v-\frac{EI_z}{GA_y}\left(k_y v''\right)+\frac{\rho I_z}{GA_y}k_y\ddot{v}\right)\\ &U_v=p_y-\frac{EI_z}{GA_y}\left(p_y''\right)+\frac{\rho I_z}{GA_y}\left(\ddot{p}_y\right)-m_z' \end{aligned} \tag{1.16b}$$

$$\begin{aligned} &EI_y w''''-\rho I_y\left(\frac{Ea_z}{G}+1\right)\frac{\partial^2\ddot{w}}{\partial x^2}+\rho A\ddot{w}+\frac{EI_y}{GA_z}(Nw')-(Nw')'\\ &-\frac{\rho I_y}{GA_z}\left(\frac{\partial^2(Nw')}{\partial t^2}-\rho A\ddddot{w}\right)+\left(k_z w-\frac{EI_y}{GA_z}\left(k_z w''\right)+\frac{\rho I_y}{GA_z}k_z\ddot{w}\right)\\ &U_w=p_z-\frac{EI_y}{GA_z}\left(p_z''\right)+\frac{\rho I_y}{GA_z}(\ddot{p}_z)+m_y' \end{aligned} \tag{1.16c}$$

Equations (1.16) constitute the governing differential equations of a Timoshenko beam-column, partially supported on a tensionless Winkler foundation, subjected to nonlinear vibrations due to the combined action of time dependent axial and transverse loading. These equations are also subjected to the pertinent boundary conditions of the problem, which are given as

$$a_1 u\left(x,t\right)+\alpha_2 N\left(x,t\right)=\alpha_3 \tag{1.17}$$

$$\beta_1 v\left(x,t\right)+\beta_2 V_y\left(x,t\right)=\beta_3 \tag{1.18a}$$

$$\bar{\beta}_1\theta_z(x,t) + \bar{\beta}_2 \mathrm{M}_z(x,t) = \bar{\beta}_3 \tag{1.18b}$$

$$\gamma_1 w(x,t) + \gamma_2 V_z(x,t) = \gamma_3 \tag{1.19a}$$

$$\bar{\gamma}_1\theta_y(x,t) + \bar{\gamma}_2 \mathrm{M}_y(x,t) = \bar{\gamma}_3 \tag{1.19b}$$

at the beam ends $x = 0, l$, together with the initial conditions

$$u(x,0) = \bar{u}_0(x) \tag{1.20a}$$

$$\dot{u}(x,0) = \dot{\bar{u}}_0(x) \tag{1.20b}$$

$$v(x,0) = \bar{v}_0(x) \tag{1.21a}$$

$$\dot{v}(x,0) = \dot{\bar{v}}_0(x) \tag{1.21b}$$

$$w(x,0) = \bar{w}_0(x) \tag{1.22a}$$

$$\dot{w}(x,0) = \dot{\bar{w}}_0(x) \tag{1.22b}$$

where $\bar{u}_0(x)$, $\bar{v}_0(x)$, $\bar{w}_0(x)$, $\dot{\bar{u}}_0(x)$, $\dot{\bar{v}}_0(x)$ and $\dot{\bar{w}}_0(x)$ are prescribed functions. In (1.18) and (1.19) V_y, V_z and M_z, M_y are the reactions and bending moments with respect to y, z, respectively, which together with the angles of rotation due to bending θ_y, θ_z are given by the following relations

$$V_y = Nv' - EI_z v''' - \frac{EI_z}{GA_y}\left[p_y' - U_v k_y v' + (Nv')'' - \rho A \frac{\partial \ddot{v}}{\partial x}\right] + \rho I_z \ddot{\theta}_z \tag{1.23a}$$

$$V_z = Nw' - EI_y w''' - \frac{EI_y}{GA_z}\left[p_z' - U_w k_z w' + (Nw')'' - \rho A \frac{\partial \ddot{w}}{\partial x}\right] - \rho I_y \ddot{\theta}_y \tag{1.23b}$$

$$M_z = EI_z v'' + \frac{EI_z}{GA_y}\left[p_y - U_v k_y v + (Nv')' - \rho A \ddot{v}\right] \tag{1.23c}$$

$$M_y = -EI_y w'' - \frac{EI_y}{GA_z}\left[p_z - U_w k_z w + (Nw')' - \rho A \ddot{w}\right] \tag{1.23d}$$

$$\begin{aligned}\theta_y = {} & \frac{EI_y}{G^2 A_z^2}\left(\rho A \frac{\partial \ddot{w}}{\partial x} - p_z' + U_w k_z w' - (Nw')''\right) - \\ & -\frac{1}{GA_z}\left(EI_y w''' + \rho I_y \ddot{\theta}_y + GA_z w'\right)\end{aligned} \tag{1.23e}$$

$$\begin{aligned}\theta_z = {} & \frac{EI_z}{G^2 A_y^2}\left(p_y' - U_v k_y v' + (Nv')'' - \rho A \frac{\partial \ddot{v}}{\partial x}\right) + \\ & +\frac{1}{GA_y}\left(EI_z v''' - \rho I_z \ddot{\theta}_z + GA_y v'\right)\end{aligned} \tag{1.23f}$$

Finally, $\alpha_k, \bar{\alpha}_k, \beta_k, \bar{\beta}_k, \gamma_k, \bar{\gamma}_k$ $(k = 1,2,3)$ are functions specified at the beam ends $x = 0, l$. Equations (1.17), (1.18) and (1.19) describe the most general nonlinear boundary conditions associated with the problem at hand and can include elastic support or restraint. It is apparent that all types of the conventional boundary conditions (clamped, simply supported, free or guided edge) can be derived from these equations by specifying appropriately these functions (e.g. for a clamped edge it is $\alpha_1 = \beta_1 = \gamma_1 = 1$, $\overline{\beta}_1 = \overline{\gamma}_1 = 1$, $\alpha_2 = \alpha_3 = \beta_2 = \beta_3 = \gamma_2 = \gamma_3 = \overline{\beta}_2 = \overline{\beta}_3 = \overline{\gamma}_2 = \overline{\gamma}_3 = 0$).

The solution of the initial boundary value problem given from (1.16), subjected to the boundary conditions (1.17), (1.18) and (1.19) and the initial conditions (1.20), (1.21) and (1.22) which represents the nonlinear flexural dynamic analysis of a Timoshenko beam-column, partially supported on a tensionless Winkler foundation, presumes the evaluation of the shear deformation coefficients a_y, a_z, corresponding to the principal coordinate system Cyz. These coefficients are established equating the approximate formula of the shear strain energy per unit length [24]

$$U_{\text{appr.}} = \frac{a_y Q_y^2}{2AG} + \frac{a_z Q_z^2}{2AG} \tag{1.24}$$

with the exact one given from

$$U_{\text{exact}} = \int_\Omega \frac{(\tau_{xz})^2 + (\tau_{xy})^2}{2G} d\Omega \tag{1.25}$$

and are obtained as [28]

$$a_y = \frac{1}{\kappa_y} = \frac{A}{\Delta^2} \int_\Omega [(\nabla\Theta) - \boldsymbol{e}] \cdot [(\nabla\Theta) - \boldsymbol{e}]\, d\Omega \tag{1.26a}$$

$$a_z = \frac{1}{\kappa_z} = \frac{A}{\Delta^2} \int_\Omega [(\nabla\Phi) - \boldsymbol{d}] \cdot [(\nabla\Phi) - \boldsymbol{d}]\, d\Omega \tag{1.26b}$$

where $(\tau_{xz})_j$, $(\tau_{xy})_j$ are the transverse (direct) shear stress components, $(\nabla) \equiv \boldsymbol{i}_y (\partial/\partial y) + \boldsymbol{i}_z (\partial/\partial z)$ is a symbolic vector with $\boldsymbol{i}_y$, $\boldsymbol{i}_z$ the unit vectors along y and z axes, respectively, Δ is given from

$$\Delta = 2(1+\nu)\, \mathrm{I}_y \mathrm{I}_z \tag{1.27}$$

ν is the Poisson ratio of the cross section material, $\boldsymbol{e}$ and $\boldsymbol{d}$ are vectors defined as

$$\boldsymbol{e} = \left(\nu I_y \frac{y^2 - z^2}{2}\right) \boldsymbol{i}_y + \nu I_y yz \boldsymbol{i}_z \tag{1.28a}$$

$$\boldsymbol{d} = \nu I_z yz \boldsymbol{i}_y - \left(\nu I_z \frac{y^2 - z^2}{2}\right) \boldsymbol{i}_z \tag{1.28b}$$

and $\Theta(y,z)$, $\Phi(y,z)$ are stress functions, which are evaluated from the solution of the following Neumann type boundary value problems Sapountzakis and Mokos [28]

$$\nabla^2 \Theta = -2I_y y \text{ in } \Omega \tag{1.29a}$$

$$\frac{\partial \Theta}{\partial n} = \boldsymbol{n} \cdot \boldsymbol{e} \text{ on } \Gamma = \bigcup_{j=1}^{K+1} \Gamma_j \tag{1.29b}$$

$$\nabla^2 \Phi = -2I_z z \text{ in } \Omega \tag{1.30a}$$

$$\frac{\partial \Phi}{\partial n} = \boldsymbol{n} \cdot \boldsymbol{d} \text{ on } \Gamma = \bigcup_{j=1}^{K+1} \Gamma_j \tag{1.30b}$$

where $\boldsymbol{n}$ is the outward normal vector to the boundary Γ. In case of negligible shear deformations $a_z = a_y = 0$. It is also worth here noting that the boundary conditions (1.29b) and (1.30b) have been derived from the physical consideration that the traction vector in the direction of the normal vector $\boldsymbol{n}$ vanishes on the free surface of the beam.

1.3 Integral Representation – Numerical Solution

According to the precedent analysis, the nonlinear flexural dynamic analysis of Timoshenko beam-columns, partially supported on a tensionless Winkler foundation, undergoing moderate large deflections reduces in establishing the displacement components $u(x,t)$ and $v(x,t)$, $w(x,t)$ having continuous derivatives up to the second order and up to the fourth order with respect to x, respectively, and also having derivatives up to the second order with respect to t (ignoring the inertia terms of the fourth order [29]. Moreover, these displacement components must satisfy the coupled governing differential equations (1.16) inside the beam, the boundary conditions (1.17), (1.18) and (1.19) at the beam ends $x = 0, l$ and the initial conditions (1.20), (1.21) and (1.22). Equations (1.16) are solved using the Analog Equation Method [17] as it is developed for hyperbolic differential equations in [30].

1.3.1 For the Transverse Displacements v, w

Let $v(x,t)$, $w(x,t)$ be the sought solution of the aforementioned boundary value problem. Setting as $u_2(x,t) = v(x,t)$, $u_3(x,t) = w(x,t)$ and differentiating these functions four times with respect to x yields

$$\frac{\partial^4 u_i}{\partial x^4} = q_i(x,t) \; (i = 2, 3) \tag{1.31}$$

Equations (1.31) are quasi-static, that is the time variable appears as a parameter. They indicate that the solution of (1.16b) and (1.16c) can be established by solving

(1.31) under the same boundary conditions (1.18) and (1.19), provided that the fictitious load distributions $q_i\,(x,t)$ $(i = 2,\ 3)$ are first established. These distributions can be determined using BEM as follows.

Following the procedure presented in [30] and employing the constant element assumption for the load distributions q_i along the L internal beam elements (as the numerical implementation becomes very simple and the obtained results are of high accuracy), the integral representations of the displacement components u_i $(i = 2,\ 3)$ and their first derivatives with respect to x when applied for the beam ends $(0,l)$, together with the boundary conditions (1.18) and (1.19) are employed to express the unknown boundary quantities $u_i\,(\zeta,t)$, $u_{i,x}\,(\zeta,t)$, $u_{i,xx}\,(\zeta,t)$ and $u_{i,xxx}\,(\zeta,t)$ $(\zeta = 0, l)$ in terms of q_i as

$$\begin{bmatrix} \mathbf{D}_{11} & 0 & \mathbf{D}_{13} & \mathbf{D}_{14} \\ \mathbf{D}_{21} & \mathbf{D}_{22} & \mathbf{D}_{23} & 0 \\ \mathbf{E}_{31} & \mathbf{E}_{32} & \mathbf{E}_{33} & \mathbf{E}_{34} \\ \mathbf{E}_{41} & \mathbf{E}_{42} & \mathbf{E}_{43} & 0 \end{bmatrix} \begin{Bmatrix} \hat{\mathbf{u}}_{2,xxx} \\ \hat{\mathbf{u}}_{2,xx} \\ \hat{\mathbf{u}}_{2,x} \\ \hat{\mathbf{u}}_{2} \end{Bmatrix} = \begin{Bmatrix} \boldsymbol{\beta}_3 \\ \bar{\boldsymbol{\beta}}_3 \\ 0 \\ 0 \end{Bmatrix} + \begin{Bmatrix} 0 \\ 0 \\ \mathbf{F}_3 \\ \mathbf{F}_4 \end{Bmatrix} \mathbf{q}_2 \tag{1.32a}$$

$$\begin{bmatrix} \mathbf{G}_{11} & 0 & \mathbf{G}_{13} & \mathbf{G}_{14} \\ \mathbf{G}_{21} & \mathbf{G}_{22} & \mathbf{G}_{23} & 0 \\ \mathbf{E}_{31} & \mathbf{E}_{32} & \mathbf{E}_{33} & \mathbf{E}_{34} \\ \mathbf{E}_{41} & \mathbf{E}_{42} & \mathbf{E}_{43} & 0 \end{bmatrix} \begin{Bmatrix} \hat{\mathbf{u}}_{3,xxx} \\ \hat{\mathbf{u}}_{3,xx} \\ \hat{\mathbf{u}}_{3,x} \\ \hat{\mathbf{u}}_{3} \end{Bmatrix} = \begin{Bmatrix} \boldsymbol{\gamma}_3 \\ \bar{\boldsymbol{\gamma}}_3 \\ 0 \\ 0 \end{Bmatrix} + \begin{Bmatrix} 0 \\ 0 \\ \mathbf{F}_3 \\ \mathbf{F}_4 \end{Bmatrix} \mathbf{q}_3 \tag{1.32b}$$

where $\mathbf{D}_{11}$, $\mathbf{D}_{13}$, $\mathbf{D}_{14}$, $\mathbf{D}_{21}$, $\mathbf{D}_{22}$, $\mathbf{D}_{23}$, $\mathbf{G}_{11}$, $\mathbf{G}_{13}$, $\mathbf{G}_{14}$, $\mathbf{G}_{21}$, $\mathbf{G}_{22}$, $\mathbf{G}_{23}$ are 2×2 known square matrices including the values of the functions $\beta_j, \bar{\beta}_j, \gamma_j, \bar{\gamma}_j$ $(j = 1, 2)$ of (1.18) and (1.19); $\boldsymbol{\beta}_3$, $\bar{\boldsymbol{\beta}}_3$, $\boldsymbol{\gamma}_3$, $\bar{\boldsymbol{\gamma}}_3$ are 2×1 known column matrices including the boundary values of the functions $\beta_3, \bar{\beta}_3, \gamma_3, \bar{\gamma}_3$ of (1.18) and (1.19); $\mathbf{E}_{jk}$, $(j = 3, 4,$ $k = 1, 2, 3, 4)$ are square 2×2 known coefficient matrices and $\mathbf{F}_j$ $(j = 3, 4)$ are $2 \times L$ rectangular known matrices originating from the integration of kernels on the axis of the beam. Moreover,

$$\hat{\mathbf{u}}_i = \{u_i\,(0,t) \quad u_i\,(l,t)\}^T \tag{1.33a}$$

$$\hat{\mathbf{u}}_{i,x} = \left\{ \frac{\partial u_i\,(0,t)}{\partial x} \quad \frac{\partial u_i\,(l,t)}{\partial x} \right\}^T \tag{1.33b}$$

$$\hat{\mathbf{u}}_{i,xx} = \left\{ \frac{\partial^2 u_i\,(0,t)}{\partial x^2} \quad \frac{\partial^2 u_i\,(l,t)}{\partial x^2} \right\}^T \tag{1.33c}$$

$$\hat{\mathbf{u}}_{i,xxx} = \left\{ \frac{\partial^3 u_i\,(0,t)}{\partial x^3} \quad \frac{\partial^3 u_i\,(l,t)}{\partial x^3} \right\}^T \tag{1.33d}$$

are vectors including the two unknown boundary values of the respective boundary quantities and $\mathbf{q}_i = \left\{q_1^i \; q_2^i \; \dots \; q_L^i\right\}^T$ $(i = 2,\ 3)$ is the vector including the L unknown nodal values of the fictitious load.

Discretization of the integral representations of the displacement components $u_i\,(i = 2, 3)$ and their derivatives with respect to x, after elimination of the boundary quantities employing (1.32), gives

$$\mathbf{u}_i = \mathbf{T}_i \mathbf{q}_i + \mathbf{t}_i \; i = 2,3 \tag{1.34a}$$

$$\mathbf{u}_{i,x} = \mathbf{T}_{ix} \mathbf{q}_i + \mathbf{t}_{ix} \; i = 2,3 \tag{1.34b}$$

$$\mathbf{u}_{i,xx} = \mathbf{T}_{ixx} \mathbf{q}_i + \mathbf{t}_{ixx} \; i = 2,3 \tag{1.34c}$$

$$\mathbf{u}_{i,xxx} = \mathbf{T}_{ixxx} \mathbf{q}_i + \mathbf{t}_{ixxx} \; i = 2,3 \tag{1.34d}$$

$$\mathbf{u}_{i,xxxx} = \mathbf{q}_i \; i = 2,3 \tag{1.34e}$$

where $\mathbf{u}_i$, $\mathbf{u}_{i,x}$, $\mathbf{u}_{i,xx}$, $\mathbf{u}_{i,xxx}$, $\mathbf{u}_{i,xxxx}$ are vectors including the values of $u_i\,(x,t)$ and their derivatives at the L nodal points, $\mathbf{T}_i$, $\mathbf{T}_{ix}$, $\mathbf{T}_{ixx}$, $\mathbf{T}_{ixxx}$ are known $L \times L$ matrices and $\mathbf{t}_i$, $\mathbf{t}_{ix}$, $\mathbf{t}_{ixx}$, $\mathbf{t}_{ixxx}$ are known $L \times 1$ matrices.

In the conventional BEM, the load vectors $\mathbf{q}_i$ are known and (1.34) are used to evaluate $u_i\,(x,t)$ and their derivatives at the L nodal points. This, however, can not be done here since $\mathbf{q}_i$ are unknown. For this purpose, $2L$ additional equations are derived, which permit the establishment of $\mathbf{q}_i$. These equations result by applying (1.16b) and (1.16c) to the L collocation points, which after ignoring the inertia terms of the fourth order arising from coupling of shear deformations and rotary inertia Thomson [29], lead to the formulation of the following set of $2L$ simultaneous equations

$$\mathbf{M}_2 \ddot{\mathbf{q}}_2 + \mathbf{S}_2 \dot{\mathbf{q}}_2 + \mathbf{K}_2 \mathbf{q}_2 = \mathbf{f}_2 \tag{1.35a}$$

$$\mathbf{M}_3 \ddot{\mathbf{q}}_3 + \mathbf{S}_3 \dot{\mathbf{q}}_3 + \mathbf{K}_3 \mathbf{q}_3 = \mathbf{f}_3 \tag{1.35b}$$

where the $\mathbf{M}_2$, $\mathbf{M}_3$, $\mathbf{S}_2$, $\mathbf{S}_3$, $\mathbf{K}_2$, $\mathbf{K}_3$ $L \times L$ matrices and the $\mathbf{f}_2$, $\mathbf{f}_3$ $L \times 1$ vectors are given as

$$\mathbf{M}_2 = \rho A \mathbf{T}_2 - \rho \mathrm{I}_z \left(\frac{Ea_y}{G} + 1 \right) \mathbf{T}_{2xx} - \frac{\rho I_z}{GA_y} \left(\mathbf{N}_x \mathbf{T}_{2x} + \mathbf{N} \mathbf{T}_{2xx} - \bar{\mathbf{K}}_y \mathbf{T}_2 \right) \tag{1.36a}$$

$$\mathbf{S}_2 = -\frac{2\rho I_z}{GA_y} \left(\mathbf{N}_{xt} \mathbf{T}_{2x} + \mathbf{N}_t \mathbf{T}_{2xx} \right) \tag{1.36b}$$

$$\begin{aligned} \mathbf{K}_2 = {} & \mathbf{EI}_z - \mathbf{N}_x \mathbf{T}_{2x} - \mathbf{N}\mathbf{T}_{2xx} - \frac{\rho I_z}{GA_y} \left(\mathbf{N}_{xtt} \mathbf{T}_{2x} + \mathbf{N}_{tt} \mathbf{T}_{2xx} \right) + \bar{\mathbf{K}}_y \mathbf{T}_2 + \\ & + \frac{EI_z}{GA_y} \left(\mathbf{N}_{xxx} \mathbf{T}_{2x} + 3\mathbf{N}_{xx} \mathbf{T}_{2xx} + 3\mathbf{N}_x \mathbf{T}_{2xxx} + \mathbf{N} - \bar{\mathbf{K}}_y \mathbf{T}_{2xx} \right) \end{aligned} \tag{1.36c}$$

$$\begin{aligned} \mathbf{f}_2 = {} & \mathbf{p}_y - \bar{\mathbf{K}}_y \mathbf{t}_2 - \frac{EI_z}{GA_y} \left[\mathbf{p}_{y,xx} - \bar{\mathbf{K}}_y \mathbf{t}_{2xx} \right] + \frac{\rho I_z}{GA_y} \mathbf{p}_{y,tt} - \mathbf{m}_{z,x} + \mathbf{N}_x \mathbf{t}_{2x} + \mathbf{N}\mathbf{t}_{2xx} - \\ & - \frac{EI_z}{GA_y} \left(\mathbf{N}_{xxx} \mathbf{t}_{2x} + 3\mathbf{N}_{xx} \mathbf{t}_{2xx} + 3\mathbf{N}_x \mathbf{t}_{2xxx} \right) + \frac{\rho I_z}{GA_y} \left(\mathbf{N}_{xtt} \mathbf{t}_{2x} + \mathbf{N}_{tt} \mathbf{t}_{2xx} \right) \end{aligned} \tag{1.36d}$$

$$\mathbf{M}_3 = \rho A \mathbf{T}_3 - \rho \mathrm{I}_y \left(\frac{Ea_z}{G} + 1 \right) \mathbf{T}_{3xx} - \frac{\rho I_y}{GA_z} \left(\mathbf{N}_x \mathbf{T}_{3x} + \mathbf{N} \mathbf{T}_{3xx} - \bar{\mathbf{K}}_z \mathbf{T}_3 \right) \tag{1.36e}$$

$$\mathbf{S}_3 = -\frac{2\rho I_y}{GA_z}\left(\mathbf{N}_{xt}\mathbf{T}_{3x} + \mathbf{N}_t\mathbf{T}_{3xx}\right) \tag{1.36f}$$

$$\begin{aligned}\mathbf{K}_3 = \mathbf{EI}_y - \mathbf{N}_x\mathbf{T}_{3x} - \mathbf{NT}_{3xx} - \frac{\rho I_y}{GA_z}\left(\mathbf{N}_{xtt}\mathbf{T}_{3x} + \mathbf{N}_{tt}\mathbf{T}_{3xx}\right) + \bar{\mathbf{K}}_z\mathbf{T}_3 + \\ + \frac{EI_y}{GA_z}\left(\mathbf{N}_{xxx}\mathbf{T}_{3x} + 3\mathbf{N}_{xx}\mathbf{T}_{3xx} + 3\mathbf{N}_x\mathbf{T}_{3xxx} + \mathbf{N} - \bar{\mathbf{K}}_z\mathbf{T}_{3xx}\right)\end{aligned} \tag{1.36g}$$

$$\begin{aligned}\mathbf{f}_3 = \mathbf{p}_z - \bar{\mathbf{K}}_z\mathbf{t}_3 - \frac{EI_y}{GA_z}\left[\mathbf{p}_{z,xx} - \bar{\mathbf{K}}_z\mathbf{t}_{3xx}\right] + \frac{\rho I_y}{GA_z}\mathbf{p}_{z,tt} + \mathbf{m}_{y,x} + \mathbf{N}_x\mathbf{t}_{3x} + \mathbf{Nt}_{3xx} - \\ - \frac{EI_y}{GA_z}\left(\mathbf{N}_{xxx}\mathbf{t}_{3x} + 3\mathbf{N}_{xx}\mathbf{t}_{3xx} + 3\mathbf{N}_x\mathbf{t}_{3xxx}\right) + \frac{\rho I_y}{GA_z}\left(\mathbf{N}_{xtt}\mathbf{t}_{3x} + \mathbf{N}_{tt}\mathbf{t}_{3xx}\right)\end{aligned} \tag{1.36h}$$

where $\mathbf{N}$, $\mathbf{N}_{km}(k,m = x,t)$ are $L \times L$ diagonal matrices containing the values of the axial force and its derivatives with respect to k and m parameters at the L nodal points, $\mathbf{I}$ is the unit matrix, $\mathbf{EI}_y$, $\mathbf{EI}_z$ are $L \times L$ diagonal matrices including the values of the corresponding quantities at the aforementioned points, while $\mathbf{p}_y$, $\mathbf{p}_{y,xx}$, $\mathbf{p}_{y,tt}$, $\mathbf{p}_z$, $\mathbf{p}_{z,xx}$, $\mathbf{p}_{z,tt}$, $\mathbf{m}_{y,x}$ and $\mathbf{m}_{z,x}$ are $L \times 1$ vectors containing the values of the external loading and its derivatives at these points. Finally $\bar{\mathbf{K}}_y$, $\bar{\mathbf{K}}_z$ are diagonal matrices whose diagonal elements are given as

$$\left(\bar{\mathbf{K}}_y\right)_{i,i} = \left(k_y\right)_i (U_v)_i \tag{1.37a}$$

$$\left(\bar{\mathbf{K}}_z\right)_{i,i} = (k_z)_i (U_w)_i \tag{1.37b}$$

where $\left(k_y\right)_i$, $(k_z)_i$, $(U_v)_i$, $(U_w)_i$ are the values of the corresponding moduli of subgrade or the unit step function at the i-th nodal point.

1.3.2 For the Axial Displacement u

Let $u_1 = u(x,t)$ be the sought solution of the boundary value problem described by (1.16a) and (1.17). Differentiating this function two times yields

$$\frac{\partial^2 u_1}{\partial x^2} = q_x(x,t) \tag{1.38}$$

Equation (1.38) indicates that the solution of the original problem can be obtained as the axial displacement of a beam with unit axial rigidity subjected to an axial fictitious load $q_x(x,t)$ under the same boundary conditions. The fictitious load is unknown. Following the same procedure as in 1.3.1, the discretized counterpart of the integral representations of the displacement component u_1 and its first derivative with respect to x when applied to all nodal points in the interior of the beam yields

$$\mathbf{u}_1 = \mathbf{T}_1\mathbf{q}_1 + \mathbf{t}_1 \tag{1.39a}$$

$$\mathbf{u}_{1,x} = \mathbf{T}_{1x}\mathbf{q}_1 + \mathbf{t}_{1x} \tag{1.39b}$$

where $\mathbf{T}_1$, $\mathbf{T}_{1x}$ are known $L \times L$ matrices, similar with those mentioned before for the displacements u_2, u_3. Application of (1.16a) to the L collocation points, after employing (1.34) and (1.39) leads to the formulation of the following system of L equations with respect to $\mathbf{q_1}$, $\mathbf{q_2}$ and $\mathbf{q_3}$ fictitious load vectors

$$\begin{aligned}\left(\mathbf{EA} - \bar{\mathbf{K}}_x\mathbf{T}_1\right)\mathbf{q_1} - \boldsymbol{\rho}\mathbf{A}\mathbf{T}_1\ddot{\mathbf{q}}_1 = -\mathbf{p_x} - \mathbf{EA}\left[\left(\mathbf{T}_{2xx}\mathbf{q}_2 + \mathbf{t}_{2xx}\right)\right]_{dg.}\left(\mathbf{T}_{2x}\mathbf{q}_2 + \mathbf{t}_{2x}\right) - \\ -\mathbf{EA}\left[\left(\mathbf{T}_{3xx}\mathbf{q}_3 + \mathbf{t}_{3xx}\right)\right]_{dg.}\left(\mathbf{T}_{3x}\mathbf{q}_3 + \mathbf{t}_{3x}\right) + \bar{\mathbf{K}}_x\mathbf{t}_1\end{aligned} \tag{1.40}$$

where $\mathbf{EA}$, $\boldsymbol{\rho}\mathbf{A}$ are $L \times L$ diagonal matrices including the values of the corresponding quantities at the L nodal points and $\bar{\mathbf{K}}_x$ is a diagonal matrix similar with those of (1.37) whose diagonal elements are given as

$$\left(\bar{\mathbf{K}}_x\right)_{i,i} = (k_x)_i\,(U_u)_i \tag{1.41}$$

Moreover, substituting (1.34) and (1.40) in (1.11a) the discretized counterpart of the axial force at the neutral axis of the beam is given as

$$\begin{aligned}\mathbf{N} = \mathbf{EA}\left(\mathbf{T}_{1x}\mathbf{q}_1 + \mathbf{t}_{1x}\right) + \frac{1}{2}\mathbf{EA}\left[\left(\mathbf{T}_{2x}\mathbf{q}_2 + \mathbf{t}_{2x}\right)\right]_{dg}\left(\mathbf{T}_{2x}\mathbf{q}_2 + \mathbf{t}_{2x}\right) + \\ +\frac{1}{2}\mathbf{EA}\left[\left(\mathbf{T}_{3x}\mathbf{q}_3 + \mathbf{t}_{3x}\right)\right]_{dg}\left(\mathbf{T}_{3x}\mathbf{q}_3 + \mathbf{t}_{3x}\right)\end{aligned} \tag{1.42}$$

Equations (1.35a), (1.35b), (1.40) and (1.42) constitute a nonlinear coupled system of equations with respect to $\mathbf{q_1}$, $\mathbf{q_2}$, $\mathbf{q_3}$ and $\mathbf{N}$ quantities. The solution of this system is accomplished iteratively by employing the average acceleration method in combination with the modified Newton Raphson method [18, 19].

1.4 For the Stress Functions Θ (*y*,*z*) and Θ (*y*,*z*)

The evaluation of the stress functions $\Theta(y,z)$ and $\Phi(y,z)$ is accomplished using BEM as this is presented in Sapountzakis and Mokos [28]. Moreover, since the nonlinear the nonlinear flexural dynamic problem of Timoshenko beam-columns is solved by the BEM, the domain integrals for the evaluation of the area, the bending moments of inertia (1.12) and the shear deformation coefficients (1.26) have to be converted to boundary line integrals, in order to maintain the pure boundary character of the method. This can be achieved using integration by parts, the Gauss theorem and the Green identity. Thus, the moments, the product of inertia and the cross section area can be written as

$$I_y = \int_{\Gamma}\left(yz^2 n_y\right) ds \tag{1.43a}$$

$$I_z = \int_\Gamma \left(zy^2 n_z\right) ds \tag{1.43b}$$

$$A = \frac{1}{2} \int_\Gamma \left(y n_y + z n_z\right) ds \tag{1.43c}$$

while the shear deformation coefficients a_y and a_z are obtained from the relations

$$a_y = \frac{A}{\Delta^2} \left((4v+2) I_y I_{\Theta y} + \frac{1}{4} v^2 I_{yy}^2 I_{ed} - I_{\Theta e} \right) \tag{1.44a}$$

$$a_z = \frac{A}{\Delta^2} \left((4v+2) I_z I_{\Phi z} + \frac{1}{4} v^2 I_z^2 I_{ed} - I_{\Phi d} \right) \tag{1.44b}$$

where

$$I_{\Theta e} = \int_\Gamma \Theta \left(\boldsymbol{n} \cdot \boldsymbol{e}\right) ds \tag{1.45a}$$

$$I_{\Phi d} = \int_\Gamma \Phi \left(\boldsymbol{n} \cdot \boldsymbol{d}\right) ds \tag{1.45b}$$

$$I_{ed} = \int_\Gamma \left(y^4 z n_z + z^4 y n_y + \frac{2}{3} y^2 z^3 n_z \right) ds \tag{1.45c}$$

$$I_{\Theta y} = \frac{1}{6} \int_\Gamma \left[-2 I_{yy} y^4 z n_z + \left(3\Theta n_y - y\left(n \cdot e\right)\right) y^2 \right] ds \tag{1.45d}$$

$$I_{\Phi z} = \frac{1}{6} \int_\Gamma \left[-2 I_{zz} z^4 y n_y + \left(3\Phi n_z - z\left(n \cdot d\right)\right) z^2 \right] ds \tag{1.45e}$$

1.5 Numerical Examples

On the basis of the analytical and numerical procedures presented in the previous sections, a computer program has been written and representative examples have been studied to demonstrate the efficiency, wherever possible the accuracy and the range of applications of the developed method. In all the examples treated the results have been obtained using $L = 41$ nodal points along the beam-column and a time step of $\Delta t = 1.0\mu$s.

1.5.1 Example 1

For comparison reasons, the linear dynamic analysis of a simply supported uniform beam-column of length $l = 6.096$ m ($E = 24.82$ GPa, $\rho = 3,387 \text{Kg/m}^3$, $v = 0.3$, $I = 143.9 \times 10^{-5} \text{m}^4$) resting on a homogeneous (either bilateral or unilateral) elastic foundation with modulus of subgrade reaction $k_z = 16.55 \text{MN/m}^2$, as this is shown in Fig. 1.2 is examined. The free vibrations case of this example was analyzed by

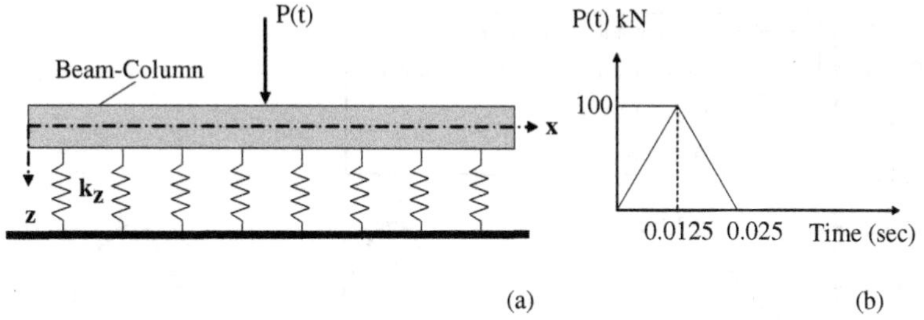

Fig. 1.2 Prismatic beam-column on elastic foundation (**a**) subjected to a triangular impulsive load (**b**)

Table 1.1 First five natural frequencies (Hz) of the simply supported beam of example 1

Modes	Timoshenko et al. [31]	Lai et al. [32]	Thambiratnam and Zhuge [33]	Friswell et al. [34]	Calim [35]	ANSYS [36]	Present study
1	32.9032	32.9049	32.9033	32.8980	32.8633	32.8624	32.7946
2	56.8135	56.8220	56.8193	56.8080	56.5972	56.5891	56.5476
3	112.908	111.973	111.961	111.900	110.759	110.739	110.722
4	–	–	–	193.760	189.939	189.901	189.489
5	–	–	–	–	222.078	222.043	222.077

Timoshenko et al. [31]. Lai et al. [32], Thambiratnam and Zhuge [33] and Friswell et al. [34] while the forced vibrations one by Calim [35]. The beam-column is subjected to a triangular impulsive load of amplitude $P_{l/2} = 100\,k\text{N}$ at its midpoint, as this is shown in Fig. 1.2.

In Table 1.1, the evaluated first five natural frequencies of the beam-column resting on the bilateral elastic foundation are presented as compared with those obtained from the literature. In Figs. 1.3, 1.4 and 1.5 the time history of the transverse displacement $w\,(l/2)$ at the beam-column's midpoint, of the bending moment $M_y\,(l/2)$ at the same point and of the shear force $Q_z\,(l)$ at the right supported end, respectively are presented either for the bilateral or the unilateral elastic foundation model and compared with those obtained from a complementary functions method and a FEM solution demonstrating the accuracy of the results of the proposed method. Moreover, in Table 1.2 the extreme values of the displacement $w\,(l/2)$ and of the soil reaction $p_{sz}(l/2)$ at the beam-column's midpoint are also presented for both cases of bilateral and unilateral soil reaction.

1.5.2 Example 2

In order to illustrate the importance of the nonlinear analysis and the influence of the shear deformation effect in flexural vibrations, a clamped beam-column of length $l = 4.90$ m, having a hollow rectangular cross section ($E = 210$ GPa, $v = 0.3$,

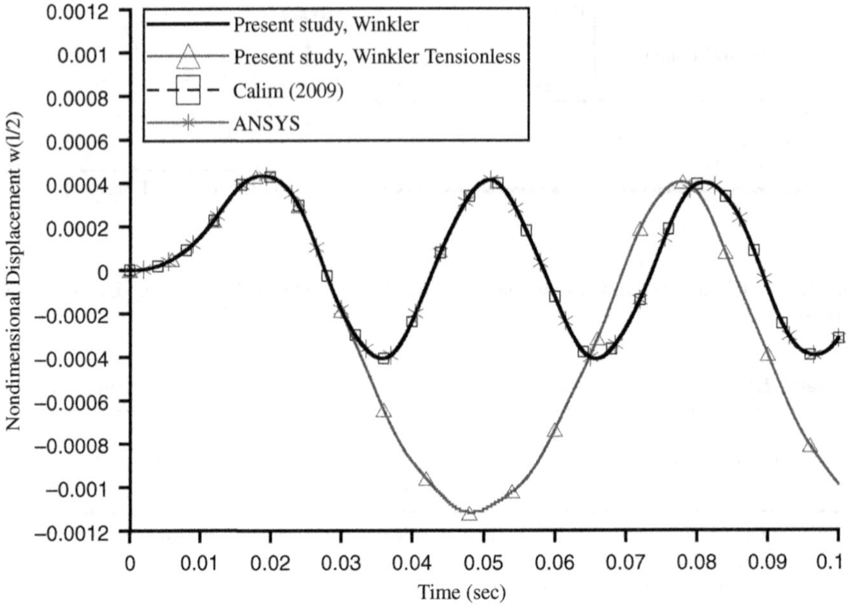

Fig. 1.3 Time history of the transverse displacement $w\,(l/2)$ at the midpoint of the beam-column of example 1

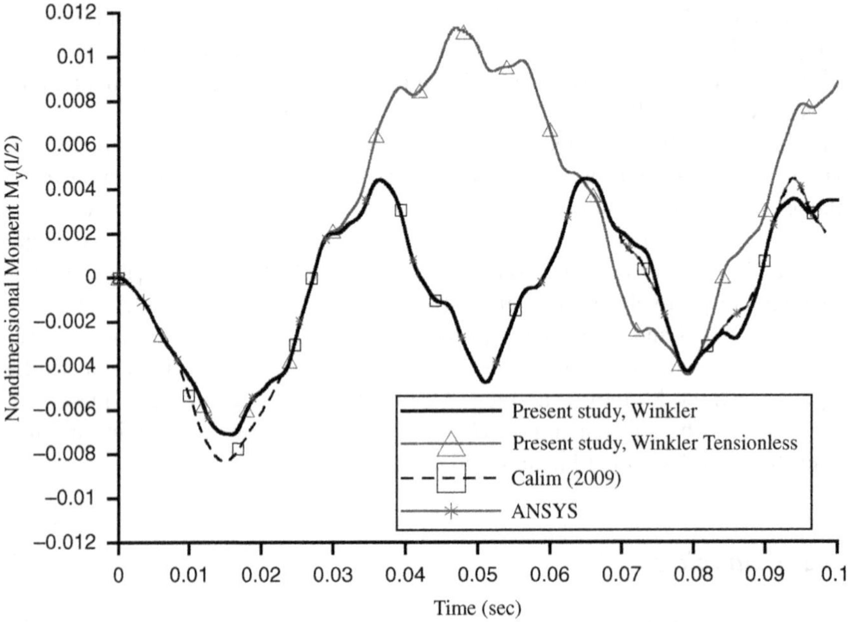

Fig. 1.4 Time history of the bending moment $M_y\,(l/2)$ at the midpoint of the beam-column of example 1

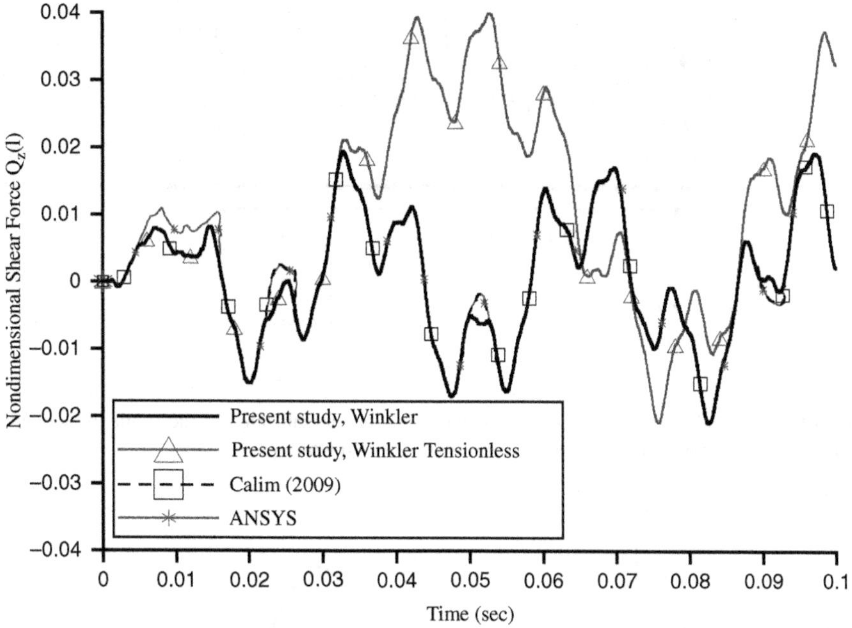

Fig. 1.5 Time history of the shear force Q_z (l) at the right supported end of the beam-column of example 1

Table 1.2 Extreme values of the displacement w $(l/2)$ and the foundation reaction $p_{sz}(l/2)$ of the beam-column of example 1

	Winkler		Winkler tensionless	
	w $(l/2)$ m	$p_{sz}(l/2)$ kN	w $(l/2)$ m	$p_{sz}(l/2)$ kN
Max	$2.63\times1`0^{-3}$	43.6	2.63×10^{-3}	43.6
Min	-2.50×10^{-3}	−41.5	-6.80×10^{-3}	0.00

$a_z = 3.664, a_y = 1.766$, $\rho = 7.85$ tn/m^3) resting on a homogeneous (either bilateral or unilateral) elastic foundation of stiffness k_z, as this is shown in Fig. 1.6 is examined.

In Figs. 1.7 and 1.8 the deflection w along the beam-column subjected to a suddenly applied consecrated bending moment $M_y = 200$ kNm at its midpoint is presented at the time instant $t = 1.6 \cdot 10^{-2}$s for various values of the stiffness k_z for the cases of bilateral and unilateral soil reaction, respectively. The influence of both the foundation stiffness parameter k_z and the unilateral character of the soil reaction are easily verified. Moreover, in Fig. 1.9 the time history of the central transverse deflection w $(l/2)$ and in Table 1.3 the maximum central deflection $w_{\max}$ (m) and the period T_z (s) of the first cycle of motion of the beam-column additionally subjected

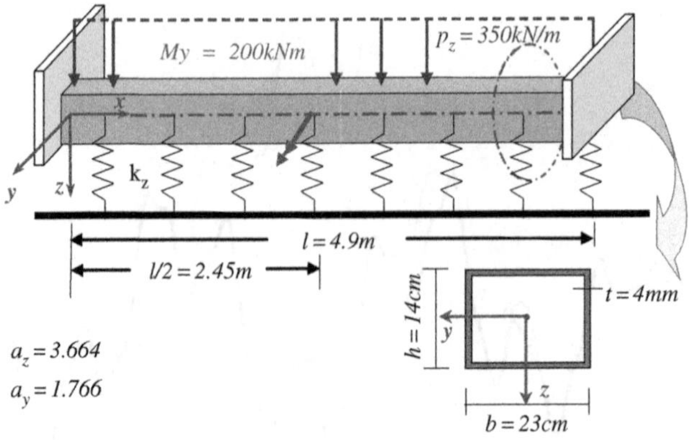

Fig. 1.6 Clamped beam of hollow rectangular cross section subjected to the suddenly applied concentrated bending moment M_y and uniformly distributed load p_z

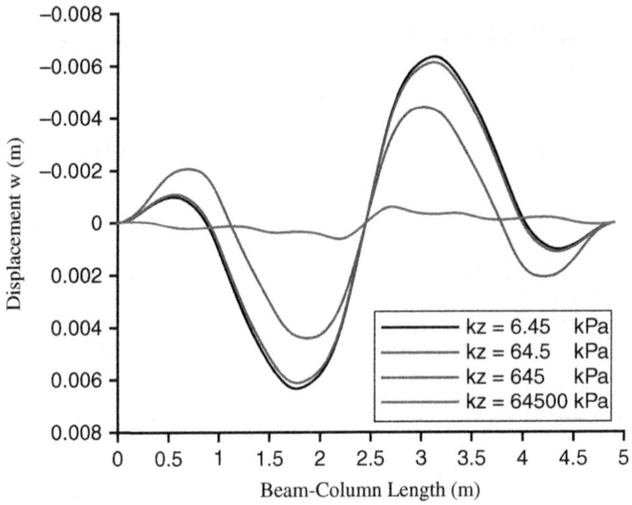

Fig. 1.7 Deflection w along the beam-column of example 2, for various stiffness k_z values of the bilateral Winkler springs

to a uniformly distributed load $p_z = 350$ kN/m (Fig. 1.6) is presented for a unilateral subgrade model with $k_z = 645.0$ kPa, performing either a linear or a nonlinear analysis and taking into account or ignoring both shear deformation effect and rotary inertia. From the obtained results, the discrepancy between the linear and the nonlinear analysis is not negligible and should not be ignored, while the significant influence of the shear deformation effect increasing both central transverse displacement and the obtained period of the first cycle of motion is remarked in both linear and nonlinear analysis.

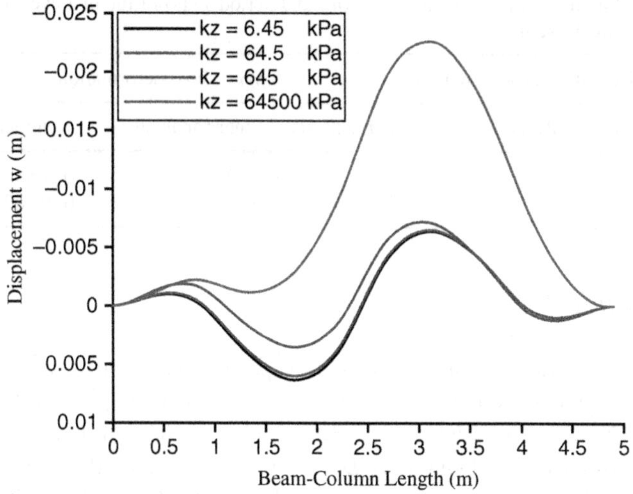

Fig. 1.8 Deflection w along the beam-column of example 2, for various stiffness k_z values of the unilateral Winkler springs

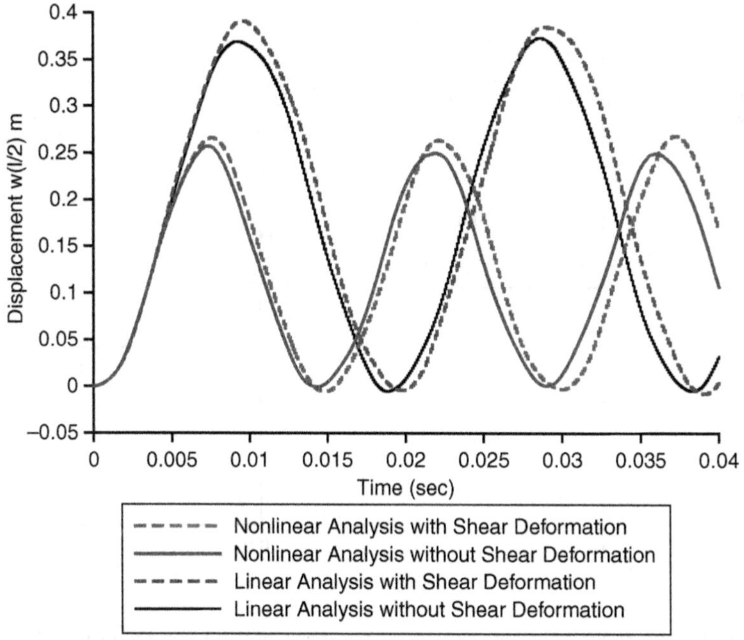

Fig. 1.9 Time history of the central deflection $w\,(l/2)$ of the beam-column of example 2, for a unilateral subgrade model with $k_z = 645.0\,\text{kPa}$

Table 1.3 Maximum central deflection w_{max} (m) and period T_z (s) of the first cycle of motion of the clamped beam of example 2

	Without shear deformation		With shear deformation	
	Linear analysis	Nonlinear analysis	Linear analysis	Nonlinear analysis
w_{max}	0.3729	0.2572	0.3914	0.2688
T_z	0.01890	0.01482	0.01973	0.01607

1.5.3 Example 3

To demonstrate the range of applications of the proposed method, a partially embedded pile in a homogeneous elastic Winker foundation with spring stiffness $k_z = k_y = 85.0\ \text{MN/m}^2$, of total length $l = 15.0$ m ($l_{\text{free}} = 6.20$ m, $l_{\text{embed}} = 8.80$ m), of circular cross section of diameter $D = 1.0$ m ($E = 29$ GPa, $A = 0.785\ \text{m}^2$, $v = 0.2$, $I_y = I_z = 0.049\ \text{m}^4$, $a_y = a_z = 1.172$), as this is shown in Fig. 1.10 has been studied. According to its boundary conditions, the pile end at the elastic foundation is clamped, while the other end is free according to its displacements and blocked according to its rotations. The pile is subjected to a suddenly applied concentrated axial load $P_x(0, t) = 1.0$ MN, ($t \geq 0.0$) and to a uniformly distributed transverse load $p_y\ (t) = 500$ kN/m, ($t \geq 0.0$) acting to the free part of the length of the pile.

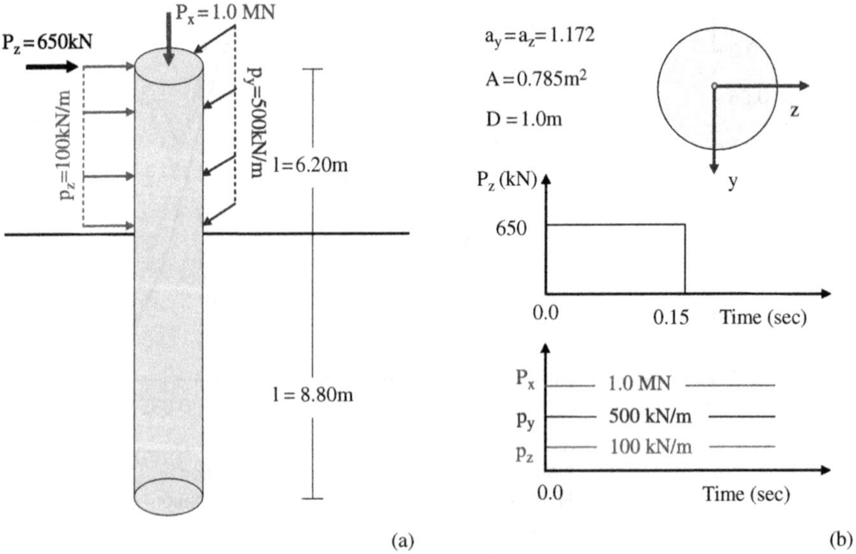

Fig. 1.10 Pile of circular cross section in axial – flexural loading subjected to rectangular impulsive concentrated load P_z, concentrated axial load P_x and to uniformly distributed loading p_y, p_z

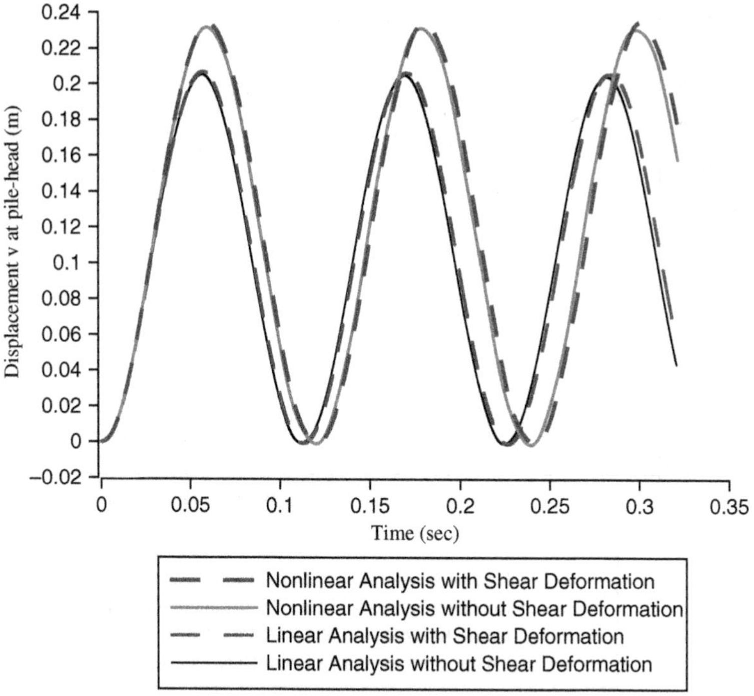

Fig. 1.11 Time history of the transverse displacement v_{top} of the head of the pile of example 3

In Figs. 1.11 and 1.12 the time history of the head displacement v_{top} of the pile and the deflection v along the pile at the time instant $t = 7.0 \cdot 10^{-2}$s are presented, respectively performing either a linear or a nonlinear analysis, taking into account or ignoring both rotary inertia and shear deformation effect. Moreover, in Table 1.4 the maximum value of the head displacement $(v_{top})_{max}$ and the period T_y of the first-cycle of motion are presented for the aforementioned cases. Moreover, the examined pile additionally to the aforementioned loading is also subjected to a uniformly distributed load $p_z = 100$ kN/m at its free length and to a suddenly applied concentrated load $P_z(t) = 650$ kN, $(0.0 \leq t \leq 0.15)$ acting at its top, as this is shown in Fig. 1.10.

In Fig. 1.13 the time history of the head displacement of the pile w_{top} performing either a linear or a nonlinear analysis is presented taking into account or ignoring shear deformation effect. Finally, in Table 1.5 the maximum values of the head displacements $(v_{top})_{max}$, $(w_{top})_{max}$ and the periods T_y, T_z(μs) of the first cycle are presented for the same cases of analysis. At this point it is worth noting that the small discrepancy of the head deflections $(v_{top})_{max}$ between the Tables 1.4, 1.5 is due to the coupling effect of the transverse displacements in y, z directions in the nonlinear analysis.

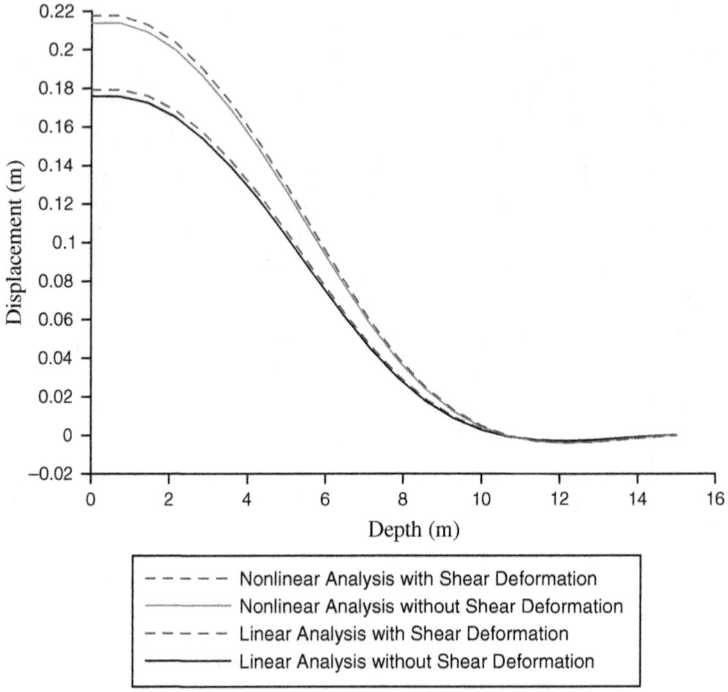

Fig. 1.12 Displacement v at time instant $t = 7.0 \times 10^{-2}$ along the pile of example 3

Table 1.4 Maximum head deflection $(v_{top})_{max}$ and period T_y (s) of the first cycle of motion of the pile of example 3

	Without shear deformation		With shear deformation	
	Linear analysis	Nonlinear analysis	Linear analysis	Nonlinear analysis
$(v_{top})_{max}$	0.2054	0.2321	0.2070	0.2351
T_y	0.1105	0.1180	0.1111	0.1201

1.5.4 Example 4

In this example, a fully embedded in stiff cohesive soil with non constant stiffness free head pile of length $l = 8.0$ m of a hollow circular cross section ($E = 210$ GPa, $a_y = a_z = 2.226$, $\rho = 7.85\text{tn/m}^3$, $v = 0.3$), as this is shown in Fig. 1.14 is examined. The pile is subjected to a concentrated axial $P_x(t) = 500\,\text{kN}$, $(t \geq 0.0)$ and transverse $P_z(t) = 750 \cdot \cos(\omega t)$ kN loading acting at the free pile head, where $\omega = 614.329$ rad/s is the first natural frequency of the pile-soil system. In Figs. 1.15 and 1.16 the time history of the pile head displacement w_{top} and the deflection w along the pile at the time instant $t = 4 \cdot 10^{-2}$ s are presented, respectively, performing either a linear or a nonlinear analysis and taking into account or ignoring both rotary

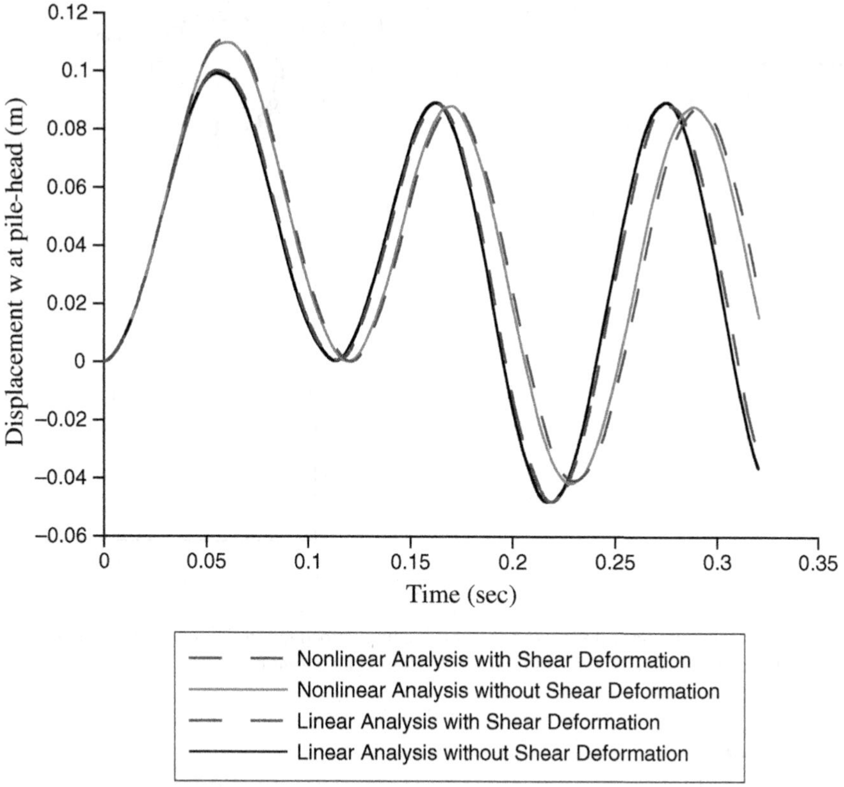

Fig. 1.13 Time history of the transverse displacement w_{top} of the head of the pile of example 3

Table 1.5 Maximum head displacements $(v_{top})_{max}$, $(w_{top})_{max}$ and periods T_y, T_z (s) of the first cycle of the examined pile of example 3

	Without shear deformation		With shear deformation	
	Linear analysis	Nonlinear analysis	Linear analysis	Nonlinear analysis
$(v_{top})_{max}$	0.2054	0.2320	0.2070	0.2353
$(w_{top})_{max}$	0.0992	0.1109	0.1002	0.1111
T_y	0.1105	0.1172	0.1111	0.1192
T_z	0.1133	0.1179	0.1143	0.1215

inertia and shear deformation effect. The discrepancy between linear and nonlinear analysis in the resonance case is remarkable and is justified from the varying with time stiffness of the pile-soil system in the case of a nonlinear analysis. Moreover, to demonstrate this discrepancy, in Fig. 1.17 the time history of the head displacement w_{top} of the pile is presented for the time interval 0.0–0.05 s.

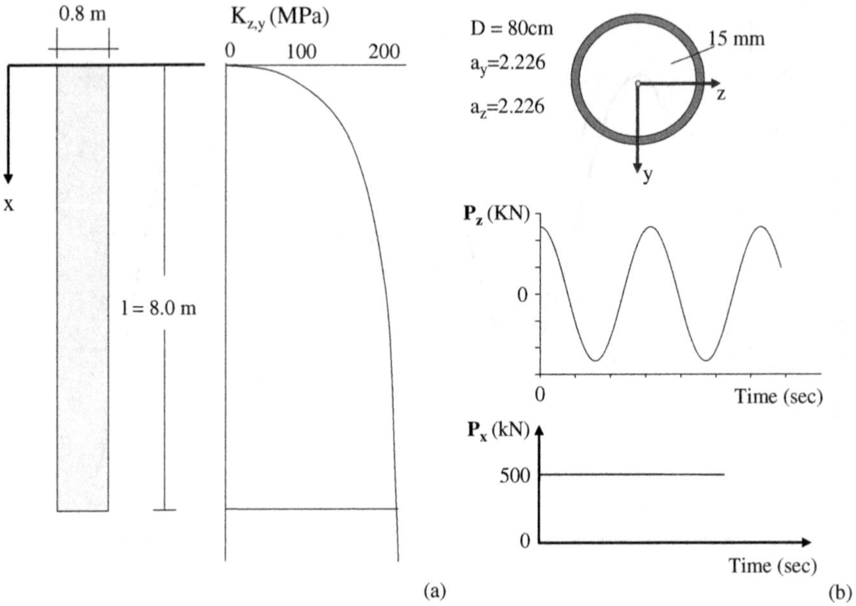

Fig. 1.14 Hollow circular pile in axial-flexural loading embedded in non constant stiffness soil, subjected to concentrated axial P_x and transverse P_z loading at pile-head

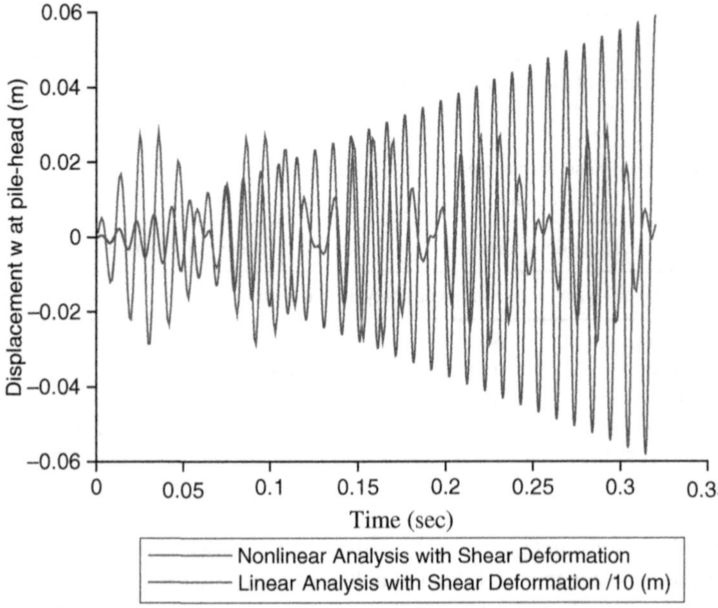

Fig. 1.15 Time history of the transverse displacement w_{top} of the head of the pile of example 4 (for graphic purposes displacements coming from linear analysis are divided by 10)

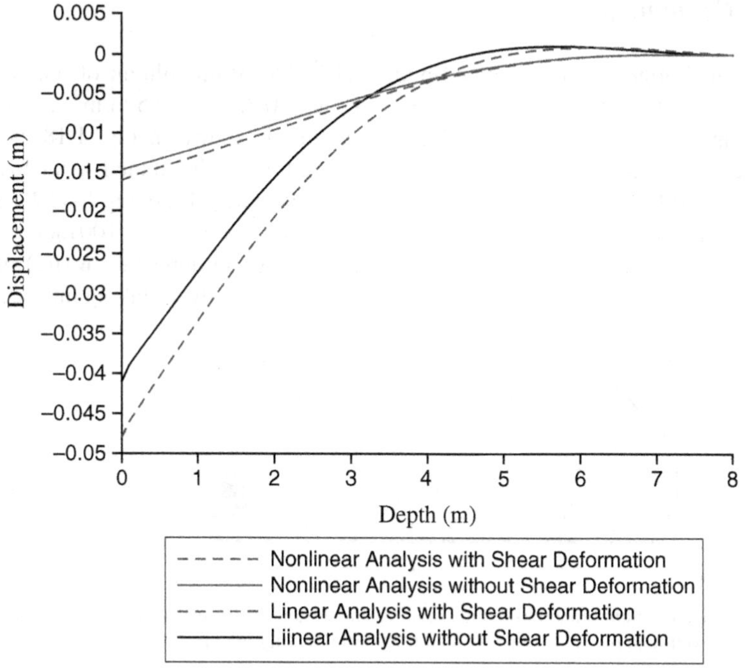

Fig. 1.16 Displacement w at time instant $t = 4.0 \times 10^{-2}$s along the pile of example 4

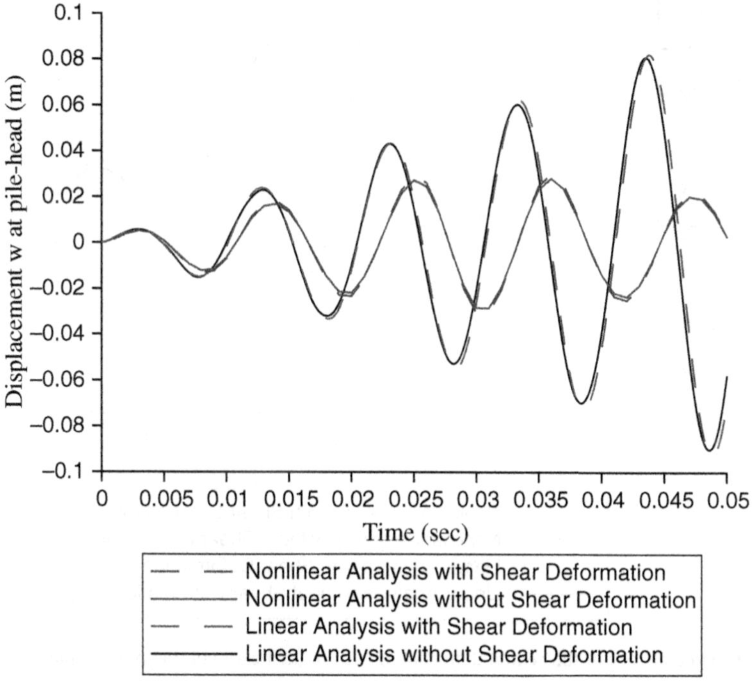

Fig. 1.17 Time history of the transverse displacement w_{top} of the head of the pile of example 4

1.5.5 Example 5

As the final application a clamped-pinned HEA320 beam-column of length $l = 6.5$ m ($E = 210$ GPa, $a_y = 1.475, a_z = 4.512$, $v = 0.3$, $\rho = 7.85$ tn/m^3) resting on a constant stiffness soil of $k_z = 1.2$ MPa, as this is shown in Fig. 1.18, is considered. The beam-column is subjected to a uniformly distributed axial loading $p_x(t) = 500$ kN, ($t \geq 0.0$) and to a transverse concentrated moving load $P_z(t) = 10 \cdot \sin(\omega t)$ MN with constant velocity of $v = 65$ m/s, where $\omega = 100$ rad/s.

In Fig. 1.19 the time history of the central transverse deflection $w(l/2)$ of the beam-column, performing either a linear or a nonlinear analysis, taking into account

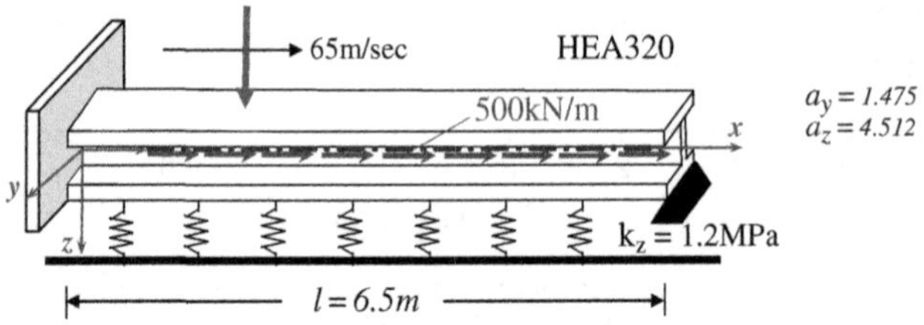

Fig. 1.18 Beam-column in axial-flexural loading on constant stiffness soil subjected to uniformly distributed axial load p_x and to concentrated transverse moving load P_z

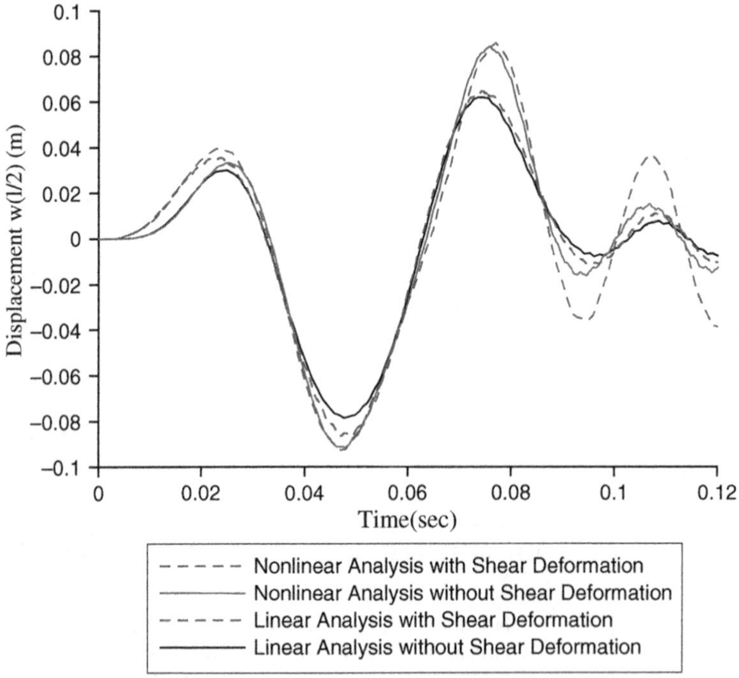

Fig. 1.19 Time history of the central deflection $w(l/2)$ of the beam-column of example 5

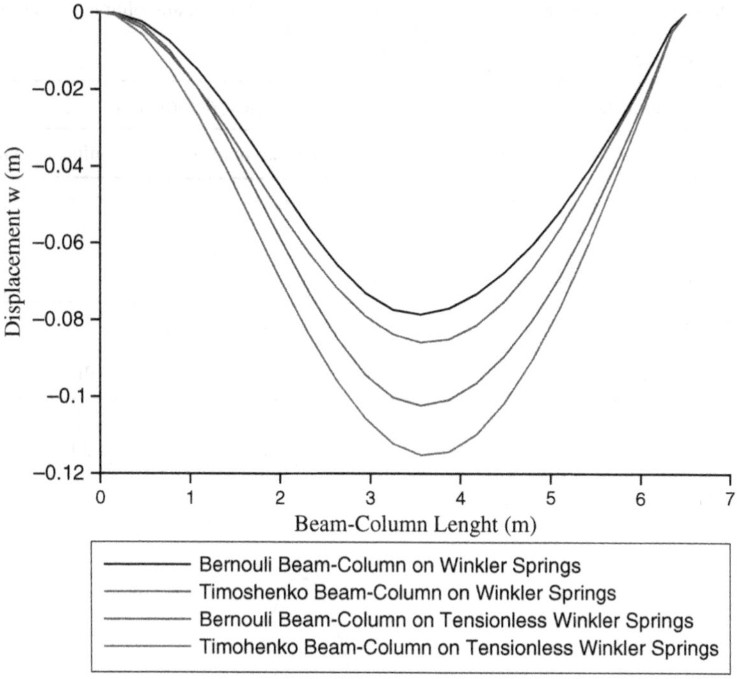

Fig. 1.20 Displacements w at time instant $t = 0.05$ s along the beam – column of example 5

or ignoring both rotary inertia and shear deformation effect is presented. In Fig. 1.20 the displacement w at the time instant $t = 0.05$ s along the beam-column either for conventional Winkler or for tensionless Winkler soil are presented performing a nonlinear analysis and taking into account or ignoring shear deformation effect. Finally, in Table 1.6 the deflection extreme values $w_{\max}$, $w_{\min}$ and in Fig. 1.21 the effect of the frequency ω of the concentrated moving load to the maximum deflection $w_{\max}$ are presented for all of the aforementioned cases.

1.6 Concluding Remarks

A boundary element method is developed for the nonlinear dynamic analysis of beam-columns of arbitrary doubly symmetric simply or multiply connected constant cross section, partially supported on tensionless Winkler foundation, undergoing moderate large deflections under general boundary conditions, taking into account the effects of shear deformation and rotary inertia. The beam-column is subjected to the combined action of arbitrarily distributed or concentrated transverse loading and bending moments in both directions as well as to axial loading. The proposed model takes into account the coupling effects of bending and shear deformations

Table 1.6 Extreme values of the displacement $w \cdot 10^{-2}$ (m) of the beam-column of example 5

	Winkler			
	Without shear deformation		With shear deformation	
	Linear analysis	Nonlinear analysis	Linear analysis	Nonlinear analysis
w_{max}	6.21	8.43	6.45	8.58
w_{min}	−7.87	−9.15	−8.17	−9.26
	Tensionless winkler			
	Without shear deformation		With shear deformation	
	Linear analysis	Nonlinear analysis	Linear analysis	Nonlinear analysis
w_{max}	5.92	6.84	6.67	6.98
w_{min}	−10.18	−10.96	−11.46	12.34

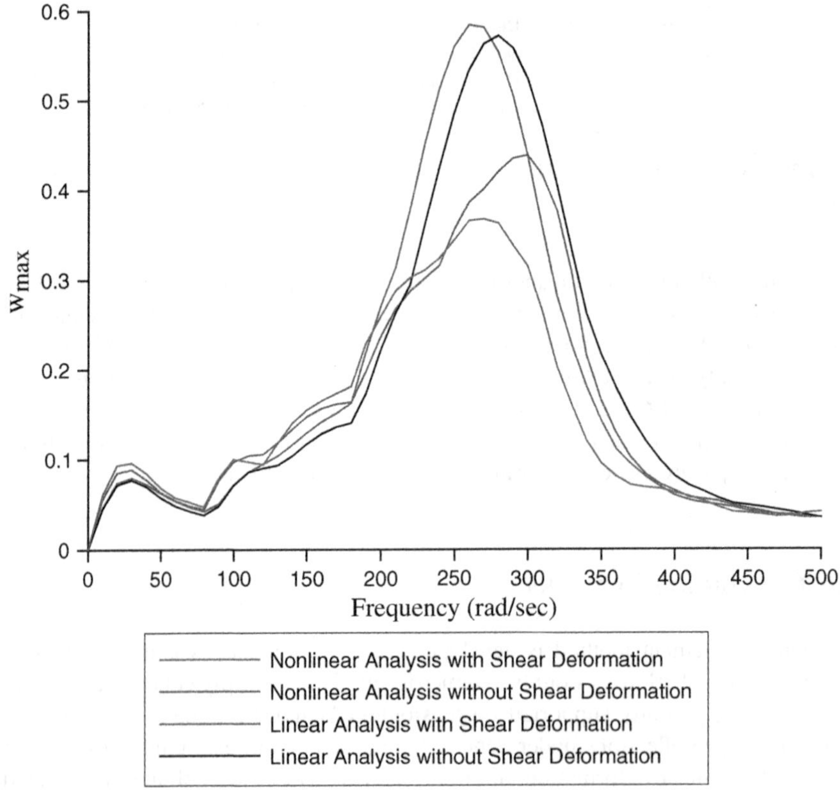

Fig. 1.21 Frequency effect to the maximum deflection of the beam – column of example 5

along the member as well as the shear forces along the span induced by the applied axial loading. The main conclusions that can be drawn from this investigation are

1. The numerical technique presented in this investigation is well suited for computer aided analysis for beams of arbitrary simply or multiply connected doubly symmetric cross section.
2. In some cases, the effect of shear deformation is significant, especially for low beam slenderness values, increasing both the maximum transverse displacements and the calculated periods of the first cycle of motion.
3. The discrepancy between the results of the linear and the nonlinear analysis is remarkable.
4. The effect of both the tensionless character and the stiffness of the soil is important.
5. The discrepancy in the response of a beam-column in the resonance case, performing a linear or a nonlinear analysis is remarkable.
6. The coupling effect of the transverse displacements in both directions in the nonlinear analysis influences these displacements.

Acknowledgements The work of this paper was conducted from the "DARE" project, financially supported by a European Research Council (ERC) Advanced Grant under the "Ideas" Programme in Support of Frontier Research [Grant Agreement 228254].

References

1. Hetenyi, M.: Beams and plates on elastic foundations and related problems. Appl. Mech. Rev. **19**, 95–102 (1966)
2. Rades, M.: Dynamic analysis of an inertial foundation model. Int. J. Solids Struct. **8**, 1353–1372 (1972)
3. Wang, T., Stephens, J.: Natural frequencies of timoshenko beams on Pasternak foundation. J. Sound Vib. **51**(2), 149–155 (1977)
4. De Rosa, M.: Free vibrations of Timoshenko beams on two-parameter elastic foundation. Comput. Struct. **57**(1), 151–156 (1995)
5. El-Mously, M.: Fundamental frequencies of Timoshenko beams mounted on Pasternak foundation. J. Sound Vib. **228**(2), 452–457 (1999)
6. Zhang, Y., et al.: Vibration and buckling of a double-beam system under compressive axial loading. J. Sound Vib. **318**, 341–352 (2008)
7. Coskun, I.: The response of a finite beam on a tensionless pasternak foundation subjected to a harmonic load. Eur. J. Mech. A/Solids **212**, 151–161 (2003)
8. Matsunaga, H.: Vibration and buckling of deep beam-columns on two-parameters elastic foundation. J. Sound Vib. **228**(2), 359–376 (1999)
9. Coskun, I.: Non-linear vibrations of a beam resting on a tensionless Winkler foundation. J. Sound Vib. **236**(3). 401–411 (2000)
10. Chen, C.: DQEM vibration analyses of non-prismatic shear deformable beams resting on elastic foundations. J. Sound Vib. **255**(5), 989–999 (2002)
11. Chen, W., et al.: A mixed method for bending and free vibration of beams resting on a Pasternak elastic foundation. Appl. Math. Model. **28**, 877–890 (2004)
12. Kim, S., Cho, Y.H.: Vibration and dynamic buckling of shear beam-columns on elastic foundation under moving harmonic loads. Int. J. Solids Struct. **43**, 393–412 (2006)

13. Yokoyama, T.: Parametric instability of Timoshenko beams resting on an elastic foundation. Comput. Struct. **28**(2), 207–216 (1988)
14. Karamanlidis, D., Prakash, V.: Buckling and vibration analysis of flexible beams resting on an elastic half-space. Earthquake Eng. Struct. Dynam. **16**, 1103–1114 (1988)
15. Arboleda, M. et al.: Timoshenko beam-column with generalized end conditions on elastic foundation: dynamic-stiffness matrix and load vector. J. Sound Vib. **310**, 1057–1079 (2007)
16. Lewandowski, R.: Nonlinear free vibrations of multispan beams on elastic supports. Comput. Struct. **32**(2), 305–312 (1989)
17. Katsikadelis, J.: The analog equation method. A boundary-only integral equation method for nonlinear static and dynamic problems in general bodies. Theor. Appl. Mech. **27**, 13–38 (2002)
18. Chang, S.: Studies of Newmark method for solving nonlinear systems: (I) basic analysis. J. Chinese Inst. Eng. **27**(5), 651–662 (2004)
19. Isaacson, E., Keller, H.B.: Analysis of Numerical Methods. Wiley, New York, NY (1966)
20. Timoshenko, S., Goodier, J.: Theory of Elasticity, 3rd edn. McGraw-Hill, New York, NY (1984)
21. Cowper, G.: The shear coefficient in Timoshenko's beam theory. J. Appl. Mech. ASME. **33**(2), 335–340 (1966)
22. Schramm, U., et al.: On the shear deformation coefficient in beam theory. Finite Elem. Anal. Des. **16**, 141–162 (1994)
23. Schramm, U., et al.: Beam stiffness matrix based on the elasticity equations. Int. J. Numer. Methods Eng. **40**, 211–232 (1997)
24. Stephen, N.: Timoshenko's shear coefficient from a beam subjected to gravity loading. ASME J. Appl. Mech. **47**, 121–127 (1980)
25. Hutchinson, J.R.: Shear coefficients for Timoshenko beam theory. ASME J. Appl. Mech. **68**, 87–92 (2001)
26. Ramm, E., Hofmann, T.: Stabtragwerke: Der Ingenieurbau. In: Mehlhorn, G. (ed.) Band Baustatik/Baudynamik. Ernst & Sohn, Berlin (1995)
27. Rothert, H., Gensichen, V.: Nichtlineare Stabstatik. Springer, Berlin (1987)
28. Sapountzakis, E., Mokos, V.: A BEM solution to transverse shear loading of beams. Comput. Mech. **36**, 384–397 (2005)
29. Thomson, W.: Theory of Vibration with Applications. Prentice–Hall, Englewood Cliffs, NJ (1981)
30. Sapountzakis, E., Katsikadelis, J.: Analysis of plates reinforced with beams. Comput. Mech. **26**, 66–74 (2000)
31. Timoshenko, S., et al.: Vibration Problems in Engineering, 4th edn. Wiley, New York, NY (1974)
32. Lai, Y., et al.: Dynamic response of beams on elastic foundation. Int. J. Struct. Eng. **118**, 853–858 (1992)
33. Thambiratnam, D., Zhuge, Y.: Free vibration analysis of beams on elastic foundation. Comput. Struct. **60**(6), 971–980 (1996)
34. Friswell, M., et al.: Vibration analysis of beams with non-local foundation using the finite element method. Int. J. Numer. Methods Eng. **71**(11), 1365–1386 (2007)
35. Calim, F.: Dynamic analysis of beams on viscoelastic foundation. Eur. J. Mech. A/Solids **28**, 469–476 (2009)
36. ANSYS Swanson Analysis System, Inc., 201 Johnson Road, Houston, PA 15342–1300, USA

Chapter 2
Mechanics of Viscoelastic Plates Made of FGMs

H. Altenbach and V.A. Eremeyev

Abstract Considering the viscoelastic behavior of polymer foams a new plate theory based on the direct approach is introduced and applied to plates composed of functionally graded materials (FGM). The governing two-dimensional equations are formulated for a deformable surface, the viscoelastic effective stiffness parameters are identified assuming linear viscoelastic material behavior. The material properties are changing in the thickness direction. Solving some problems of the global structural analysis it will be demonstrated that in some cases the results significantly differ from the results based on the Kirchhoff-type theory. The aim of this paper is to extend the results of the analysis given in (ZAMM 88:332–341, 2008; Acta Mech 204:137–154, 2009; Key Eng Mater 399:63–70, 2009) related to the case of general linear viscoelastic behaviour and to discuss how the effective viscoelastic properties reflect the properties in the thickness direction.

Keywords Plates · Viscoelasticity · Functionally graded material · Foam

2.1 Introduction

Foams are a very perspective class of materials for modern engineering applications. Metallic and polymeric foams are applied as a material for lightweight structures which are used in civil engineering, in the automotive or aerospace industries since they combine low weight, high specific strength, and excellent possibilities to absorb energy, see [4, 5]. The technical realization is mostly performed as sandwich panels (plates or shells with hard and stiff face sheets and a core layer made of foam). From the mechanical point of view, foam is a very complex material, and it can be modeled with the help of three- or two-dimensional theories. Below the non-homogeneous foam is presented as a functionally graded material (FGM)

H. Altenbach (✉)
Lehrstuhl für Technische Mechanik, Zentrum für Ingenieurwissenschaften,
Martin-Luther-Universität Halle-Wittenberg, D-06099, Halle (Saale), Germany
e-mail: holm.altenbach@iw.uni-halle.de

J. Murín et al. (eds.), *Computational Modelling and Advanced Simulations*,
Computational Methods in Applied Sciences 24, DOI 10.1007/978-94-007-0317-9_2,

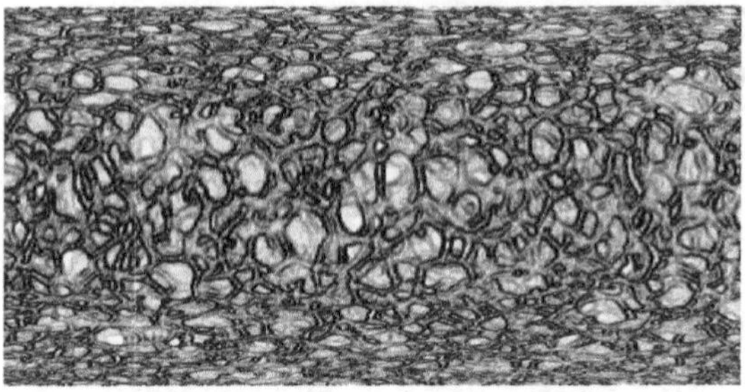

Fig. 2.1 Non-homogeneous foam

with "smeared" mechanical properties. By this way the changing over the thickness direction properties are substituted by effective stiffness parameters.

The analysis of plates and shell structures can be performed applying theories deduced by various approaches. Here we present a new theory of viscoelastic plates with changing properties in the thickness direction based on the direct approach in the plate theory and extended by the effective properties concept. We consider plates made of polymer foams with highly non-homogeneous structure through the thickness (see, for example Fig. 2.1). From the direct approach point of view a plate or a shell is modeled as a material surface each particle of which has five degrees of freedom (three displacements and two rotations, the rotation about the normal to plate is not considered as a kinematically independent variable, which corresponds to ignoring drilling moments effects). Such a model can be accepted in the case of plates with constant or slow changing thickness. For the linear variant of such theory the identification of the elastic stiffness tensors considering changing properties is proposed in [6–8], see also [9, 10]. Some extensions of the proposed theory of plates to the case of viscoelastic materials are given in [11] and applied to FGMs in [1–3].

Let us note that functionally graded plates and shells are investigated in many papers, see [12–17] among others. Different theories based on kinematical hypotheses or some mathematical treatments of the three-dimensional equations are presented. The suggested theories can be classified mostly as first order shear deformation theories or higher order theories. The first ones are based on the improvement of the strains introducing both independent translations and rotations of the points of the mid-plane of the plate while the second class of theories is very popular in the computational community. Both types of theories have advantages and disadvantages. Within the proposed theories the elastic, thermoelastic, and magneto- or electroelastic behaviour of the plate material are taken into account. In general, the viscoelastic properties of FGM plates are less considered see [1–3] and [18].

Here we focus our attention on the viscoelastic behavior. The dependence of the relaxation functions on the plate through-the-thickness structure as well as on the bulk material viscoelastic properties is analyzed. We show that for FGM plates the effective viscoelastic properties significantly depend on the bulk properties as functions of the thickness coordinate z. In particular, we discuss the influence of the plate geometry on the spectrum of the relaxation time. As a special case the viscoelastic behavior of a sandwich plate with a core made of a FGM is considered.

2.2 Governing Equations of a 5-Parametric Plate Theory

Let us consider the geometrically and physically linear theory of plates. The equations of motion are formulated as the Euler's laws of dynamics

$$\nabla \cdot \mathbf{T} + \mathbf{q} = \rho \ddot{\mathbf{u}} + \rho \boldsymbol{\Theta}_1 \cdot \ddot{\boldsymbol{\varphi}}, \tag{2.1}$$

$$\nabla \cdot \mathbf{M} + \mathbf{T}_{\times} + \mathbf{m} = \rho \boldsymbol{\Theta}_1^T . \ddot{\mathbf{u}} + \rho \boldsymbol{\Theta}_2 \cdot \ddot{\boldsymbol{\varphi}}. \tag{2.2}$$

Here $\mathbf{u}$ and $\boldsymbol{\varphi}$ are the vectors of displacements and rotations,

$$\mathbf{u} = u_1 \mathbf{e}_1 + u_2 \mathbf{e}_2 + w \mathbf{n}, \;\; \boldsymbol{\varphi} = -\varphi_2 \mathbf{e}_1 + \varphi_1 \mathbf{e}_2, \tag{2.3}$$

$\mathbf{n}$ is the unit outer normal vector at the plate surface, $\mathbf{e}_1$, $\mathbf{e}_2$ are orthonormal vectors in the tangent plane, $\mathbf{T}$ and $\mathbf{M}$ are the tensors of forces and moments,

$$\mathbf{T} = T_1 \mathbf{e}_1 \mathbf{e}_1 + T_2 \mathbf{e}_2 \mathbf{e}_2 + T_{12}(\mathbf{e}_1 \mathbf{e}_2 + \mathbf{e}_2 \mathbf{e}_1) + T_{1n} \mathbf{e}_1 \mathbf{n} + T_{2n} \mathbf{e}_2 \mathbf{n}, \tag{2.4}$$

$$\mathbf{M} = M_1 \mathbf{e}_1 \mathbf{e}_2 - M_2 \mathbf{e}_2 \mathbf{e}_1 - M_{12}(\mathbf{e}_1 \mathbf{e}_1 - \mathbf{e}_2 \mathbf{e}_2), \tag{2.5}$$

$\mathbf{q}$ and $\mathbf{m}$ are the vectors of surface loads (forces and moments), $\mathbf{T}_{\times}$ is the vector invariant of the force tensor, ∇ is the nabla (Hamilton) operator, $\boldsymbol{\Theta}_1$ and $\boldsymbol{\Theta}_2$ are the first and the second tensor of inertia, ρ is the density (effective property of the deformable surface), the upper index T denotes transposed, and $(\cdots)^{\cdot}$ the time derivative, respectively.

In the case of orthotropic behavior the following constitutive equations for the stress resultants are valid:

- In-plane forces

$$\mathbf{T} \cdot \mathbf{a} = \int\limits_{-\infty}^{t} \mathbf{A}(t-\tau) \cdot \cdot \dot{\boldsymbol{\mu}} \, d\tau + \int\limits_{-\infty}^{t} \mathbf{B}(t-\tau) \cdot \cdot \dot{\boldsymbol{\kappa}} \, d\tau, \tag{2.6}$$

- Transverse shear forces

$$\mathbf{T} \cdot \mathbf{n} = \int\limits_{-\infty}^{t} \boldsymbol{\Gamma}(t-\tau) \cdot \cdot \dot{\boldsymbol{\gamma}} \, d\tau, \tag{2.7}$$

- Moments

$$\mathbf{M}^T = \int\limits_{-\infty}^{t} \dot{\boldsymbol{\mu}} \cdot \cdot \mathbf{B}(t-\tau) \, d\tau + \int\limits_{-\infty}^{t} \mathbf{C}(t-\tau) \cdot \cdot \dot{\boldsymbol{\kappa}} \, d\tau, \tag{2.8}$$

where $\mathbf{A}$, $\mathbf{B}$, $\mathbf{C}$ are fourth-order tensors, and $\boldsymbol{\Gamma}$ is a second-order tensor. They describe the relaxation functions of the plate, $\boldsymbol{\mu}$, $\boldsymbol{\kappa}$, and $\boldsymbol{\gamma}$ are the tensor of in-plane strains, the tensor of the out-of-plane strains, and the vector of transverse shear strains, respectively. They are given by the relations

$$2\boldsymbol{\mu} = \nabla\mathbf{u} + (\nabla\mathbf{u})^T, \boldsymbol{\kappa} = \nabla\boldsymbol{\varphi}, \boldsymbol{\gamma} = \nabla\mathbf{u} - \mathbf{a}\times\boldsymbol{\varphi}. \tag{2.9}$$

Here $\mathbf{a}$ is the first metric tensor. The component form of (2.9) is given by

$$\boldsymbol{\mu} = \mu_{11}\mathbf{e}_1\mathbf{e}_1 + \mu_{22}\mathbf{e}_2\mathbf{e}_2 + \mu_{12}(\mathbf{e}_1\mathbf{e}_2 + \mathbf{e}_2\mathbf{e}_1),$$

$$\boldsymbol{\kappa} = \kappa_{11}\mathbf{e}_1\mathbf{e}_2 - \kappa_{12}\mathbf{e}_1\mathbf{e}_1 + \kappa_{21}\mathbf{e}_2\mathbf{e}_2 - \kappa_{22}\mathbf{e}_2\mathbf{e}_1,$$

$$\boldsymbol{\gamma} = \gamma_1\mathbf{e}_1 + \gamma_2\mathbf{e}_2.$$

In Cartesian coordinates the constitutive equations are

$$T_{\alpha\beta} = \int_{-\infty}^{t} A_{\alpha\beta\gamma\delta}(t-\tau)\,\dot{\mu}_{\gamma\delta}(\tau)\,d\tau + \int_{-\infty}^{t} B_{\alpha\beta\gamma\delta}(t-\tau)\,\dot{\kappa}_{\gamma\delta}(\tau)d\tau,$$

$$T_{\alpha 3} = \int_{-\infty}^{t} \Gamma_{\alpha\beta}(t-\tau)\,\dot{\gamma}_{\beta}(\tau)d\tau,$$

$$M_{\alpha\beta} = \int_{-\infty}^{t} B_{\alpha\beta\gamma\delta}(t-\tau)\,\dot{\mu}_{\gamma\delta}(\tau)\,d\tau + \int_{-\infty}^{t} C_{\alpha\beta\gamma\delta}(t-\tau)\,\dot{\kappa}_{\gamma\delta}(\tau)\,d\tau,$$

with

$$\begin{aligned} \mu_{11} &= u_{1,1},\ \mu_{22} = u_{2,2},\ 2\mu_{12} = u_{1,2} + u_{2,1},\\ \kappa_{11} &= \varphi_{1,1},\ \kappa_{22} = \varphi_{2,2},\ \kappa_{12} = \varphi_{2,1},\ \kappa_{21} = \varphi_{1,2},\\ \gamma_1 &= w_{,1}+\varphi_1,\ \gamma_2 = w_{,2}+\varphi_2, \end{aligned}$$

where $T_{\alpha\beta}$ are the in-plane forces, $T_{\alpha 3}$ are the transverse shear forces, $M_{\alpha\beta}$ are the moments, $A_{\alpha\beta\gamma\delta}$, $B_{\alpha\beta\gamma\delta}$, $C_{\alpha\beta\gamma\delta}$ and $\Gamma_{\alpha\beta}$ are the relaxation functions for the plate, $\alpha,\beta,\gamma,\delta = 1,\ 2$, μ_{11}, μ_{22} are the in-plane normal strains, μ_{12} is the in-plane shear strain, γ_α are the transverse shear strains, κ_{11}, κ_{22} are the bending deformations and κ_{12} is the twist deformation.

For the orthotropic material behavior the effective relaxation tensors have the form

$$\begin{aligned} \mathbf{A} &= A_{11}\mathbf{a}_1\mathbf{a}_1 + A_{12}(\mathbf{a}_1\mathbf{a}_2 + \mathbf{a}_2\mathbf{a}_1) + A_{22}\mathbf{a}_2\mathbf{a}_2 + A_{44}\mathbf{a}_4\mathbf{a}_4,\\ \mathbf{B} &= B_{13}\mathbf{a}_1\mathbf{a}_3 + B_{14}\mathbf{a}_1\mathbf{a}_4 + B_{23}\mathbf{a}_2\mathbf{a}_3 + B_{24}\mathbf{a}_2\mathbf{a}_4 + B_{42}\mathbf{a}_4\mathbf{a}_2,\\ \mathbf{C} &= C_{22}\mathbf{a}_2\mathbf{a}_2 + C_{33}\mathbf{a}_3\mathbf{a}_3 + C_{34}(\mathbf{a}_3\mathbf{a}_4 + \mathbf{a}_4\mathbf{a}_3) + C_{44}\mathbf{a}_4\mathbf{a}_4,\\ \boldsymbol{\Gamma} &= \Gamma_1\mathbf{a}_1 + \Gamma_2\mathbf{a}_2 \end{aligned}$$

with $(\mathbf{a}_1;\mathbf{a}_2) = \mathbf{e}_1\mathbf{e}_1 \pm \mathbf{e}_2\mathbf{e}_2$, $(\mathbf{a}_3;\mathbf{a}_4) = \mathbf{e}_1\mathbf{e}_2 \mp \mathbf{e}_2\mathbf{e}_1$.

In the case of isotropic and symmetric over the thickness plates the effective relaxation tensors have a reduced structure

$$\mathbf{A} = A_{11}\mathbf{a}_1\mathbf{a}_1 + A_{22}(\mathbf{a}_2\mathbf{a}_2 + \mathbf{a}_4\mathbf{a}_4),\ \mathbf{B} = \mathbf{0},$$

$$\mathbf{C} = C_{22}\mathbf{a}_2\mathbf{a}_2 + C_{44}(\mathbf{a}_3\mathbf{a}_3 + \mathbf{a}_4\mathbf{a}_4),\ \boldsymbol{\Gamma} = \boldsymbol{\Gamma}\mathbf{a}.$$

2.3 Effective Properties

Using the same technique as for the elastic plates (see, for example [10]) below we compute the viscoelastic stiffness tensor components. Let us consider the three-dimensional viscoelastic constitutive equations

$$\boldsymbol{\sigma} = \int_{-\infty}^{t} \mathbf{R}(t-\tau)\cdot\cdot\dot{\boldsymbol{\varepsilon}}\, d\tau, \tag{2.10}$$

or in the inverse form

$$\boldsymbol{\varepsilon} = \int_{-\infty}^{t} \mathbf{J}(t-\tau)\cdot\cdot\dot{\boldsymbol{\sigma}}\, d\tau, \tag{2.11}$$

where $\boldsymbol{\sigma}$ and $\boldsymbol{\varepsilon}$ are the stress and strain tensors, $\mathbf{R}$ and $\mathbf{J}$ are the 4th order tensors of relaxation and creep functions, respectively.

Further we consider two cases:

- *Case 1.* Homogeneous plates – all properties are constant (no dependency of the thickness coordinate z).
- *Case 2.* Inhomogeneous plates (sandwich, multilayered, functionally graded) – all properties are functions of z only.

These means that in the general case $\mathbf{R}$ and $\mathbf{J}$ depend on the thickness coordinate z and on the time t.

Using the Laplace transform

$$\overline{f}(s) = \int_{-\infty}^{t} f(t)e^{-st}dt,$$

of a function $f(t)$one can write (2.10) and (2.11) as follows

$$\overline{\boldsymbol{\sigma}} = s\overline{\mathbf{R}}\cdot\cdot\overline{\boldsymbol{\varepsilon}}, \overline{\boldsymbol{\varepsilon}} = s\overline{\mathbf{J}}\cdot\cdot\overline{\boldsymbol{\sigma}}. \tag{2.12}$$

Using the correspondence principle (the analogy between (2.12) and the Hooke's law) we can extend the identification procedure presented in [6–8] to the Laplace mapping of the effective relaxation or creep functions, see [1, 3]. For the orthotropic viscoelastic material the in-plane and the out-of-plane stiffness tensor components are

$$
\begin{aligned}
(\overline{A}_{11};\overline{B}_{13};\overline{C}_{33}) &= \frac{1}{4}\left\langle\frac{\overline{E}_1+\overline{E}_2+2\overline{E}_1\overline{\nu}_{12}}{1-\overline{\nu}_{12}\overline{\nu}_{21}}\left(1;z;z^2\right)\right\rangle,\\
(\overline{A}_{12};-\overline{B}_{23}=\overline{B}_{14};\overline{C}_{34}) &= \frac{1}{4}\left\langle\frac{\overline{E}_1-\overline{E}_2}{1-\overline{\nu}_{12}\overline{\nu}_{21}}\left(1;z;z^2\right)\right\rangle,\\
(\overline{A}_{22};-\overline{B}_{24};\overline{C}_{44}) &= \frac{1}{4}\left\langle\frac{\overline{E}_1+\overline{E}_2-2\overline{E}_1\overline{\nu}_{21}}{1-\overline{\nu}_{12}\overline{\nu}_{21}}\left(1;z;z^2\right)\right\rangle,\\
(\overline{A}_{44};\overline{B}_{42};\overline{C}_{22}) &= \frac{1}{4}\left\langle\overline{G}_{12}\left(1;z;z^2\right)\right\rangle,
\end{aligned}
$$

while the transverse shear relaxation tensor components are

$$
\begin{aligned}
\overline{\Gamma}_1 &= \frac{1}{2}\left(\lambda^2+\eta^2\right)\frac{\overline{A}_{44}\overline{C}_{22}-\overline{B}_{42}^2}{\overline{A}_{44}},\\
\overline{\Gamma}_2 &= \frac{1}{2}\left(\lambda^2-\eta^2\right)\frac{\overline{A}_{44}\overline{C}_{22}-\overline{B}_{42}^2}{\overline{A}_{44}},
\end{aligned}
$$

where λ and η are the minimal nonzero eigen-values following from the Sturm-Liouville problems

$$
\begin{aligned}
&\frac{d}{dz}\left(\overline{G}_{2n}\frac{dZ}{dz}\right)+\lambda^2\overline{G}_{12}Z=0, \qquad \left.\frac{dZ}{dz}\right|_{|z|=h/2}=0,\\
&\frac{d}{dz}\left(\overline{G}_{1n}\frac{d\tilde{Z}}{dz}\right)+\lambda^2\overline{G}_{12}\tilde{Z}=0, \qquad \left.\frac{d\tilde{Z}}{dz}\right|_{|z|=h/2}=0.
\end{aligned}
$$

Here h is the plate thickness and $\langle(\cdots)\rangle$ denotes integration over the thickness.

In the case of isotropic material behaviour these formulas are simplified, see [1] for details. The non-zero components of the relaxation tensors are given by

- the in-plane relaxation functions

$$
\overline{A}_{11}=\frac{1}{2}\left\langle\frac{\overline{E}}{1-\overline{\nu}}\right\rangle,\ \overline{A}_{22}=\frac{1}{2}\left\langle\frac{\overline{E}}{1+\overline{\nu}}\right\rangle=\overline{A}_{44}=\langle\overline{G}\rangle,
$$

- the coupling relaxation functions

$$
\begin{aligned}
\overline{B}_{13} &= -\frac{1}{2}\left\langle\frac{\overline{E}_1+\overline{E}_2+2\overline{E}_1\overline{\nu}_{21}}{1-\overline{\nu}_{12}\overline{\nu}_{21}}z\right\rangle,\\
\overline{B}_{24} &= \frac{1}{2}\left\langle\frac{\overline{E}}{1+\overline{\nu}}z\right\rangle=-\overline{B}_{42}=\langle\overline{G}z\rangle,
\end{aligned}
$$

- the out-of-plane relaxation functions

$$\overline{C}_{33} = \frac{1}{2}\left\langle \frac{\overline{E}}{1-\overline{\nu}} z^2 \right\rangle,$$

$$\overline{C}_{22} = \frac{1}{2}\left\langle \frac{\overline{E}}{1+\overline{\nu}} z^2 \right\rangle = \overline{C}_{44} = \left\langle \overline{G} z^2 \right\rangle,$$

- the transverse shear relaxation function

$$\overline{\Gamma}_1 = \overline{\Gamma} = \lambda^2 \frac{\overline{A}_{44}\overline{C}_{22} - \overline{B}_{42}^2}{\overline{A}_{44}},$$

with λ following from

$$\frac{d}{dz}\left(\overline{G}\frac{dZ}{dz}\right) + \lambda^2 \overline{G} Z = 0, \qquad \left.\frac{dZ}{dz}\right|_{|z|=h/2} = 0. \tag{2.13}$$

For the plate which is symmetrically to the mid-plane the relation $\mathbf{B} = \mathbf{0}$ holds true. The relaxation functions of the isotropic viscoelastic plate with symmetric cross-section were considered in [1]. Note that for isotropic viscoelastic material we introduced three functions$\overline{E}(s)$, $\overline{G}(s)$ and $\overline{\nu}(s)$. They are interlinked by the formula

$$\overline{E}(s) = 2\overline{G}(s)\left(1 + \overline{\nu}(s)\right). \tag{2.14}$$

Following [19, 20] we use (2.14) as the definition of the Poisson's ratio for isotropic viscoelastic material.

In the theory of viscoelasticity of solids the assumption $\nu(t) = \nu = \text{const}$ is often used. It is fulfilled in many applications (see arguments in [21–23] concerning $\nu(t) \approx \text{const}$), for example, $\nu = 0.5$ for an incompressible viscoelastic material. In the general case, ν is a function of t. $\nu(t)$ is assumed to be an increasing function of t [23–25] or non-monotonous function of t, see [19, 20]. The latter case may be realized for cellular materials or foams. Further we consider the influence of $\nu(t)$ on the deflection of viscoelastic plate and its effective relaxation functions.

2.4 Example of Effective Properties

2.4.1 Homogeneous Plate

The simplest test for the correctness of the estimated stiffness properties is the homogeneous isotropic plate. The basic geometrical property is the thickness h. The plate is symmetrically with respect to the mid-plane. All material properties are constant over the thickness, i.e. they do not depend on the thickness coordinate z.

For the sake of simplicity, at first let us first consider the case $\nu(t) = \nu = \text{const}$. This means that the following relation is held true: $E(t) = 2G(t)(1+\nu)$. The non-zero components of the classical tensors are

$$A_{11}(t) = \frac{E(t)h}{2(1-\nu)}, A_{22}(t) = \frac{E(t)h}{2(1+\nu)} = G(t)h,$$

$$C_{33}(t) = \frac{E(t)h^3}{24(1-\nu)}, C_{22}(t) = \frac{E(t)h^3}{24(1+\nu)} = \frac{G(t)h^3}{12}.$$

The bending stiffness D results in

$$D(t) = \frac{E(t)h^3}{12(1-\nu^2)}.$$

The transverse shear relaxation function follows from (2.13). The solution of (2.13) yields the smallest eigen-value $\lambda = \pi/h$ which does not depend on s. Finally, one obtains

$$\Gamma(t) = \frac{\pi^2}{12}G(t)h. \tag{2.15}$$

$\pi^2/12$ is a factor similar to the shear correction factor which was first introduced by Timoshenko in the theory of beams [26]. Here this factor is a result of the non-classical establishments of the transverse shear stiffness. Comparing this value with the Mindlin's estimate $\pi^2/12$ and the Reissner's estimate 5/6 one concludes that the direct approach yields in the same value like in Mindlin's theory (note that Mindlin's shear correction is based on the solution of a dynamic problem, here was used the solution of a quasi-static problem), see [27–30]. The Reissner's value slightly differs. The graphs of $D(t)$ are given in Fig. 2.2 for two values of ν, i.e. for $\nu = 0.1;\ 0.4$. Let us note that in this case $D(t)$ and $\Gamma(t)$ demonstrate the same spectrum of relaxation times as the bulk material.

At second, let us consider the general case $\nu = \nu(t)$. Using the convolution theorem [21–25] in this case D is reconstructed from

$$\overline{D}(s) = \frac{\overline{E}(s)h^3}{12[1-\overline{\nu}^2(s)]}$$

as follows

$$D(t) = \int_{-\infty}^{t} \frac{E(t-\tau)h^3}{12[1-\nu^2(\tau)]}\,d\tau.$$

Using the initial and the final value theorems [21–25]

$$f(0) = \lim_{s\to\infty}\overline{f}(s), \quad f(\infty) \equiv \lim_{t\to\infty} f(t) = \lim_{s\to 0}\overline{f}(s)$$

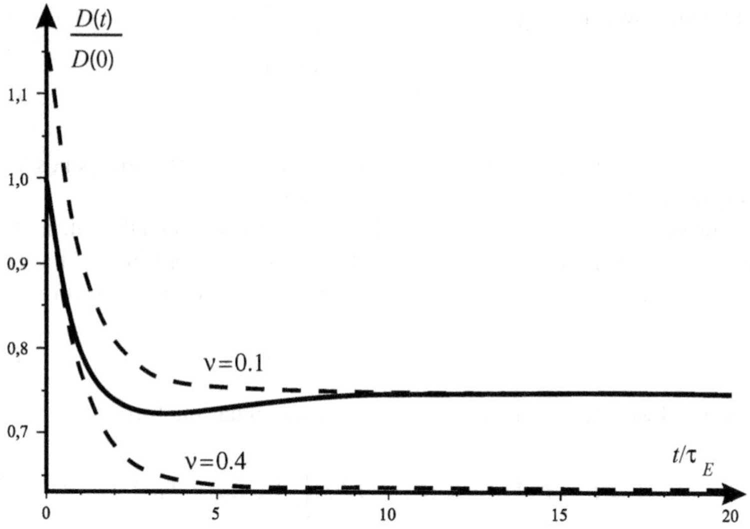

Fig. 2.2 Dimensionless bending stiffness in dependence on time: General case – *solid line*; Constant Poisson's ratio – *dashed lines*

we establish that

$$D(0) = \frac{E(0)h^3}{12[1-\nu^2(0)]}, D(\infty) = \frac{E(\infty)h^3}{12[1-\nu^2(\infty)]}.$$

where the values

$$\nu(0) = \frac{E(0)}{2G(0)} - 1, \nu(\infty) = \frac{E(\infty)}{2G(\infty)} - 1$$

may be considered as the Poisson's ratio in the initial and the relaxed state, respectively. In the general case $\nu = \nu(t)$ the relaxation function $D(t)$ is a non-monotonous function of t, while $D(t)$ is a monotonous decreasing function for constant Poisson's ratio, see Fig. 2.2. Here $\nu(0) = 0.4$, $\nu(\infty) = 0.1$. This means that in general case $D(t)$ and $\Gamma(t)$ demonstrate the spectrum of relaxation times which is not coincident with the spectrum of bulk material.

2.4.2 FGM Plate

In this section we consider small deformations of a FGM plate made of a viscoelastic polymer foam. For the panel made of a porous polymer foam the distribution of the pores over the thickness can be inhomogeneous (see, for example, Fig. 2.1). Let us introduce ρ_s as the density of the bulk material and ρ_p as the minimum value of the density of the foam. For the description of the symmetric distribution of the porosity we assume the power law [2]

$$V(z) = \alpha + (1-\alpha)\left|\frac{2z}{h}\right|^n, \tag{2.16}$$

where $\alpha = \rho_p/\rho_s$ is the minimal relative density. $n = 0$ corresponds to the homogeneous plate described in the previous paragraph.

The properties of the foam strongly depend on the porosity and the cell structure. For the polymer foam in [4, 5] the modification of the standard linear viscoelastic solid is proposed. For the open-cell foam the constitutive law has the form

$$\dot{\sigma} + \tau_E\sigma = C_1 V^2(z)\,[E_\infty\tau_E\varepsilon + E_0\dot{\varepsilon}], \tag{2.17}$$

while for the closed-cell foam the constitutive equation has the form

$$\dot{\sigma} + \tau_E\sigma = C_2\left[\phi^2\, V^2(z) + (1-\phi)V(z)\right]\,[E_\infty\tau_E\varepsilon + E_0\dot{\varepsilon}]\ . \tag{2.18}$$

Here $C_1 \approx 1$, $C_2 \approx 1$, ϕ describes the relative volume of the solid polymer concentrated near the cell ribs. Usually, $\phi = 0.6\ldots0.7$. E_∞, E_0, τ_E are material constants of the polymer used in manufacturing of the foam.

From (2.17) and (2.18) one can see that the corresponding relaxation function is given by the relations

$$E = E(z,t) = E(t)k(z), \tag{2.19}$$

where $E(t)$ is defined by

$$E(t) = E_\infty + (E_0 - E_\infty)e^{-t/\tau_E},$$

while

$$k(z) = C_1 V^2(z)$$

for open-cell foam and

$$k(z) = C_2\left[\phi^2\, V^2(z) + (1-\phi)V(z)\right]$$

for closed-cell foam, respectively. Analogous to (2.19) the following relation can be established for the shear relaxation function

$$G = G(z,t) = G(t)m(z). \tag{2.20}$$

Equations (2.19) and (2.20) state that the viscoelastic properties of the foam, for example, the time of relaxation do not depend on the porosity distribution. Note that representations (2.19) and (2.20) are only simple assumptions for spatial non-homogeneous foams.

Using experimental data presented in [5] one can assume $\nu(t) = \nu = \text{const}$. In this case we have the relations [3]

$$A_{11} = A_{22}\frac{1+\nu}{1-\nu}, C_{33} = C_{22}\frac{1+\nu}{1-\nu}. \tag{2.21}$$

A_{22} and C_{22} are given by the relations

$$A_{22} = h\left[\alpha^2 + \frac{2\alpha(1-\alpha)}{n+1} + \frac{(1-\alpha)^2}{2n+1}\right]G(t), \tag{2.22}$$

$$C_{22} = \frac{h^3}{12}\left[\alpha^2 + \frac{6\alpha(1-\alpha)}{n+3} + \frac{3(1-\alpha)^2}{2n+3}\right]G(t), \tag{2.23}$$

for the open-cell foam, and

$$\begin{aligned} A_{22} = h\Big\{&\phi^2\left[\alpha^2 + \frac{2\alpha(1-\alpha)}{n+1} + \frac{(1-\alpha)^2}{2n+1}\right] \\ &+(1-\phi)\left[\alpha + \frac{1-\alpha}{n+1}\right]\Big\}G(t), \end{aligned} \tag{2.24}$$

$$\begin{aligned} C_{22} = \frac{h^3}{12}\Big\{&\phi^2\left[\alpha^2 + \frac{6\alpha(1-\alpha)}{n+3} + \frac{3(1-\alpha)^2}{2n+3}\right] \\ &+(1-\phi)\left[\alpha + \frac{3(1-\alpha)}{n+3}\right]\Big\}G(t) \end{aligned} \tag{2.25}$$

for the closed-cell foam, respectively. Here we assume that $C_1 = C_2 = 1$, and that ϕ does not depend on z.

From (2.21), (2.22), (2.23), (2.24) and (2.25) it is easy to see that the classical relaxation functions differ only by factors from the shear relaxation function. Note that one can easily extend (2.17) and (2.18) to the case of general constitutive equations, used in the linear viscoelasticity [13–15]. Thus taking into account the assumption that ν = const, one can calculate the classical effective stiffness relaxation functions for general viscoelastic constitutive equations multiplying the shear relaxation function $G(t)$ with the corresponding factor similar to (2.22), (2.23), (2.24) and (2.25). In the general situation and taking into account other viscoelastic phenomena, for example, the filtration of a fluid in the saturated foam, the effective stiffness relaxation functions may be more complex than for the pure solid polymer discussed here.

Finally, we should mention that in the case of constant Poisson's ratio and with the assumption (2.19) and (2.20) the determination of the effective in-plane, bending and transverse shear stiffness tensors of a symmetric FGM viscoelastic plate made of a polymer foam can be realized by the same method as for elastic plates [9–12]. The relaxation functions for viscoelastic FGM plates can be found from the values of the corresponding effective stiffness of an elastic FGM plate by multiplication with the normalized shear relaxation function of the polymer solid.

2.4.3 Sandwich Plate

Sandwich structures with a core made of foam have various applications in the engineering. Let us consider a sandwich plate with the following geometry: h_c is the core thickness and h_f the thickness of the face sheets ($h_f << h_c$). The material properties of the core and the face sheets are given by the relaxation functions $E_c(t)$, $E_f(t)$, $G_c(t)$ and $G_f(t)$ with $E_c(t) << E_f(t)$ and $G_c(t) << G_f(t)$. We have again a symmetry with respect to the mid-plane that means $\mathbf{B} = \mathbf{0}$. With the thickness $h = h_c + h_f$ one gets

$$\overline{A}_{11} = \frac{1}{2}\left(\frac{\overline{E}_f h_f}{1-\overline{\nu}_f} + \frac{\overline{E}_c h_c}{1-\overline{\nu}_c}\right),$$

$$\overline{A}_{11} = \frac{1}{2}\left(\frac{\overline{E}_f h_f}{1+\overline{\nu}_f} + \frac{\overline{E}_c h_c}{1+\overline{\nu}_c}\right) = \overline{A}_{44},$$

$$\overline{C}_{33} = \frac{1}{24}\left[\frac{\overline{E}_f(h^3-h_c^3)}{1-\overline{\nu}_f} + \frac{\overline{E}_c h_c^3}{1-\overline{\nu}_c}\right],$$

$$\overline{C}_{44} = \frac{1}{24}\left[\frac{\overline{E}_f(h^3-h_c^3)}{1+\overline{\nu}_f} + \frac{\overline{E}_c h_c^3}{1+\overline{\nu}_c}\right] = \overline{C}_{22},$$

The bending relaxation function results in

$$\overline{D} = \frac{1}{12}\left[\frac{\overline{E}_f(h^3-h_c^3)}{1-\overline{\nu}_f^2} + \frac{\overline{E}_c h_c^3}{1-\overline{\nu}_c^2}\right].$$

Let us consider the latter relation in detail. For the sake of simplicity we assume that the Poisson ratios are constant. Then we obtain the relation

$$D(t) = \frac{1}{12}\left[\frac{E_f(t)(h^3-h_c^3)}{1-\nu_f^2} + \frac{E_c(t) h_c^3}{1-\nu_c^2}\right].$$

Considering the simplest form for the bulk relaxation functions, i.e. the standard visoelastic body model with

$$E_f(t) = E_\infty^f + (E_0^f - E_\infty^f)e^{-t/\tau_E^f},$$

$$E_c(t) = E_\infty^c + (E_0^c - E_\infty^c)e^{-t/\tau_E^c},$$

then we immediately obtain that D has two times of relaxation τ_E^f and τ_E^c.

A typical sandwich structure has a very weak core. In this case the bending relaxation function and the transverse shear relaxation function can be approximated by

$$\overline{D} = \frac{1}{4}\frac{\overline{E}_f h^2 h_f}{1-\overline{\nu}_f^2}, \overline{\Gamma} = \overline{G}_c h,$$

see the details given in [7–9] for the elastic case. Let us note that for such approximation the bending relaxation function D is determined by the viscoelastic properties of the faces while the transverse shear relaxation function Γ depends on the viscoelastic behaviour of the core only. For example, if the faces are made of an elastic material and the core shows the viscoelastic behavior then D is constant while Γ is a function of time

$$D = \frac{1}{4}\frac{E_f h^2 h_f}{1-\nu_f^2}, \Gamma(t) = G_c(t)h.$$

Using the technique presented here one may consider more complicated cases, for example, the case of the sandwich plate with a core made of FGM viscoelastic material or the laminate plate made of viscoelastic laminae.

2.5 Bending of a Symmetric Isotropic Plate

Using [3] and the Laplace transform, one can reduce (2.1) and (2.2) to

$$s\overline{D}_{\text{eff}}\Delta\Delta\overline{w} = \overline{q}_n - \frac{\overline{D}_{\text{eff}}}{\overline{\Gamma}}\Delta\overline{q}_n, \tag{2.26}$$

where w is the deflection, and q_n is the transverse load. Here we consider the symmetry of the material properties with respect to the mid-plane and $\mathbf{m} = \mathbf{0}$. Using assumption $\nu = \text{const}$ we transform (2.26) to the form

$$s\overline{D}_{\text{eff}}\Delta\Delta\overline{w} = \overline{q}_n - \frac{2}{\lambda^2(1-\nu)}\Delta\overline{q}_n. \tag{2.27}$$

Let us consider a rectangular plate $x \in [0, a]$, $y \in [0, b]$, where a and b are the length and width of the plate, simple support boundary conditions, and the sinusoidal load

$$q_n = Q(t)\sin\frac{\pi x}{a}\sin\frac{\pi y}{b}. \tag{2.28}$$

Then the solution of (2.27) has the form

$$\overline{w} = \frac{K}{\eta^4 h^3}\frac{\hat{Q}}{s\overline{G}(s)}\sin\frac{\pi x}{a}\sin\frac{\pi y}{b}, \tag{2.29}$$

where

$$K = 1 + \frac{2\eta^2}{(1-\nu)\lambda^2}, \quad \hat{Q} = \frac{\overline{Q}h^3}{D^0_{\text{eff}}}, \quad \eta = \left(\frac{\pi}{a}\right)^2 + \left(\frac{\pi}{b}\right)^2,$$

$$D^0_{\text{eff}} = \frac{C_{22} + C_{33}}{G(t)}.$$

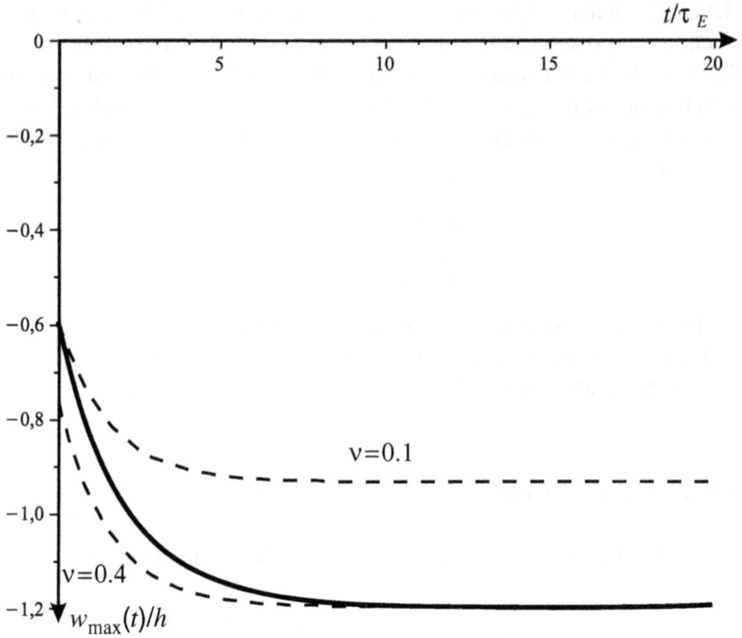

Fig. 2.3 Dimensionless maximal deflection in dependence on time: General case – *solid line*; Constant Poisson's ratio – *dashed lines*

The factor K determines the maximum deflection. For the Kirchhoff's plate theory$K = 1$, for the homogeneous plate modeled in the sense of Mindlin's plate theory $K = 1 + \frac{2\eta}{\pi^2(1-\nu)}$. For the FGM plate the bounds for K are given in [3].

For the viscoelastic plate both the qualitative and the quantitative influence of the shear stiffness is the same as in [2]. For example, let us consider an open-cell foam and following values $\nu = 0.3$, $a = b$, $h = 0.05a$, $\alpha = 0.9$. Using the calculation of [2] we obtain the following values of λ: $\lambda = 0.83\,h$ for $n = 2$, $\lambda = 0.82\,h$ for $n = 5$. The corresponding values of K are given by $K \approx 1.2 \quad (n = 2)$, $K \approx 1.21 \quad (n = 5)$. That means that for the functionally graded plates the influence of transverse shear stiffness may be significant. As well as for elastic FGM plates for the cases of other types of boundary conditions the influence of the structure of viscoelastic plate on the deflection may be greater than for the used simple support type boundary conditions. A numerical example concerning the maximal deflection vs. time is given in Fig. 2.3.

2.6 Conclusions

Here we presented the new model of the linear viscoelastic plates made of such FGM as polymer foam with the non-homogeneous distribution of porosity. The so-called direct approach is applied to the statement of the boundary value problem of the viscoelastic plates. Within this approach the plate is considered as a deformable surface. The balance laws and the constitutive equations are formulated as for 2D continuum. The procedure of the identification of the viscoelastic material properties is described and the example of the bending of a FGM viscoelastic plate is given. The given examples of effective relaxation functions in the cases of homogeneous, sandwich and FGM plates show that any viscoelastic plate considered as a 2D viscoelastic continuum has more complicated viscoelastic properties than the bulk material. These properties depend on the bulk properties and the through-the-thickness structure of plate, in general.

Acknowledgements The second author was supported by the RFBR with the grants No. 09-01-00459, 09-01-00849 and by the DFG grant No. AL 341/33-1.

References

1. Altenbach, H., Eremeyev, V.A.: Analysis of the viscoelastic behavior of plates made of functionally graded materials. Zeitschrift für Angewandte Mathematik und Mechanik (ZAMM). **88**, 332–341 (2008)
2. Altenbach, H., Eremeyev, V.A.: On the bending of viscoelastic plates made of polymer foams. Acta Mech. **204**, 137–154 (2009)
3. Altenbach, H., Eremeyev, V.A.: On the time dependent behaviour of FGM plates. Key Eng. Mater. **399**, 63–70 (2009)
4. Gibson, L.J., Ashby, M.F.: Cellular Solids: Structure and Properties, 2nd edn, Cambridge Solid State Science Series. Cambridge University Press, Cambridge (1997)
5. Mills, N.: Polymer Foams Handbook Engineering and Biomechanics Applications and Design. Guide Butterworth-Heinemann, Boston, MA (2007)
6. Altenbach, H.: Definition of elastic moduli for plates made from thickness-uneven anisotropic material. Mech. Solids **22**, 135–141 (1987)
7. Altenbach, H.: An alternative determination of transverse shear stiffnesses for sandwich and laminated plates. Int. J. Solids Struct. **37**, 3503–3520 (2000)
8. Altenbach, H.: On the determination of transverse shear stiffnesses of orthotropic plates. ZAMP. **51**, 629–649 (2000)
9. Altenbach, H., Eremeyev, V.A.: Direct approach based analysis of plates composed of functionally graded materials. Arch. Appl. Mech. **78**, 775–794 (2008)
10. Altenbach, H., Eremeyev, V.A.: Eigen-vibrations of plates made of functionally graded material. CMC: Comput. Mater. Continua **9**, 153–178 (2009)
11. Altenbach, H.: Eine direkt formulierte lineare Theorie für viskoelasische Platten und Schalen. Ingenieur-Arch. **58**, 215–228 (1988)
12. Praveen, G.N., Reddy, J.N.: Nonlinear transient thermoelastic analysis of functionally graded ceramic-metal plates. Int. J. Solids Struct. **35**, 4457–4476 (1998)
13. He, X.Q., Ng, T.Y., Sivashanker, S., Liew, K.M.: Active control of FGM plates with integrated piezoelectric sensors and actuators. Int. J. Solids Struct. **38,** 1641–1655 (2001)
14. Yang, J., Shen, H.S.: Dynamic response of initially stressed functionally graded rectangular thin plates. Compos. Struct. **54**, 497–508 (2001)
15. Javaheri, R., Eslami, M.R.: Thermal buckling of functionally graded plates. AIAA J. **40,** 162–169 (2002)
16. Arciniega, R., Reddy, J.: Large deformation analysis of functionally graded shells. Int. J. Solids Struct. **44**, 2036–2052 (2007)
17. Batra, R.C.: Higher order shear and normal deformable theory for functionally graded incompressible linear elastic plates. Thin-Walled Struct. **45**, 974–982 (2007)
18. Zhang, N.-H., Wang, M.-L.: Thermoviscoelastic deformations of functionally graded thin plates. Eur. J. Mech. A-Solids **26**, 872–886 (2007)

19. Lakes, R.S.: The time-dependent Poisson's ratio of viscoelastic materials can increase or decrease. Cell. Polym. **11**, 466–469 (1992)
20. Lakes, R.S., Wineman, A.: On Poisson's ratio in linearly viscoelastic solids. J Elast. **85**, 45–63 (2006)
21. Christensen, R.M.: Theory of Viscoelasticity An Introduction. Academic Press, New York (1971)
22. Drozdov, A.D.: Finite Elasticity and Viscoelasticity. World Scientific, Singapore (1996)
23. Tschoegl, N.W.: The Phenomenological Theory of Linear Viscoelastic Behavior. An Introduction. Springer, Berlin (1989)
24. Brinson, H.F., Brinson, C.L.: Polymer Engineering Science and Viscoelasticity. An Introduction. Springer, New York, NY (2008)
25. Riande, E., et al. (eds.): Polymer Viscoelasticity: Stress and Strain in Practice. Marcel Dekker, New York, NY (2000)
26. Timoshenko, S.P.: On the correction for shear of the differential equation for transverse vibrations of prismatic bars. Phil. Mag. Ser. **41**, 744–746 (1921)
27. Reissner, E.: On the theory of bending of elastic plates. J. Math. Phys. **23**, 184–194 (1944)
28. Reissner, E.: On bending of elastic plates. Q. Appl. Math. **5**, 55–68 (1947)
29. Mindlin, R.D.: Influence of rotatory inertia and shear on flexural motions of isotropic elastic plates. Trans. ASME J. Appl. Mech. **18**, 31–38 (1951)
30. Reissner, E.: Reflection on the theory of elastic plates. Appl. Mech. Rev. **38**, 1453–1464 (1985)

Chapter 3
Indirect Trefftz Method for Solving Cauchy Problem of Linear Piezoelectricity

G. Dziatkiewicz and P. Fedeliński

Abstract In this chapter, the indirect Trefftz collocation method is applied to solve the inverse Cauchy problem of linear piezoelectricity. The static analysis of coupled fields in 2D piezoelectric solids is considered. The piezoelectric material is modelled as a homogenous, linear elastic, linear dielectric and transversal isotropic. The T-functions are obtained by the Stroh formalism. The inverse Cauchy problem in linear plane piezoelectricity is formulated. The Cauchy problem is solved using the singular value decomposition (SVD) regularisation technique with L-curve method.

Keywords Cauchy problem · Indirect Trefftz method · Piezoelectricity · Singular value decomposition

3.1 Introduction

Piezoelectric materials are widely used as sensors and actuators in "smart structures" and micro-electro-mechanical systems (MEMS). Other applications are ultrasonic transducers and electromechanical filters. An analysis of piezoelectric devices requires a solution of coupled electrical and mechanical partial differential equations. In this paper, the indirect Trefftz collocation method (TCM) is implemented to solve both the direct and the inverse Cauchy problems in linear plane piezoelectricity. In most boundary value problems, the governing equations have to be solved with the appropriate boundary conditions [1, 2]. These problems are called direct. However, when the boundary conditions are incomplete on a certain boundary part, the boundary value problems are generally ill-posed, then the existence, uniqueness and stability of the solution is not always guaranteed [3, 4]. These problems are inverse problems. In this paper the Cauchy problem is considered, which is solved by the indirect Trefftz collocation regularization method with the SVD technique [3].

G. Dziatkiewicz (✉)
Department of Strength of Materials and Computational Mechanics, Silesian University of Technology, Konarskiego 18A, 44-100, Gliwice, Poland
e-mail: grzegorz.dziatkiewicz@polsl.pl

J. Murín et al. (eds.), *Computational Modelling and Advanced Simulations*, Computational Methods in Applied Sciences 24, DOI 10.1007/978-94-007-0317-9_3,

To solve both direct and inverse problems of piezoelectricity, the TCM computer code is developed. Numerical examples are given and good results are obtained. When the inverse problem is analyzed, numerical solutions are compared with the solution obtained from the direct problem.

3.2 Governing Equations of Linear Piezoelectricity

Consider a piezoelectric material which occupies a domain $\Omega \subset R^2$, with a boundary surface $\Gamma = \partial\Omega$. The basic equations governing the elastic and electric fields in a piezoelectric material in the static case are: equilibrium equations, constitutive equations and coupled field equations.

The equilibrium equations are both elastostatics and electrostatics equations [1, 2]:

$$\begin{aligned} \sigma_{ij,i} + f_j &= 0 \\ D_{i,i} - q &= 0 \end{aligned} \tag{3.1}$$

where: σ_{ij} denotes the stress tensor and f_j is the body force vector per unit volume; D_i is the electric displacement vector and q denotes the intrinsic electric charge per unit volume. The piezoelectric materials are dielectrics, which possess no free electric charges, thus q is equal to zero.

The constitutive equations of piezoelectricity are given by [1, 2]:

$$\begin{aligned} \sigma_{ij} &= C_{ijkl}s_{kl} - e_{lij}E_l \\ D_i &= e_{ikl}s_{kl} + \varepsilon_{il}E_l \end{aligned} \tag{3.2}$$

where s_{kl} is the strain tensor, E_l – electric field. The tensors C_{ijkl}, e_{lij}, ε_{il} denote elastic moduli, measured in a constant electric field, piezoelectric constants and dielectric constants, measured at constant strains, respectively. The first equation in the system (3.2) describes a converse piezoelectric effect, and the second one – a direct effect. These equations describe a linear Voigt model of the piezoelectric material.

In linear elasticity, the strain tensor is given by:

$$s_{kl} = \frac{1}{2}\left(u_{k,l} + u_{l.k}\right). \tag{3.3}$$

In (3.3) u_k is the displacement vector.

The static electric field is irrotational, so the electric field vector can be described using the electric potential [1]:

$$E_l = -\phi_{,l} \tag{3.4}$$

where φ denotes the electric potential.

After some mathematical operations, using (3.1), (3.2), (3.3) and (3.4), the coupled field equations of static piezoelectricity can be expressed by the following system of partial differential equations [1, 2]:

$$\begin{aligned} C_{ijkl}u_{k,li} + e_{lij}\phi_{,li} &= -f_j \\ e_{ikl}u_{k,li} - \varepsilon_{il}\phi_{,li} &= 0 \end{aligned} . \tag{3.5}$$

In (3.5) the intrinsic electric charge is neglected.

To get the classical boundary – value problem formulation, (3.5) must be completed with the boundary conditions. First, boundary quantities are defined as:

$$\begin{aligned} t_j &= \sigma_{ij}n_i \\ \omega &= D_i n_i \end{aligned} . \tag{3.6}$$

In (3.6) t_j denotes the tractions, ω is the charge flux density and n_i is the unit outward normal vector. Boundary conditions, in the direct problem, are both mechanical and electrical and they can be applied in the form:

$$\begin{aligned} &\Gamma_t : t_i = \bar{t}_i \; ; \Gamma_u : u_i = \bar{u}_i \\ &\Gamma_\phi : \phi = \bar{\phi} \; ; \Gamma_\omega : \omega = \bar{\omega} \end{aligned} \tag{3.7}$$

where Γ_t, Γ_u, Γ_φ and Γ_ω denote parts of the boundary Γ where tractions, displacements, potentials and charge flux densities are prescribed. These parts of the boundaries fulfil the following relations:

$$\begin{aligned} &\Gamma = \Gamma_t \cup \Gamma_u = \Gamma_\phi \cup \Gamma_\omega \\ &\Gamma_t \cap \Gamma_u = \emptyset \\ &\Gamma_\phi \cap \Gamma_\omega = \emptyset \end{aligned} \tag{3.8}$$

The coupled field equations with boundary conditions formulate the direct boundary-value problem of linear piezoelectricity.

To simplify the notation of equations, generalized quantities are introduced [1]:

$$\mathbf{U}_K = \begin{Bmatrix} u_k \\ \phi \end{Bmatrix}; \; \mathbf{T}_J = \begin{Bmatrix} t_j \\ \omega \end{Bmatrix}; \; \mathbf{B}_J = \begin{Bmatrix} b_j \\ 0 \end{Bmatrix} \tag{3.9}$$

where $\mathbf{U}_K$, $\mathbf{T}_J$ and $\mathbf{B}_J$ are generalized displacement, traction and body force vector, respectively.

Then, the coupled field equations are given by the operator equation [1]:

$$\mathbf{L}_{JK}\mathbf{U}_K = -\mathbf{B}_J \tag{3.10}$$

where $\mathbf{L}_{JK}$ is the 2D elliptic operator of static piezoelectricity.

The most popular piezoelectric materials are ceramics, which are called PZT. These piezoelectrics are solids, which belong to the hexagonal symmetry class of the crystals. These crystalic solids have anisotropic physical properties (mechanical and electrical), therefore in the present work homogeneous, transversal isotropic, linear elastic and linear dielectric model of the piezoelectric material is chosen. For this model, the operator L_{JK}, in the two dimensional case, has a form:

$$\mathbf{L}_{JK} = \begin{bmatrix} c_{11}\partial_{11} + c_{44}\partial_{33} & (c_{13} + c_{44})\,\partial_{13} & (e_{15} + e_{31})\,\partial_{13} \\ & c_{44}\partial_{11} + c_{33}\partial_{33} & e_{15}\partial_{11} + e_{33}\partial_{33} \\ sym & & -\varepsilon_{11}\partial_{11} - \varepsilon_{33}\partial_{33} \end{bmatrix} \tag{3.11}$$

where ∂_{ij} is a differential operator (differentiation with respect to the spatial coordinates). The coefficients c_{ij}, e_{ij} and ε_{ij} are the values of the elastic, piezoelectric and dielectric constants, respectively, which are written by using the contracted matrix notation.

3.3 Indirect Trefftz Collocation Method

In the indirect Trefftz method, the solution of the boundary-value problem, in arbitrary chosen point P, is approximated by the series of the T-complete functions [5, 6]:

$$\begin{aligned} U(P_i) &\simeq \tilde{U}(P_i) = \sum_{i=1}^{N} c_i U_i^* = \mathbf{c}^{\mathbf{T}}\mathbf{U}^*(P) \\ T(P_i) &\simeq \tilde{T}(P_i) = \sum_{i=1}^{N} c_i T_i^* = \mathbf{c}^{\mathbf{T}}\mathbf{T}^*(P) \end{aligned} \tag{3.12}$$

where quantities with star denote the T-complete functions of the Trefftz method. The vector **c** contains unknown series coefficients.

These functions satisfy the system of the governing equations, i.e. the homogenous system of the elliptic differential equations of the linear piezoelectricity.

The superposition of the T-complete functions satisfies the governing equations, but does not satisfy the mechanical and electric boundary conditions. This problem leads to the minimization problem of the boundary residuals [7, 8]:

$$\begin{aligned} P_i \in \Gamma_U &: \mathrm{R}_1 = \tilde{U} - \bar{U} = \mathbf{c}^{\mathbf{T}}\mathbf{U}^*(P_i) - \bar{U}(P_i) \neq 0 \\ P_i \in \Gamma_T &: \mathrm{R}_2 = \tilde{T} - \bar{T} = \mathbf{c}^{\mathbf{T}}\mathbf{T}^*(P_i) - \bar{T}(P_i) \neq 0 \end{aligned} . \tag{3.13}$$

The unknowns are the coefficients of the superposition of the T-complete functions, which define the searched mechanical and electric fields. The collocation method assumes that the residuals vanish at the boundary points [8]:

$$\begin{aligned} P_i \in \Gamma_U &: \mathrm{R}_1 = \tilde{U} - \bar{U} = \mathbf{c}^{\mathbf{T}}\mathbf{U}^*(P_i) - \bar{U}(P_i) = 0 \quad (i = 1, \ldots, M_1) \\ P_i \in \Gamma_T &: \mathrm{R}_2 = \tilde{T} - \bar{T} = \mathbf{c}^{\mathbf{T}}\mathbf{T}^*(P_i) - \bar{T}(P_i) = 0 \quad (i = 1, \ldots, M_2) \end{aligned} \tag{3.14}$$

where M_1 and M_2 denote the number of the boundary collocation nodes, where Dirichlet and Neumann boundary conditions are prescribed, respectively. The matrix form of the system (3.14) is given below:

$$\begin{bmatrix} U^*_{11} & U^*_{12} & \cdots & U^*_{1N} \\ U^*_{21} & U^*_{22} & \cdots & U^*_{2N} \\ \cdots & \cdots & \cdots & \cdots \\ U^*_{M_1 1} & U^*_{M_1 2} & \cdots & U^*_{M_1 N} \\ \mathrm{T}^*_{11} & \mathrm{T}^*_{12} & \cdots & \mathrm{T}^*_{1N} \\ \mathrm{T}^*_{21} & \mathrm{T}^*_{22} & \cdots & \mathrm{T}^*_{2N} \\ \cdots & \cdots & \cdots & \cdots \\ \mathrm{T}^*_{M_2 1} & \mathrm{T}^*_{M_2 2} & \cdots & \mathrm{T}^*_{M_2 N} \end{bmatrix} \begin{bmatrix} c_1 \\ c_2 \\ \cdots \\ c_N \end{bmatrix} = \begin{bmatrix} \bar{U}_1 \\ \bar{U}_2 \\ \cdots \\ \bar{U}_{M_1} \\ \bar{T}_1 \\ \bar{T}_2 \\ \cdots \\ \bar{T}_{M_2} \end{bmatrix}. \tag{3.15}$$

When **D** is main matrix of the system, **F** is a vector of given boundary values, then system of equations in the indirect Trefftz collocation method has a form:

$$\mathbf{Dc} = \mathbf{F} \tag{3.16}$$

The indirect Trefftz collocation approach usually requires the solution of the overdetermined system of equations, which determines the unknown coefficients of the superposition of the T-complete functions [7, 8]. The matrix of the system of equations is usually rectangular, nearly singular and ill-conditioned [4]. For a system of equations with these properties, a singular value decomposition (SVD) regularization technique is one of the most popular solution. The SVD allows to regularize the solution with the minimal norm [3].

3.4 The Stroh Formalism and the T-Complete Functions

Since piezoelectric materials are anisotropic, the fundamental solutions are rather complicated, even for the transversal isotropic model of the material. To obtain the general solutions and the T-complete functions, the Stroh formalism is used [1, 6]. In this method it is assumed that the field of the generalized displacements has a form [1]:

$$\mathbf{U} = \mathbf{a}f(z) \tag{3.17}$$

where **a** is the unknown vector and $f(z)$ is an analytic complex function and z is a complex variable:

$$z = x_1 + px_3 \tag{3.18}$$

where x_1 and x_3 are the coordinates, p denotes the unknown complex constant. Putting expression (3.18) into the coupled field equations, the quadratic eigenvalue problem is obtained [1, 9]:

$$\left\{\mathbf{Q} + p(\mathbf{R} + \mathbf{R}^T) + p^2\mathbf{T}\right\}\mathbf{a} = 0 \tag{3.19}$$

where the matrices **Q**, **R** and **T** depend only on the material constants. The above equation can be transformed into the standard eigenvalue problem [6]:

$$\mathbf{N}\boldsymbol{\xi} = p\boldsymbol{\xi} \tag{3.20}$$

where:

$$\mathbf{N} = \begin{bmatrix} -\mathbf{T}^{-1}\mathbf{R}^T & \mathbf{T}^{-1} \\ \mathbf{R}\mathbf{T}^{-1}\mathbf{R}^T - \mathbf{Q} & (-\mathbf{T}^{-1}\mathbf{R}^T)^T \end{bmatrix}, \ \boldsymbol{\xi} = \begin{bmatrix} \mathbf{a} \\ \mathbf{b} \end{bmatrix} \tag{3.21}$$

where the vector **b** is equal to:

$$\mathbf{b}_J = -\frac{1}{p_J}\left(\mathbf{Q} + p_J\mathbf{R}\right)\mathbf{a}. \tag{3.22}$$

When the vector **b** is known, the generalized stress function $\boldsymbol{\Psi}$ can be introduced:

$$\boldsymbol{\Psi} = \mathbf{b}f(z). \tag{3.23}$$

The stress tensor and electric displacement vector can be written using the above quantities in a form [1, 5]:

$$\begin{aligned} \begin{Bmatrix} \sigma_{3j} \\ D_3 \end{Bmatrix} &= \boldsymbol{\Psi}_{,1} \\ \begin{Bmatrix} \sigma_{1j} \\ D_1 \end{Bmatrix} &= -\boldsymbol{\Psi}_{,3} \end{aligned} . \tag{3.24}$$

It is known, that the eigenvalue problem (3.20) or (3.21), in two-dimensional case, gives three pairs of complex conjugates eigenvalues and corresponding eigenvectors. If $p_J(J = 1,2,3)$ are distinct eigenvalues with a positive imaginary part, then general solutions are given by [1]:

$$\begin{aligned} \mathbf{U} &= 2\mathrm{R}e\left\{\mathbf{A}\left\langle\mathbf{f}(z_*)\right\rangle\mathbf{q}\right\}, \\ \boldsymbol{\Psi} &= 2\mathrm{R}e\left\{\mathbf{B}\left\langle\mathbf{f}(z_*)\right\rangle\mathbf{q}\right\}. \end{aligned} \tag{3.25}$$

where matrices **A** and **B** contain eigenvectors of the problem (3.20), $\langle\mathbf{f}(z_*)\rangle$ is diagonal matrix, which contains values of arbitrary chosen complex functions of the argument in the following form :

$$z_J = x_1 + p_J x_3 \tag{3.26}$$

where p_J are eigenvalues with positive imaginary part; q denotes arbitrary complex vector.

It can be noticed, that the general solution of the operator equation (3.10) was obtained with arbitrary constants and the form of the analytic complex function must be chosen. It is known, that the set of the T-complete functions is the subset of the general solutions [8]. The T-complete functions have special properties [8, 9]. To construct these functions the complex polynomial base is applied, as for the Laplace operator [8, 9]:

$$B_T = \left\{1, z^k, iz^k\right\} \tag{3.27}$$

where $k = 1, 2, \ldots$, and i is the imaginary unit.

Then, the T-complete functions for piezoelectric operator L_{JK} are developed using the Stroh formalism and have a form:

$$\left\{\begin{array}{l}
\phi_1 = 2\mathrm{Re}\begin{bmatrix}\mathbf{A}_{11}\\ \mathbf{A}_{21}\\ \mathbf{A}_{31}\end{bmatrix}, \phi_2 = 2\mathrm{Re}\begin{bmatrix}\mathbf{A}_{12}\\ \mathbf{A}_{22}\\ \mathbf{A}_{32}\end{bmatrix}, \phi_3 = 2\mathrm{Re}\begin{bmatrix}\mathbf{A}_{13}\\ \mathbf{A}_{23}\\ \mathbf{A}_{33}\end{bmatrix}\\
\phi_{6k-2} = 2\mathrm{Re}\begin{bmatrix}\mathbf{A}_{11}z_1^k\\ \mathbf{A}_{21}z_2^k\\ \mathbf{A}_{31}z_3^k\end{bmatrix}, \phi_{6k-1} = 2\mathrm{Re}\begin{bmatrix}\mathbf{A}_{12}z_1^k\\ \mathbf{A}_{22}z_2^k\\ \mathbf{A}_{32}z_3^k\end{bmatrix}, \phi_{6k} = 2\mathrm{Re}\begin{bmatrix}\mathbf{A}_{13}z_1^k\\ \mathbf{A}_{23}z_2^k\\ \mathbf{A}_{33}z_3^k\end{bmatrix}\\
\phi_{6k+1} = 2\mathrm{Re}\begin{bmatrix}\mathbf{A}_{11}iz_1^k\\ \mathbf{A}_{21}iz_2^k\\ \mathbf{A}_{31}iz_3^k\end{bmatrix}, \phi_{6k+2} = 2\mathrm{Re}\begin{bmatrix}\mathbf{A}_{12}iz_1^k\\ \mathbf{A}_{22}iz_2^k\\ \mathbf{A}_{32}iz_3^k\end{bmatrix}, \phi_{6k+3} = 2\mathrm{Re}\begin{bmatrix}\mathbf{A}_{13}iz_1^k\\ \mathbf{A}_{23}iz_2^k\\ \mathbf{A}_{33}iz_3^k\end{bmatrix}
\end{array}\right\}$$

$$k = 1, 2, \ldots \tag{3.28}$$

In the same way, the set of the T-complete function for the generalized tractions is constructed – the matrix B and the generalized stress function are used.

3.5 The Cauchy Inverse Problem Formulation

The TCM solutions describe the solution of the direct problem of piezoelectricity. In the direct problem the governing equation have to be solved with the appropriate boundary conditions. However, when the boundary conditions are incomplete on a certain boundary part, the boundary value problems are generally ill-posed, then the existence, uniqueness and stability of the solution is not always guaranteed [3, 4]. These problems are inverse problems. In this paper the Cauchy problem is considered, which is solved by the indirect Trefftz collocation regularization method with the SVD technique.

Consider again, a piezoelectric material in R^2 space with a smooth boundary Γ in the sense of Lapunov. Let the boundary is composed of three parts, namely Γ_1, Γ_2 and Γ_3 , and the relations between these parts of the boundary are in the form:

$$\begin{array}{l}\Gamma = \Gamma_1 \cup \Gamma_2 \cup \Gamma_3\\ \Gamma_1, \Gamma_2 \neq \emptyset\\ \Gamma_1 \cap \Gamma_2 \cap \Gamma_3 = \emptyset\end{array}. \tag{3.29}$$

Let the generalized displacements U and tractions T are both measured on the boundary Γ_2. Then, the boundary conditions on this part of the boundary Γ are given by:

$$\Gamma_2 : U = \bar{U}, T = \bar{T}. \tag{3.30}$$

If on the boundary Γ_3 the boundary conditions are prescribed in the usual sense, then on the boundary Γ_1 both generalized displacements and generalized tractions are unknown and have to be determined. The problem is called the Cauchy inverse problem of piezoelectricity. The solution of this problem does not satisfy the general conditions of well-posedness. The main problem in this case, is that the solution is unstable with respect to small perturbations in the data on boundary Γ_2, so the regularization technique must be applied [3, 4]. The most popular regularization techniques are: the singular value decomposition method [3, 4, 10], the generalized singular value decomposition method [3, 4], the Tikhonov regularization method [10], the conjugate gradient method [10]. As a regularization method in the present work, the singular value decomposition (SVD) technique is applied.

Let on the boundaries Γ_1 ,Γ_2 and Γ_3 we have N_1 and N_2 and N_3 collocation node, respectively (with the condition : $N = N_1 + N_2 + N_3$, N is a total number of boundary nodes). The algebraic system of equations (3.16) has a following form in this case:

$$\mathbf{D_{inv}c} = \mathbf{F}. \tag{3.31}$$

This is a system of $N_{\text{inv}} = 6N_1 + 3N_3$ equations with M unknowns (M denotes the number of the T-complete functions), the vector $\mathbf{F}$ contains boundary conditions prescribed on the Γ_2 and Γ_3. The vector c contains the unknown values of the coefficients of Trefftz functions series. To build this system of equations only collocation nodes from the boundary Γ_2 and Γ_3 are taken into account.

In practice, the boundary conditions can not be determined precisely. To make the measurement more realistic the vector F may contain a random noise of boundary conditions with a zero mean value. If $N_{\text{inv}} > M$, and the matrix $\mathbf{D_{inv}}$ has the $rank(\mathbf{D_{inv}}) = N_{\text{inv}}$, the least-squares solution is unique and unbiased [3]. If the matrix $\mathbf{D_{inv}}$ has a deficient rank and in the vector $\mathbf{F}$ a noise is implemented, then the solution c is unstable and the system (3.31) is strongly ill-conditioned. The SVD allows to regularize the solution with a minimal norm.

The singular value decomposition of matrix $\mathbf{D_{inv}}$ has a form [3, 4]:

$$\mathbf{D_{inv}} = \mathbf{W}\boldsymbol{\Sigma}\mathbf{V}^T \tag{3.32}$$

where $\mathbf{W} \in R^{N_{\text{inv}} \times N_{\text{inv}}}$ and $\mathbf{V} \in R^{M \times M}$ are orthogonal matrices (which contain left and right singular vectors) and the rectangular matrix Σ contains singular values of the matrix $\mathbf{D}$:

$$\boldsymbol{\Sigma} = diag\,(\sigma_1, \sigma_2, \ldots, \sigma_m) \in R^{N_{\text{inv}} \times M} \tag{3.33}$$

with the order:

$$\sigma_1 \geq \sigma_2 \geq \cdots \geq \sigma_m > 0 \tag{3.34}$$

The solution of the (3.31), using the SVD is equal to [3, 4]:

$$\mathbf{c} = \sum_{i=1}^{rank(\mathbf{D_{inv}})} \frac{\mathbf{w}_i^T \mathbf{F} \mathbf{v}_i}{\sigma_i} \tag{3.35}$$

where vectors $\mathbf{w}_i$ and $\mathbf{v}_i$ are the column vectors from the matrix $\mathbf{W}$ and $\mathbf{V}$. In numerical computations the matrix $\mathbf{D_{inv}}$ has no rank exactly equal to the mathematical rank [3, 4]. The numerical rank is smaller than the mathematical rank, because of small nonzero singular values. In (3.31) when the matrix $\mathbf{D_{inv}}$ has very small nonzero singular values, then a norm of the solution $\mathbf{c}$ is very large and noise effects will be amplified (when $\sigma_i < \mathbf{w}_i^T \mathbf{F}$). To remove this effect, the least singular values must be neglected, so the new solution takes the form:

$$\mathbf{c}^{(n)} = \sum_{i=1}^{n} \frac{\mathbf{w}_i^T \mathbf{F} \mathbf{v}_i}{\sigma_i} \tag{3.36}$$

and n is a truncation number [3, 4]. Equation (3.36) corresponds to the new least-squares problem, given by [3]:

$$\min \|\mathbf{D_n c} - \mathbf{F}\|_2 \tag{3.37}$$

$\|\cdot\|_2$ is an Euclidean norm. The solution $\mathbf{c}^{(n)}$ is also known as the truncated singular value decomposition (TSVD) solution [3, 4]. This solution is equivalent to the solution obtained by the Monroe-Penrose pseudoinverse of the matrix $\mathbf{D_{inv}}$ [3, 4]. The problem is to find the matrix $\mathbf{D_{inv}}$ with truncated singular values, which will give the solution $\mathbf{c}$ with a minimal norm and with the condition of the form:

$$\|\mathbf{D_n c} - \mathbf{F}\|_2 < err \tag{3.38}$$

where *err* denotes a certain measure of noise in the vector $\mathbf{F}$.

The truncation number is a regularization parameter in this method. To find this number several algorithms are developed, for example: discrete Picard condition, discrepancy principle, generalized cross-validation technique, L-curve algorithm [3, 4, 10] and others. In this work the L-curve algorithm is used with Triangle Method [11], to automatically detect the regularization parameter.

In Fig. 3.1 the spectrum of the singular values is presented. It is easy to see, that when the number of the singular values increases, singular values become smaller [3, 4].

In present the work the L – curve method is used to determine the truncation number. The L – curve is a parametric curve of the norm of the regularized solutions vs. the norm of the residual [3, 11], where the parameter is the truncation number. In Fig. 3.2, L – curve is presented for the case without noise in the data, a log-log scale is used, because this scales emphasize the two flat parts of the curve, where the variation of the variables are small [4].

When the noise is in the data, L – curve has a different shape, which is presented in Fig. 3.3.

The corner of the L – curve determines the minimization of the residual norm and the norm of the regularized solution [3, 11]. The corner point (optimal truncation

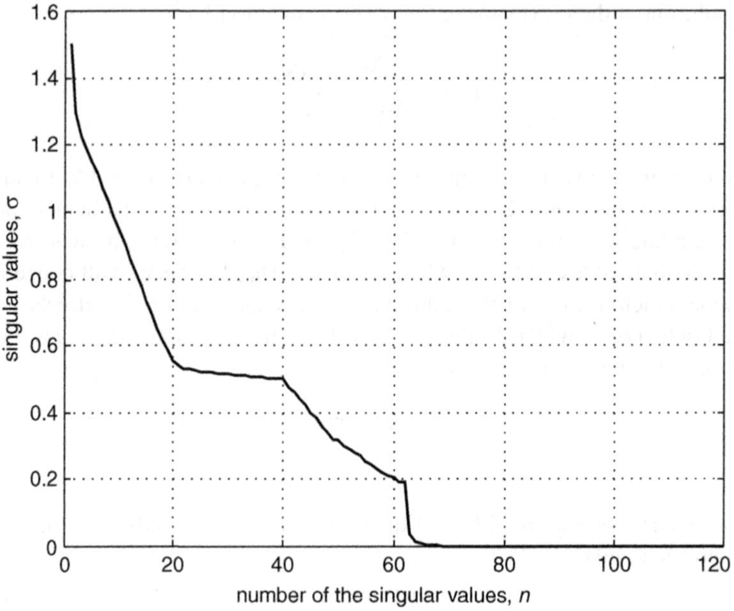

Fig. 3.1 Spectrum of the singular values of the matrix D

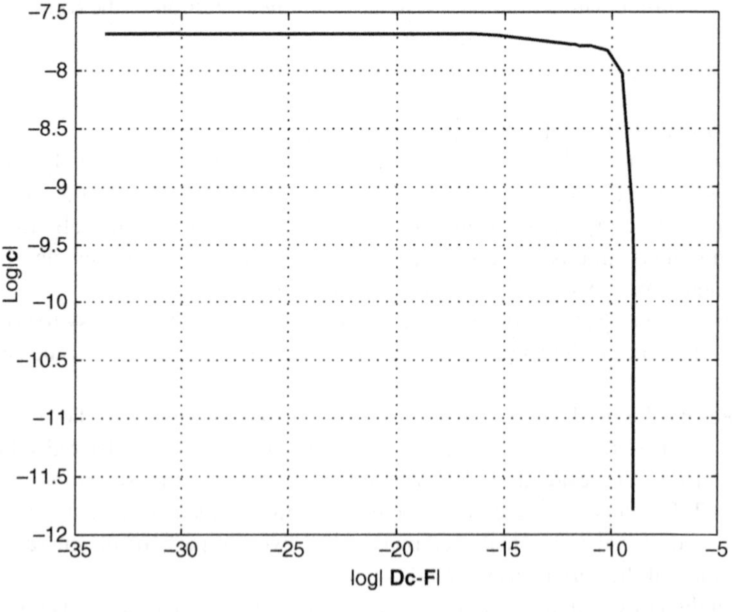

Fig. 3.2 L – curve for the data without noise

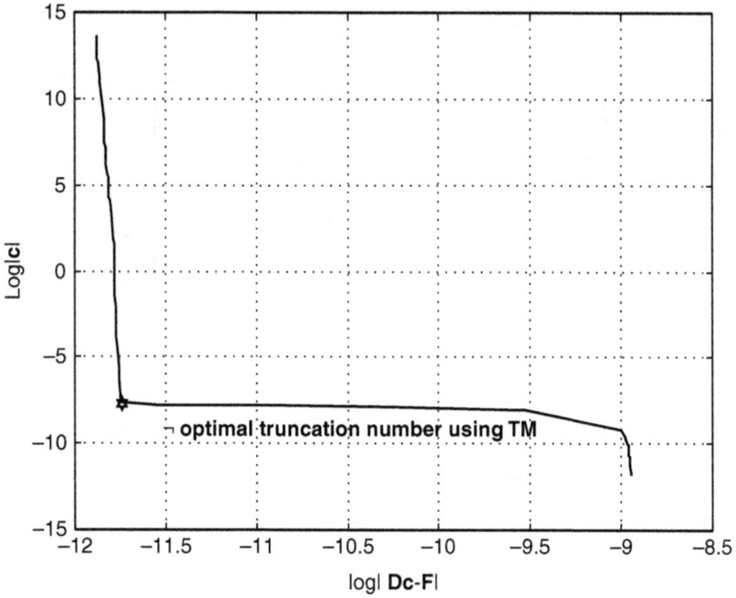

Fig. 3.3 L – curve for the data with noise r = 5%

number, which is the searched parameter) is determined automatically using the Triangle Method (TM) [11].

3.6 Numerical Examples

The piezoelectric plate, shown in Fig. 3.4 is considered. The PZT-4 ceramic material is modelled [9]. The applied voltage is $\varphi = 200$ [V]. On the vertical edges of the strip, charge flux density is equal to zero. The material properties are: elastic compliance constants: $s_{11} = 16.4 \cdot 10^{-6}$, $s_{12} = -7.22 \cdot 10^{-6}$, $s_{22} = 18.8 \cdot 10^{-6}$, $s_{33} = 47.5 \cdot 10^{-6}$ [mm^2/N]; piezoelectric moduli: , $d_{21} = -172 \cdot 10^{-9}$, $d_{22} = 374 \cdot 10^{-9}$, $d_{13} = 584 \cdot 10^{-9}$ [mm/V]; dielectric constants $\varepsilon^{\sigma}_{11} = 1.53 \cdot 10^{-9}$, $\varepsilon^{\sigma}_{22} = 1.51 \cdot 10^{-7}$ [N/V^2] are measured at constant stresses [9]. Elastic moduli, piezoelectric constants and dielectric constant measured at constant strains, can be easily obtained using the constitutive equations (3.2). The applied stresses are equal to $\sigma = 1$ [MPa]. The length and the height of the strip is equal to $L = 1$ [mm]. To discretize the boundary of the strip 80 boundary collocation nodes are applied. On the boundary part Γ_2 (nodes 21–40) both generalized displacements and tractions are known, on the part Γ_1 (nodes 61–80) the above quantities are both unknown. Nodes 1–20 and 41–60 constitute the part Γ_3, where the boundary conditions are applied as in the direct problem. The results obtained using the SVD, are compared with the solutions obtained by the direct method.

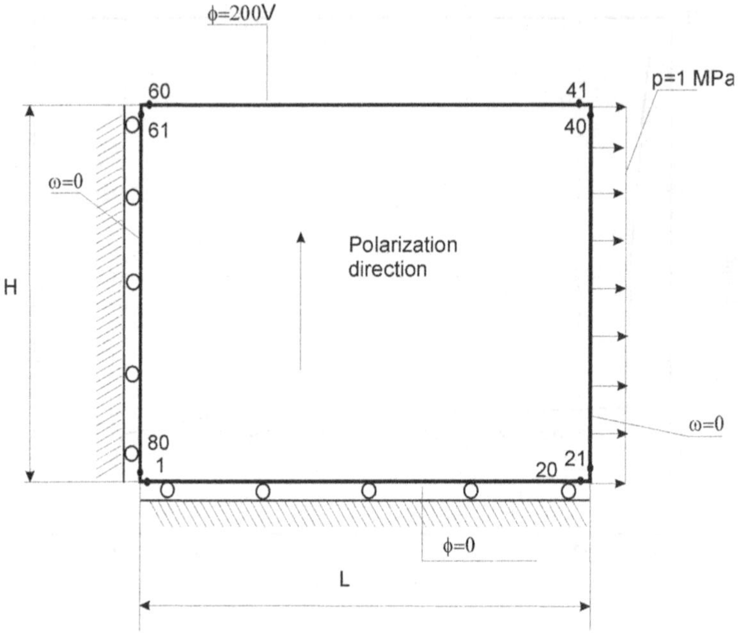

Fig. 3.4 Piezoelectric plate

The case of the perturbated boundary conditions is also considered. Quantities Q_i along the boundary Γ_2 were perturbed using an additived normal noise, with the mean equal to zero and the standard deviation equal to [3]:

$$d_i = \max_{\Gamma_2} |Q_i| \frac{r}{100} \tag{3.39}$$

where r is the percentage of the additive noise included in the input data quantities Q_i.

The error of the numerical inverse solution is calculated for a generalized boundary displacement vector **U** and generalized traction vector **T**. The percentage error is defined as a L^2 norm:

$$\varepsilon = \frac{1}{\left|f_{\max}^e\right|} \sqrt{\frac{1}{N} \sum_{i=1}^{N} (f_i^n - f_i^e)^2} \cdot 100\% \tag{3.40}$$

where: $\left|f_{\max}^e\right|$ maximal value of quantity f over N nodes obtained by the direct problem solution, f_i^n – nodal value obtained by TCM at node i, f_i^e – nodal value obtained by the direct problem solution at node i.

To investigate stability and accuracy of the present method, the noise is implemented into the data. Five levels of noise are considered $r = 1$, 2, 5, 10 and 20%. Additional quantities (the generalized displacements) on the boundary Γ_2 is perturbed in accordance with (3.39). The average errors, for 50 numerical experiments on each level of the noise are presented in Table 3.1.

Table 3.1 Errors of the inverse solution for a data with different level of noise

Noise level r (%)	L^2-norm error for generalized displacements (%)	L^2-norm error for generalized tractions (%)
0	1.13	4.38
1	1.15	4.41
2	1.20	4.47
5	1.54	6.35
10	2.12	10.94
20	3.40	12.42

It is easy to see, that the solution for a generalized displacements is better than the solution for a generalized tractions. The measure of the error shows that, if the level of noise is higher, then solution is less accurate. But, the TCM allows to obtain solutions with rather good accuracy, even for the high level of noise into data. The special form of the solution in the TCM method (coefficients, not the boundary quantities) and proper regularization technique give very efficient computational tool for solving the Cauchy inverse problems of linear piezoelectricity.

The comparison between the solution obtained using the direct and inverse formulation is shown in Figs. 3.5, 3.6, 3.7, 3.8, 3.9 and 3.10.

In Fig. 3.5 comparison between the direct and inverse z-direction (x_3) displacement solution for a noise level $r = 1\%$ is shown.

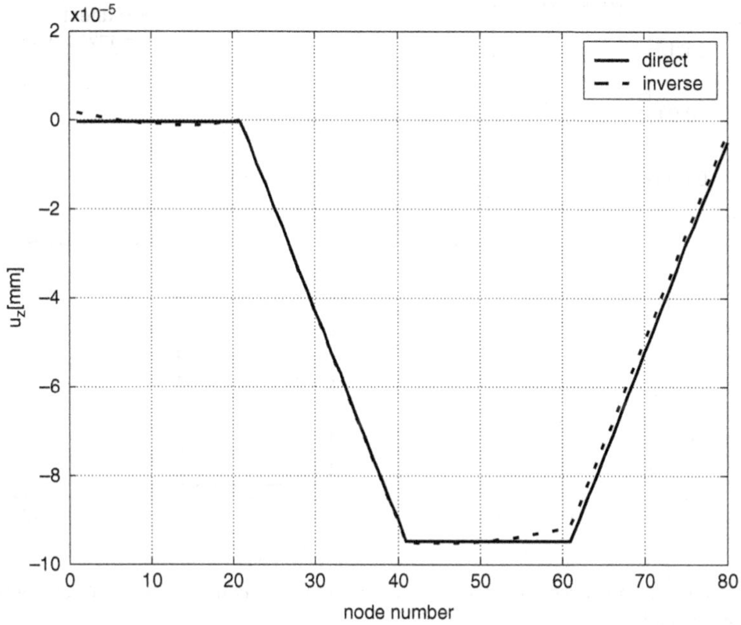

Fig. 3.5 Comparison between the direct and inverse z-direction displacement solution (data with the noise level $r = 1\%$)

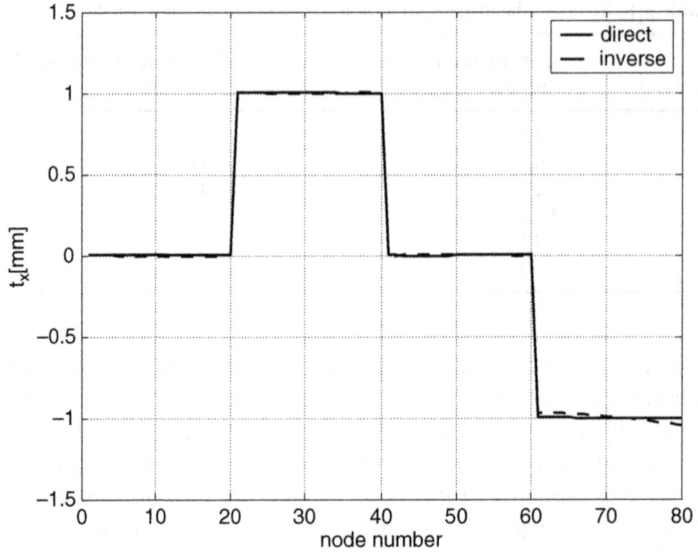

Fig. 3.6 Comparison between the direct and inverse x-direction traction solution (data with the noise level $r = 1\%$)

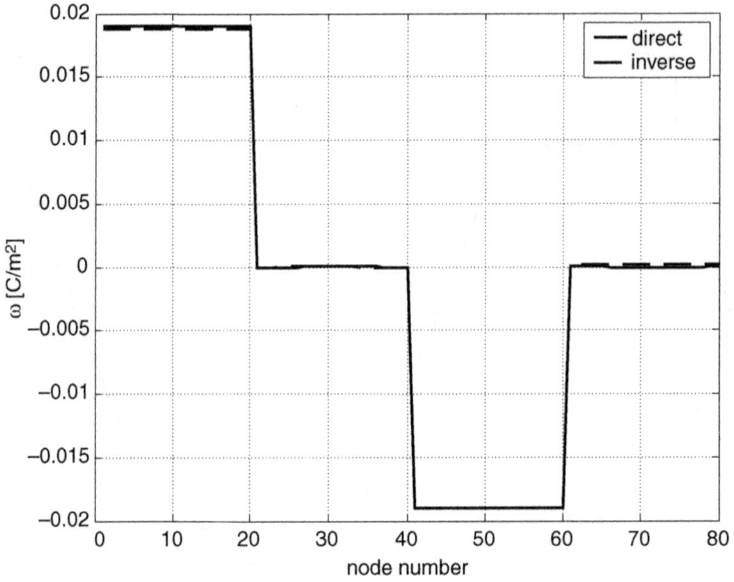

Fig. 3.7 Comparison between the direct and inverse charge flux density solution (data with the noise level $r = 1\%$)

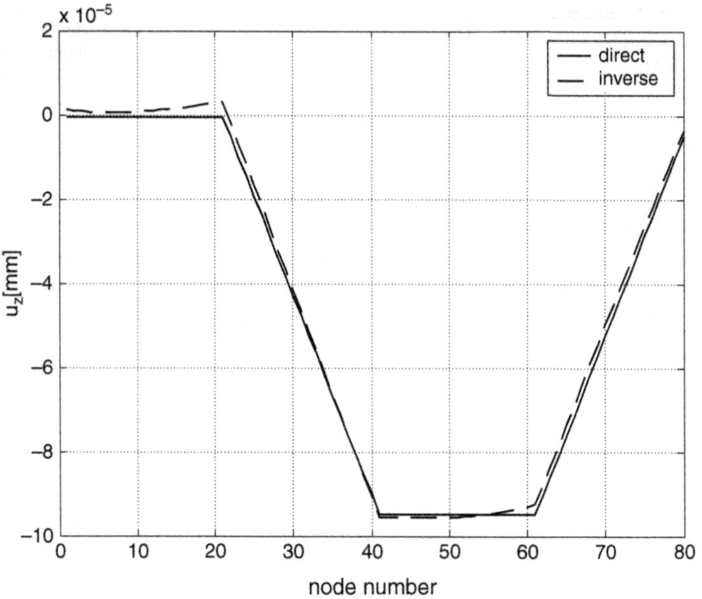

Fig. 3.8 Comparison between the direct and inverse z-direction displacement solution (data with the noise level $r = 20\%$)

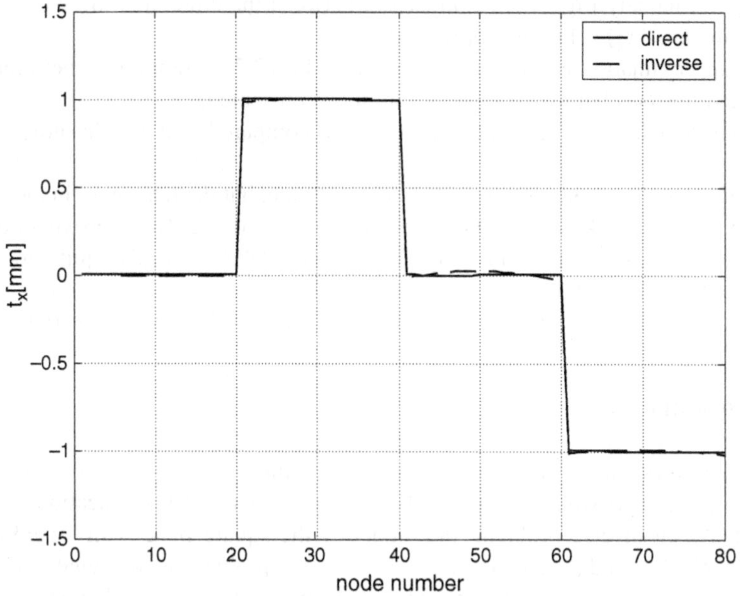

Fig. 3.9 Comparison between the direct and inverse x-direction traction solution (data with the noise level $r = 20\%$)

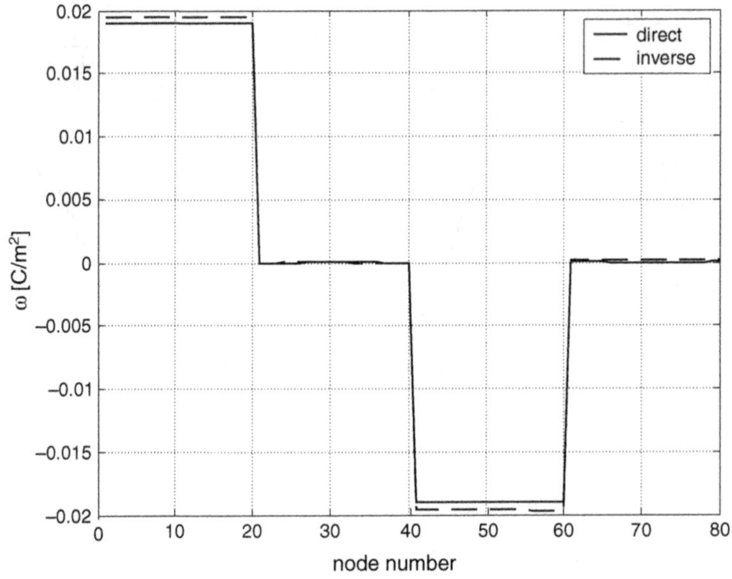

Fig. 3.10 Comparison between the direct and inverse charge flux density solution (data with the noise level $r = 20\%$)

This solution gives good approximation, but the noise level is quite low. In the next figure, namely Fig. 3.6, a comparison between the direct and inverse traction in x −direction (x_I) solution is presented.

The approximation is also quite good. In the Fig. 3.7, a comparison between the direct and inverse charge flux density solution is presented.

In the next figures the same quantities are compared, but for the noise level $r = 20\%$.

In this case, the approximation is more inaccurate, but the tendency (in the sense of least squares) is kept. For this level of the noise ($r = 20\%$) approximation of tractions in the x-direction and charge flux density solutions is also quite good, as shown in Figs. 3.9 and 3.10.

Also, it can be notice that the discrepancy between both solution is very small.

3.7 Conclusions

In present work the inverse Cauchy problem of linear piezoelectricity is considered. To solve approximately these problems the indirect Trefftz collocation method is used. As an inverse problem, the model of the square plate with mixed electromechanical boundary conditions is solved. This problem is also analyzed with perturbated data. The examples show that the truncated singular value decomposition procedure with the L − curve method (and the triangle method) allows to obtain correct solutions of the inverse problems of piezoelectricity, even for high level of noise. The L – curve method is helpful to detect the optimal value of the truncation number, which minimizes both residual and solution norm. The present regularization technique allows to obtain stable, quite accurate results, even for high level of the noise in the input data.

References

1. Pan, E.: A BEM analysis of fracture mechanics in 2D anisotropic piezoelectric solids. Eng. Anal. Bound. Elem. **23**, 67–76 (1999)
2. Jin, W.G., et al.: Trefftz indirect method for plane piezoelectricity. Int. J. Numer. Methods Eng. **63**, 139–158 (2005)
3. Marin, L., et al.: Boundary element regularization methods for solving the Cauchy problem in linear elasticity. Inverse Probl. Eng. **10**, 335–357 (2002)
4. Marin, L., Lesnic, D.: Boundary element solution for Cauchy problem in linear elasticity using singular value decomposition. Comput. Methods Appl. Mech. Eng. **191**, 3257–3270 (2002)
5. Qin, Q.-H.: Variational formulations for TFEM of piezoelectricity. Int. J. Solids Struct. **40**, 6335–6346 (2003)
6. Ting, T.C.T., Wang, Y.M.: The Stroh formalism for anisotropic materials that possess an almost extraordinary degenerate matrix N. Int. J. Solids. Struct. **34**, 401–413 (1997)
7. Portela, A., Charafi, A.: Programming Trefftz boundary elements. Adv. Eng. Soft. **28**, 509–523 (1997)
8. Xiaoping, Z., Zhen-han, Y.: Some applications of the Trefftz method in linear elliptic boundary-value problems. Adv. Eng. Soft. **24**, 133–145 (1995)
9. Sheng, N., Sze, K.Y., Cheung, Y.K.: Trefftz solutions for piezoelectricity by Lekhnitskii's formalism and boundary-collocation method. Int. J. Numer. Methods Eng. **65**, 2113–2138 (2006)
10. Ahmadian, H., Mottershead, J.E., Friswell, M.I.: Regularization methods for finite element model updating. Mech. Syst. Signal Process **12**, 47–64 (1998)
11. Castellanos, J.L., Gomez, S., Guerra, V.: The triangle method for finding the corner of the L – curve. Appl. Numer. Math. **43**, 359–373 (2002)

Chapter 4
New Phenomenological Model for Solid Foams

V. Goga

Abstract New phenomenological material model for solid foam materials is presented. This model describes the uniaxial compressive stress-strain curves of foam materials. Parameters of this model specifically control the shape of compressive stress-strain curve and they are function of foam density. Specimens of polyurethane (PUR) foam with different densities were compressed and their stress-strain curves were fitted by model. Functions of model parameters were used for predict the compressive stress-strain curve of other density foam specimen.

Keywords Foam materials · Material model · Stress-strain curve · Compression test

4.1 Introduction

Natural cellular solids like wood, corals, cancellous bones, bodies of plants and animals, etc. are using in load-bearing structures in nature. Combination of cellular structure and properties of the material from which the cellular solid is made gives to cellular solids specific properties, which were the reason for man-made cellular solids. Their low density, thermal and mechanical properties are very interesting for engineering, automotive and building industries. There are many possible applications for cellular materials ranging from light-weight construction, sound and heat insulation, heat exchangers, energy absorption systems, vibration control and acoustical scattering. Their production and utilization increase in last 30 years. It is possible to make cellular solids almost from all kind of materials at present. Metals and polymers are particularly use, but also the glass and ceramics can be fabricated into cells [1].

In cellular structures the solid material builds up an interconnected network of struts and walls. The voids which are bordered by them thus create cells. The

V. Goga (✉)
Department of Mechanics, Faculty of Electrical Engineering and Information Technology,
Slovak University of Technology, Bratislava, Slovakia
e-mail: vladimir.goga@stuba.sk

J. Murín et al. (eds.), *Computational Modelling and Advanced Simulations*,
Computational Methods in Applied Sciences 24, DOI 10.1007/978-94-007-0317-9_4,

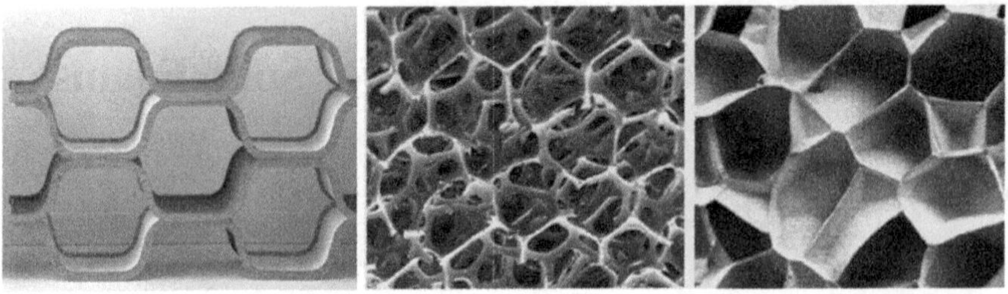

a) b) c)

Fig. 4.1 Structure of cellular solids: (**a**) honeycomb, (**b**) open-cell foam, (**c**) closed-cell foam

simplest structure of cellular materials is two-dimensional cellular solids called honeycombs (Fig. 4.1a). They are two-dimensional array of polygons, which pack to fill a plane like the hexagonal cells of the bee. More commonly are three-dimensional cellular materials. Their cells are polyhedrons, which pack in three dimensions to fill space. These materials are called foams. The structures of foams could have open or closed-cells. Open-cell foam (Fig. 4.1b) has the solid material only in cell edges (so that the cells connect through open faces). Closed-cell foam (Fig. 4.1c) has the solid material in edges and faces of cells (so that each cell is sealed off from its neighbours) [1].

Foam energy absorbing capabilities are widely used in packing and automotive industry to protect the package object from failure when it is subjected to impact and to prevent injuries to the occupants in the event of front or side collisions. Foams absorb the impact energy while keeping the stress acting on the object lower than its damage limit. One of the common features of energy absorption materials is that there is a discernible plateau in their compressive stress-strain curves. This feature means that the materials can absorb energy by deformation but keep the stress almost constant [2, 3].

Figure 4.2 shows a typical shape of the compressive stress-strain curve for foams. The curve exhibits three definite regions: linear elasticity, plateau and densification. At small strains, usually less then 5%, the behaviour is linear elastic, with a slope equal to the Young modulus of the foam. As the load increases, the foam cells begin to collapse by elastic buckling, plastic yielding or brittle crushing, depending on the mechanical properties of the cell walls. Collapse progresses at roughly constant load, giving a stress plateau, until the opposing walls in the cells meet and touch, when densification causes the stress to increase steeply. In the unloading phase, the stress varies non-linearly with the strain. The extent of each region depends on its relative density and responds to different mechanical properties [1, 2, 4]. Relative density ρ_{rel} is the most important aspect of the foam structure. It's ratio ρ^*/ρ_S, where ρ^* is the density of the foam and ρ_S that of solid of which the foam is made. It can vary from 1 to as little ass 0.01 [1, 4].

The work done per unit volume – specific energy (4.1) in deforming the foam to a given strain ε is simply the area under the stress-strain curve up to the strain ε (Fig. 4.2). Very little energy is absorbed in the linear elastic region; it is the long plateau of the stress-strain curve that allows large energy absorption at near constant

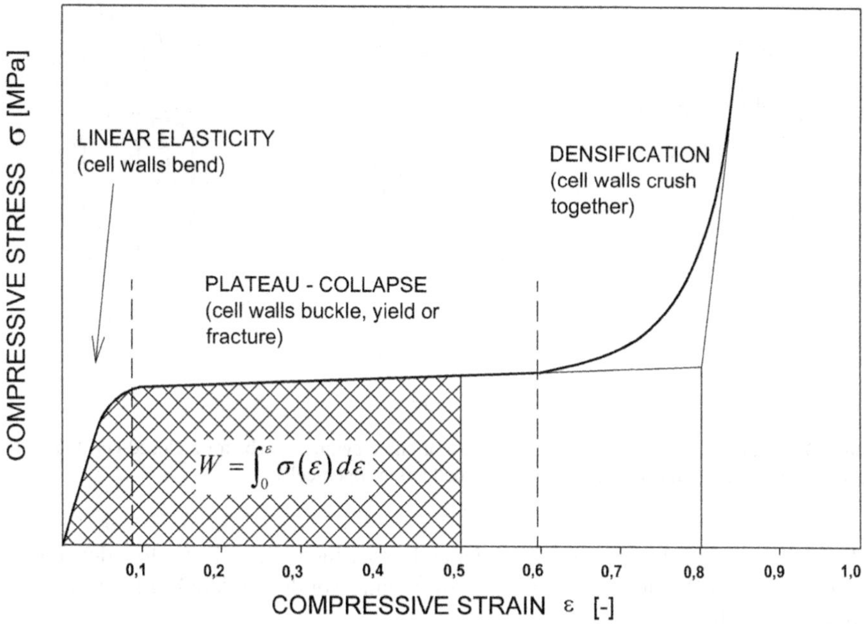

Fig. 4.2 The typical shape of the compressive stress-strain curve for solid foams

load. For calculation absorbed energy is needed to describe the stress-strain relationship. For this reason some material models and constitutive models have been researched [1, 3].

$$W = \int_0^{\varepsilon} \sigma(\varepsilon)\, d\varepsilon \tag{4.1}$$

4.2 Modeling of Cellular Solids

Cellular solids models can be divided in two categories [5]: phenomenological models and micromechanical models. The phenomenological models aim to reach the best fit with experimental mechanical behaviour without direct relationship with the physics of the phenomenon. The micromechanical models are based on the analysis of the deformation mechanisms of the micro-cell structure under loading.

4.2.1 Micromechanical Models

The most known and widely used micromechanical model of an open cell foam was put forth by Gibson and Ashby [1, 6] in which the foam is modelled as an array of cubic cells of length l, and struts of thickness t, as shown in Fig. 4.3a [1, 7]. The elastic (Young's) modulus of the cellular structure can be calculated from the elastic deflection δ of length l loaded at its midpoint by a load F, as shown in Fig. 4.3c

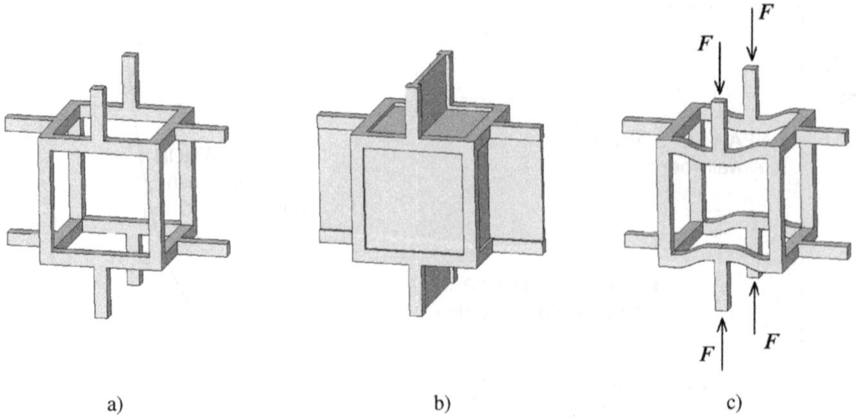

Fig 4.3 Unit cell [1]: (**a**) for an open-cell foam of cubic symmetry, (**b**) for a closed-cell foam of cubic symmetry, (**c**) shown after linear-elastic deflection (open-cell)

[1, 7]. Analysis of the mentioned model and experimental results (at low densities) indicate that the Young's modulus (E^*) of cellular solids is related to their density (ρ^*) through the relation [1, 6]:

$$\frac{E^*}{E_S} = C\left(\frac{\rho^*}{\rho_S}\right)^n \tag{4.2}$$

where E_S and ρ_S are the Young's modulus and density of the solid skeleton. The constants C and n depend on the microstructure of the solid material. Similar relations hold for the bulk and shear modulus, plastic collapse stress of plastic foam (4.6) and crushing strength of brittle foam (4.7) with possibly different values of C and n. The value of n generally lies in the range $1 \leq n \leq 4$ [8].

In deriving (4.2), it is assumed that all material of the foam is found in the struts that define the cell. In a closed cell foam (Fig. 4.3b [1, 7]), some fraction of the material resides in the cell walls or faces rather than in the struts. If the the fraction of material contained in the cell struts having thicknes t is Φ (for open cell foam $\Phi = 1$), then the fraction contained in the cell walls of thickness t_f is $1 - \Phi$ The stiffness of a closed-cell foam results then from three contributions [9]:

- The first component is strut bending, as for open cell foams.
- The second component is membrane (cell face) stretching, which arises as the result of strut flexure causing the cell walls to deform.
- The final component is the internal gas pressure of the closed-cells.

Gibson and Ashby [1, 6, 9] derived the modulus of a closed-cell foam which accounts for all three components:

$$\frac{E^*}{E_S} \approx \Phi^2\left(\frac{\rho^*}{\rho_S}\right)^2 + (1-\Phi)\frac{\rho^*}{\rho_S} + \frac{p_0\,(1-2\upsilon^*)}{E_S(1-\rho^*/\rho_S)} \tag{4.3}$$

The first term on the right describes the contribution of the cell struts to modulus while the second term accounts for the cell walls. The third term is the contribution due to the internal gas pressure, where υ^* is Poisson's ratio for the foam and p_O is a gas pressure in the cell. Nevertheless, a number of experiments have proved that most foams behave as if their cells were open because surface tension concetrates material into the cell edge during their manufacture. The faces are frequently so thin that they contribute very little to the overall stiffness and strength, and the mechanical properties of a closed-cell foam can be treated as open-cell [2]. The equations to describe the mechanical properties of cellular solids [1, 4]:

– Linear elastic region:

$$\frac{E^*}{E_S} = C_1 \left(\frac{\rho^*}{\rho_S}\right)^2, \tag{4.4}$$

– Nonlinear elastic collapse:

$$\frac{\sigma^*_{el}}{E_S} = C_2 \left(\frac{\rho^*}{\rho_S}\right)^2, \tag{4.5}$$

– Plastic collapse:

$$\frac{\sigma^*_{pl}}{\sigma_Y} = C_3 \left(\frac{\rho^*}{\rho_S}\right)^{3/2}, \tag{4.6}$$

– The crushing strength:

$$\frac{\sigma^*_f}{\sigma_f} = C_4 \left(\frac{\rho^*}{\rho_S}\right)^{3/2}, \tag{4.7}$$

where E^* is Young's modulus of foam, E_S is Young's modulus of cell wall materials, σ^*_{el} is elastic collapse stress of elastic foam, σ^*_{pl} is plastic collapse stress of plastic foam (equivalent to plateau stress), σ_Y is yielding strength of cell wall material, σ_f^* is crushing strength of brittle foam, σ_f is modulus of rupture of the cell-wall material, ρ^* and ρ_S is density of foam and cell wall material of which the foam is made, respectively, and C_1, C_2, C_3 and C_4 are constants [1, 4].

Constants $C_1 = 1$; $C_2 = 0.05$; $C_3 = 0.3$ and $C_4 = 0.65$ give a good description of a wide range of foam materials with relative density below 0.3 [1, 4].

In densification region the stress rises steeply. Starting point of densification is densification strain ε_D. It is a strain where the plateau stress starts rapidly increasing. Gibson and Ashby [1] have found that the densification strain is a linear function of the relative density both for closed- and open-cell foams:

$$\varepsilon_D = 1 - 1,4\rho_{\text{rel}}. \tag{4.8}$$

A coefficient of correction is proportional to the cube of the relative density was introduced later by Ashby in [10] in order to improve the quality of the fitting:

$$\varepsilon_D = (0,9 \div 1)\left(1 - 1,4\rho_{\text{rel}} + 0,4\rho_{\text{rel}}^3\right). \tag{4.9}$$

Chan and Xie suggested an analytical model for the densification strain [11]. They define it as the intersection of the slopes of stress-strain curves in the plateau and densification regime. According to their calculations the expression for the densification strain of closed-cell foams is similar to (4.8):

$$\varepsilon_D = 1 - \alpha_C \rho_{\text{rel}}. \tag{4.10}$$

The α_C factor indicates that the foam cannot be totally compressed, in practice, a certain amount of pores remains in the crushed material. Equation (4.10) changes to (4.11) in the case of open-cell foams:

$$\varepsilon_D = 1 - \alpha_O \left(\rho_{\text{rel}}\right)^{1/2} \tag{4.11}$$

where α_O is a constant, depending on the foam structure and the cell-edge material. The difference between (4.10) and (4.11) indicates that the amount of the poresin the crushed foams depends also on the cell-structure, and this modifies the expression for the densification strain [12]. The compression test performed by Chan and Xie shows, that α_O is approximately 1.4 ($\alpha_O \approx 1.4 - 1.7$), so (4.8) describes well the densification strain of closed-cell foams. In the case of open-cell foams (4.8) and (4.9) were found to give good results only when the fitting regime is not too wide ($\Delta\rho_{\text{rel}} < 10\%$). When the fitting regime is greater than approximately 20%, (4.11) seems to give better fit [12].

4.2.2 Phenomenological Models

Polymeric foams mechanical properties were experimental characterized by several researchers [13–16, 17]. Rusch presented one widely accepted work on this topic [18]. The relationship between compressive stress and strain was expressed as:

$$\sigma = Ef(\varepsilon), \tag{4.12}$$

where E is the initial compressive modulus of the foam and $f(\varepsilon)$ is a nonlinear strain function. Although this model represents the loading characteristics of rigid foams reasonably well, id does not consider the effect of the strain rate and temperature. Meinecke and Schwaber [19] included strain rate dependence of compressive modulus in the above equation and proposed the following model:

$$\sigma = E(\dot{\varepsilon})f(\varepsilon). \tag{4.13}$$

Nagy et al. [20] modified the above model by replacing the modulus term using a coupled strain and strain rate function as:

$$M(\varepsilon, \dot{\varepsilon}) = \left(\frac{\dot{\varepsilon}}{\dot{\varepsilon}_0}\right)^{b_1/b_2\varepsilon}, \tag{4.14}$$

where b_1 and b_2 are empirical constants and $\dot{\varepsilon}_0$ is a reference strain rate.

All the above constitutive models were verified to be applicable to foams with a given density, but can not be used to characterize the effect of density. Sherwood and Frost [21] combined the effect of temperature, density, strain and strain rate as:

$$\sigma = H\,(T)\,G\,(\rho)\,M(\varepsilon, \dot{\varepsilon})f\,(\varepsilon)\,, \tag{4.15}$$

where $H(T)$ and $G(\rho)$ are functions of temperature and initial density, respectively, and $f(\varepsilon)$ is a shape function. All these functions can be approximated from experimentally obtained data. Chou et al. [22] coupled temperature and strain rate effect in the above model as:

$$\sigma = H\,(\dot{\varepsilon}, T)\,G\,(\rho)f\,(\varepsilon)\,, \tag{4.16}$$

and verified the model on polyurethane foams at a range of temperatures, strain rates and initial densities.

For crushable and hysteretic foams, Faruque et al. [23] formulated a stain rate and temperature dependent constitutive model and implement it in an explicit dynamic finite element code. The primary feature of the model is the inclusion of strain rate effect in Young's modulus and yield strength, hardening of Young's modulus with compaction and tension cut off. The constitutive model is expressed in terms of modulus, which is a function of strain rate and volumetric strain (densification) as:

$$E = E_0\left(1 + a\ln\left(\frac{\dot{\varepsilon}_{\text{eff}}}{\dot{\varepsilon}_0}\right)\right)\left(1 + b\left(\frac{\varepsilon_v}{\varepsilon_{0v}}\right)^2\right), \tag{4.17}$$

where E_0 is the quasistatic modulus at zero volume change, a is the strain rate coefficient and b is the densification coefficient, $\dot{\varepsilon}_{eff}, \dot{\varepsilon}_0, \varepsilon_v, \varepsilon_{0v}$ are the effective strain rate, reference strain rate, volumetric strain and limiting volumetric strain beyond which the modulus is independent of volumetric strain, respectively.

For low-density polymeric foams subjected to high strain rate impact loading, Zhang et al. [24] proposed the following strain rate and temperature dependent constitutive equation and implemented it in a finite element code:

$$\sigma\,(\varepsilon) = \sigma_0\,(\varepsilon)\,H\,(T)\left(\frac{\dot{\varepsilon}}{\dot{\varepsilon}_0}\right)^{a+b\varepsilon}. \tag{4.18}$$

Liu and Subhash in [17] presented a multi-parameter phenomenological model that captures the entire nonlinear stress-strain characteristics of structural porous materials under large deformations:

$$\sigma(\varepsilon) = A\left(\frac{e^{\alpha\varepsilon} - 1}{B + e^{\beta\varepsilon}}\right) + e^{C}\left(e^{\gamma\varepsilon} - 1\right), \tag{4.19}$$

where parameters A, B, α and β are constant for a given density and strain rate. Parameters C and γ captured the densification phase.

4.3 New Phenomenological Model for Solid Foams

New phenomenological model for foam materials describes the phenomenon, which is in this case the shape of compressive stress-strain curve. The ambition was to create model with few parameters and every parameter should be depended on relative density, because mechanical properties like Young's modulus and plateau stress usually increases and densification strain (densification strain – it is strain where plateau ends and densification starts) decreases with increasing foam density.

Materials undergoing strain are often modelled with mechanical components, such as springs (restorative force component) and dashpots (damping component) – rheological models. Connecting a spring and dashpot in series yields a model of a Maxwell material (Fig. 4.4a.) while connecting a spring and damper in parallel yields a model of a Kelvin-Voight material (Fig. 4.4b.). These models have been used to modelling the behaviours (creep and stress relaxation) of a viscoelastic materials. Spring represents the elastic component of a viscoelastic material and dashpot represents the viscous component of a viscoelastic material. Different combinations of these basic models give an opportunity to modeling various behaviours of materials.

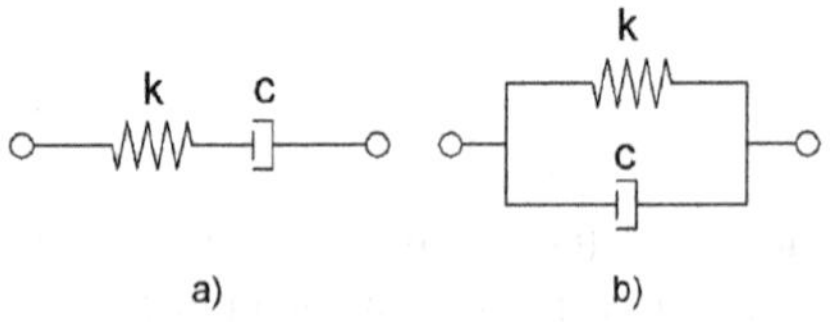

Fig. 4.4 Rheological models: (**a**) Maxwell model, (**b**) Kelvin-Voight model

4.3.1 Foam Model

New foam model consists of three systems in parallel (Fig. 4.5). The first, referred to as the Maxwell arm, contains a spring (stiffness k) and dashpot (viscosity c) in series. The other two systems contain only springs (k_P and k_D). This model can be used to accurately predict the general shape of the foam's compressive stress-strain curve.

Stress-strain response of Maxwell model describes the first and the second region of foam compression deformation, but only if the plateau stress is constant

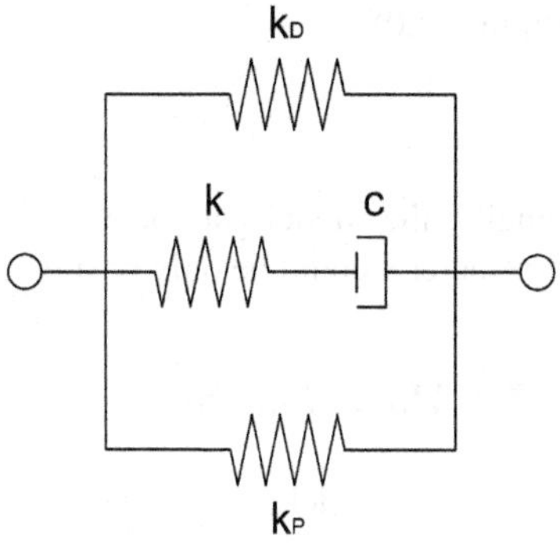

Fig. 4.5 New foam model

(Fig. 4.6). If the plateau stress increases or decreases, it is necessary to integrate next spring with stiffness k_P parallel with Maxwell model. Stiffness k_P represents a slope of the plateau stress. Densification region is controlled by last spring with nonlinear stiffness k_D. This spring is in parallel connection with other two components.

Stiffness coefficient k is equivalent to foam Young's modulus, damping coefficient c is equivalent to plateau stress, stiffness k_P represents a slope of the plateau stress and stiffness coefficient k_D is 2-parameters exponential function of

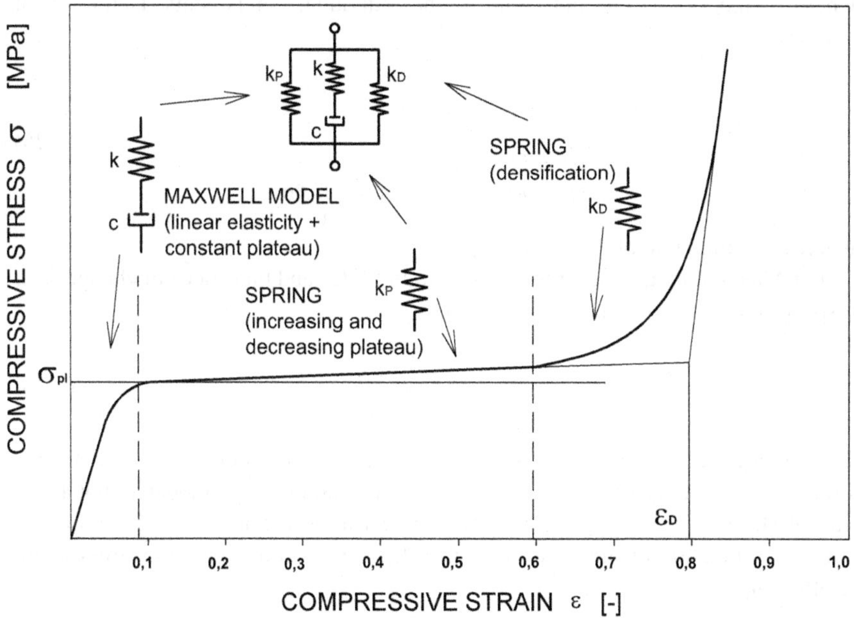

Fig. 4.6 Compressive stress-strain curve describes by foam model

strain (4.20):

$$k_D(\varepsilon) = \gamma(1 - e^{\varepsilon})^n \tag{4.20}$$

Finally, the model parameters are: k, c, k_P and γ and n from (4.20). Values of parameter n could be only positive even numbers.

4.3.2 Model Solving

In order to model this system, the following physical relations must realize:

- for series components:

$$\sigma_M = \sigma_k = \sigma_c \tag{4.21}$$

$$\varepsilon_M = \varepsilon_k + \varepsilon_c \tag{4.22}$$

- for parallel components:

$$\sigma = \sigma_M + \sigma_P + \sigma_D \tag{4.23}$$

$$\varepsilon = \varepsilon_M = \varepsilon_P = \varepsilon_D \tag{4.24}$$

where σ and ε is stress and strain of the individual components. Definition of stresses:

$$\begin{aligned} \sigma_k &= k\varepsilon_k \\ \sigma_c &= c\dot{\varepsilon}_c \\ \sigma_P &= k_P\varepsilon_P \\ \sigma_D &= k_D\varepsilon_D = \gamma\left(1 - e^{\varepsilon_D}\right)^n \varepsilon_D \end{aligned} \tag{4.25}$$

where $\dot{\varepsilon}_c$is strain rate of the dashpot.

For Maxwell model the stress (4.21), strain (4.22) and their rates of change with respect to time are governed by equation:

$$\dot{\varepsilon}_M = \frac{\dot{\sigma}_M}{k} + \frac{\sigma_M}{c} \tag{4.26}$$

The strain rate is a constant in this foam model (we consider $\dot{\varepsilon}_M = 1\,\mathrm{s}^{-1}$). This strain rate is not real rate of compression deformation. This model is not function of strain rate. Than, there is only one derivation element. It is derivation of stress as a function of time. Result of this differential equation is the stress-strain relationship:

$$\sigma_M(t) = e^{-\frac{kt}{c}}\left(-1 + e^{\frac{kt}{c}}\right) c\dot{\varepsilon}_M \tag{4.27}$$

Time t was substituted by strain ε_M :

$$t = \frac{\varepsilon_M}{\dot{\varepsilon}_M} \tag{4.28}$$

and (4.27) was changed to:

$$\sigma_M(\varepsilon_M, \dot{\varepsilon}_M) = e^{-\frac{k\varepsilon_M}{c\dot{\varepsilon}_M}} \left(-1 + e^{\frac{k\varepsilon_M}{c\dot{\varepsilon}_M}}\right) c\dot{\varepsilon}_M \tag{4.29}$$

Strain rate is constant ($\dot{\varepsilon}_M = 1\,\mathrm{s}^{-1}$) see above, therefore (4.29) is function only of the strain ε_M :

$$\sigma_M(\varepsilon_M) = e^{-\frac{k\varepsilon_M}{c}} \left(-1 + e^{\frac{k\varepsilon_M}{c}}\right) c \tag{4.30}$$

With respect to the (4.23) and (4.24) the finally equation of the foam model is:

$$\sigma(\varepsilon) = e^{-\frac{k\varepsilon}{c}} \left(-1 + e^{\frac{k\varepsilon}{c}}\right) c + (k_P + k_D)\varepsilon \tag{4.31}$$

$$\sigma(\varepsilon) = e^{-\frac{k\varepsilon}{c}} \left(-1 + e^{\frac{k\varepsilon}{c}}\right) c + \left[k_P + \gamma(1 - e^{\varepsilon})^n\right] \varepsilon \tag{4.32}$$

Equation (4.32) presents a stress-strain relationship for modelling compressive stress-strain curve. It is a phenomenological model because it is just a function for modelling the shape of foam materials stress-strain curves. Figure 4.7 shows the influence of model parameters k, c, k_P, γ and n on the shape of stress-strain curve.

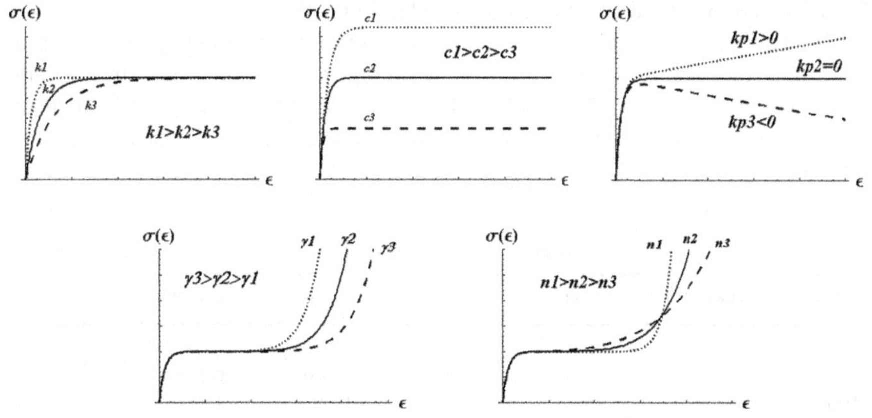

Fig. 4.7 Influence of model parameters k, c, k_P, γ and n

4.4 Compression Test of Polyurethane Foam

Uniaxial compression test was performed for open-cell rigid polyurethane (PUR) foam with densities 40, 50, 100 and 145 kg/m^3. Testing foam is used as a heat insulation and construction material (foam core of sandwich panels).

Mechanical properties were evaluated as a function of foam density using a conventional mechanical test frame. Tests were conducted at room temperature. Deformation and acting force was measured by position sensor and compression load cell. For each density was cut one specimen with dimensions 50×50×30 mm. Figure 4.8 shows stress-strain (σ-ε) curves of PUR foams.

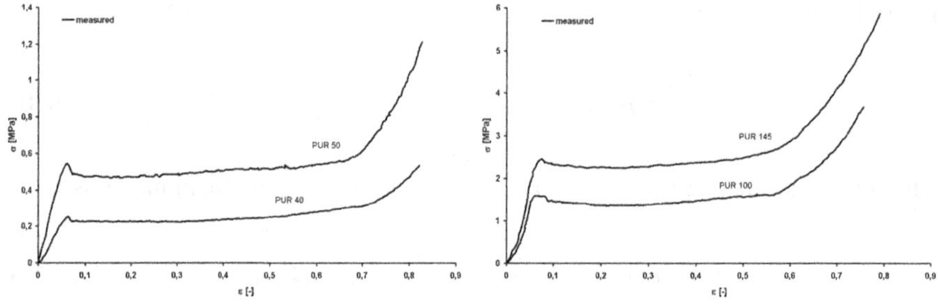

Fig. 4.8 Compressive stress-strain curves of PUR foam

4.5 Model Application

Four foam specimens with different densities were compressed. Their stress-strain curves were approximated by new foam model. Least square method was used to find a good estimation of model parameters. Calculated parameters are in Table 4.1. Modeled and measured curves are compared in Fig. 4.9.

Comparison is specific energy, what is the work done per unit volume (4.1) in deforming the foam to a given strain ε is simply the area under the stress-strain curve. Polynomial equation of the 8th degree was used to fit curves and calculate specific energies. Those results were compared with energies calculated from new foam model in Table 4.2.

Table 4.1 Model parameters

Foam density (kg/m$^{3)}$	k (MPa)	c (MPa)	k_P (MPa)	γ (MPa)	n [–]
40	6.817	0.222	0.085	0.073	6
50	16.46	0.465	0.085	0.176	6
100	45.68	1.481	0.09	1.181	6
145	60.485	2.25	0.1	2.25	4

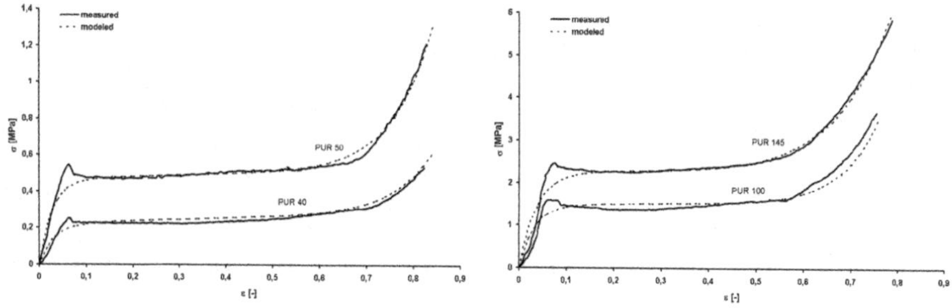

Fig. 4.9 Modeled and measured compressive stress-strain curves

Table 4.2 Specific energies

Density (kg/ m^3)	40		50		100		145	
Energy W (MJ/m^3)	Polynomial	Model	Polynomial	Model	Polynomial	Model	Polynomial	Model
ε = 20%	0.039	0.039	0.085	0.082	0.254	0.25	0.382	0.369
ΔW (%)	0		-3.5		-1.6		-3.4	
ε = 40%	0.087	0.088	0.182	0.18	0.549	0.552	0.836	0.827
ΔW (%)	1.2		-1.1		0.6		-1.1	
ε = 60%	0.145	0.142	0.286	0.283	0.856	0.87	1.337	1.336
ΔW (%)	-2.17		-1.1		1.6		-0.1	
ε = 80%	0.228	0.213	0.436	0.426	1.352	1.411	2.171	2.177
ΔW (%)	-6.6		-3		4.4		0.28	

4.6 Estimation of Stress-Strain Curve

Stress-strain curve for PUR foam with density 60 kg/m^3 was firstly modeled and then compared with measured data. Figure 4.10 shows the relationships of the model parameters on relative density except the parameter n. This parameter must be positive and even number and it is 6 and for high density foam it is 4. Solid density 1,200 kg/m^3 was used to calculate relative density [4, 7, 25].

Linear approximations of model parameters:

$$
\begin{aligned}
k &= 325,62\left(\frac{\rho^*}{\rho_S}\right)^{0,5} - 50,915 \\
c &= 12,268\left(\frac{\rho^*}{\rho_S}\right)^{0,5} - 2,0331 \\
k_P &= 1,1271\left(\frac{\rho^*}{\rho_S}\right)^{2} + 0,0831 \\
\gamma &= 61,316\left(\frac{\rho^*}{\rho_S}\right)^{1,5} - 0,3163
\end{aligned}
\tag{4.33}
$$

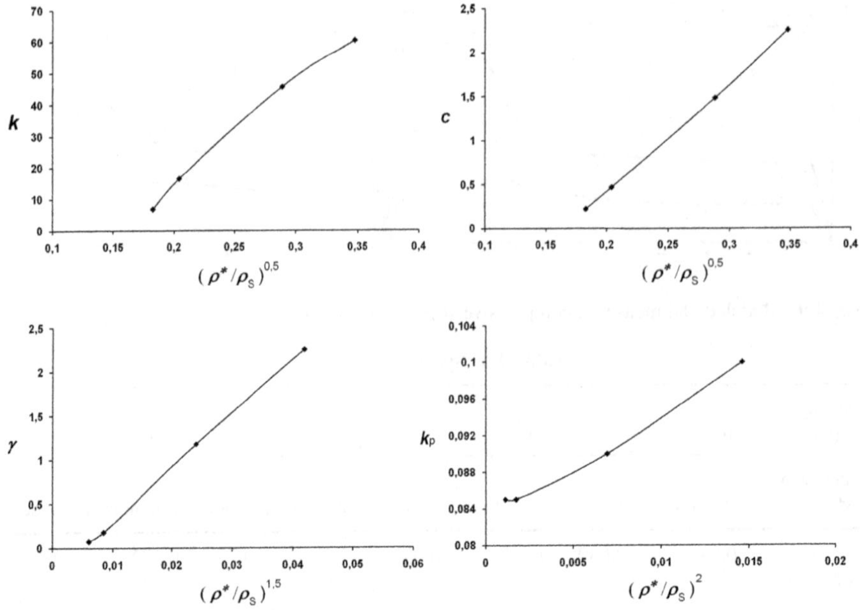

Fig. 4.10 Model parameters vs. relative density

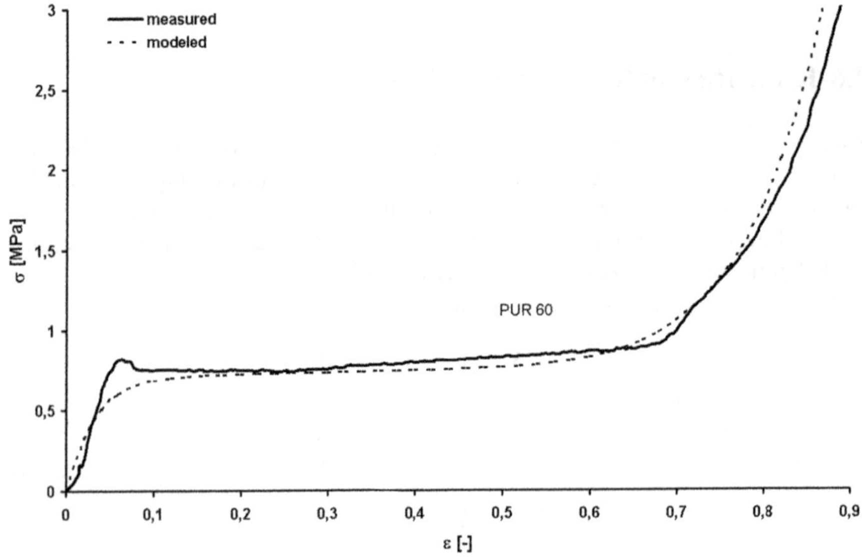

Fig. 4.11 Foam (60 kg/m^3) modeled and measured compressive stress-strain curves

Table 4.3 Specific energy – foam with density 60 kg/m^3

Density (kg/m^3)	60	
Energy W (MJ/m^3)	Polynomial	Model
$\varepsilon = 20\%$	0.131	0.121
ΔW (%)	–7.6	
$\varepsilon = 40\%$	0.281	0.268
ΔW (%)	–4.6	
$\varepsilon = 60\%$	0.435	0.423
ΔW (%)	–2.8	
$\varepsilon = 80\%$	0.609	0.649
ΔW (%)	6.6	

Calculated parameters (4.33) for PUR foam with density 60 kg/m^3:

$k = 21.9$ MPa, $c = 0.71$ MPa, $k_P = 0.086$ MPa, $\gamma = 0.369$ MPa and $n = 6$.

Modeled stress-strain curve for PUR foam with density 60 kg/m^3 was compared with measured curve in Fig. 4.11. Comparison of the specific energies for this foam is in Table 4.3. Differences of calculated energies (ΔW) are less than 10%.

4.7 Conclusion

New phenomenological foam model for solid foams in compression load case was presented. Properties of rheological models were utilized for derive model function. This model has only five parameters and each of them has certain influence on the shape of compressive stress-strain curve and they are depended on foam density. Model was used for fit the stress-strain curves of testing polyurethane foam. Four foam specimens, each of the different density, were tested. Measured stress-strain curves were fitted by polynomial equation of the 8th degree and by new model function. Polynomial equation gives the better fit but it has a lot of parameters and their relationship with the foam density is unknown. This equation was used only for comparison of the results of the calculated specific energies for new model. Thus, specific energies for different strain values were calculated by both functions. The differences of specific energies were less than 10%. New model adequately approximates measured curves.

Relationships of model parameters on relative density of used foam were evaluated and used for modelling stress-strain curve of the same foam with density, which specimen has not been compressed firstly. After compression test the modelled and measured curves were compared. Differences of specific energies were less than 10% again. Model parameters are dependent on foam density and therefore this model can be useful for predict the stress-strain curve of the foam which was firstly tested and relationships of the parameters and foam density were performed. It can reduce the number of experimental tests.

Disadvantages of this model are: it does not consider a temperature, compression strain rate, cell topology of the foam and reload case.

Acknowledgements This article has been accomplished under VEGA grant no. 1/4122/07.

References

1. Gibson, L.J., Ashby, M.F.: Cellular Solids: Structure and Properties, 2nd edn. Cambridge University Press, Cambridge (1997)
2. Han, F., Zhu, Z., Gao, J.: Compressive deformation and energy absorbing characteristic of foamd aluminum. Metallurgical Mater. Trans. A **29A**, 2497 (1998)
3. Avalle, M., Belingardi, G., Montanini, R.: Characterization of polymeric structural foams under compressive impact loading by means of energy-absorption diagram. Int. J. Impact Eng. **25**, 455–472 (2001)
4. Ashby, M.F.: The mechnical properties of cellular solids. Metall. Trans. A **14A**, 1755–1769 (1983)
5. Avalle, M., Belingardi, G., Ibba, A.: Mechanical models of cellular solids: parameters identification from experimental tests. Int. J. Impact Eng. **34**, 3–27 (2007)
6. Gibson, L.J., Ashby, M.F.: Cellular Solids: Structure and Properties. Pergamon Press, Oxford (1988)
7. Goods, S.H., Neuschwanger, C.L., Henderson, C.C., Skala, D.M.: Mechanical properties of CRETE, a poluyrethane foam. J. Appl. Polym. Sci. **68**, 1045–1055 (1998)
8. Roberts, A.P., Garboczi, E.J.: Elastic properties of model random three-dimensional open-cell solids. J. Mech. Phys. Solid **50**, 33–55 (2002)
9. Gibson, L.J., Ashby, M.F.: The mechanics of three-dimensional cellular materials. Proc. R. Soc. London **A382**, 43 (1982)
10. Ashby, M.F., Evans, A., Fleck, N.A., Gibson, L.J., Hutchinson, J.W., Wadley, H.N.G.: Metal Foams – A Design Guide. Butterworth-Heinemann, Oxford (2000)
11. Chan, K.C., Xie, L.S.: Dependecy of densifications on cell topology of metal foams. Scripta Mater. **48**, 1147–1152 (2003)
12. Kádar, Cs., Kenesei, P., Ledvai, J., Rajkovits, Zs.: Energy absorption properties of metal foams. Mater. World **6**(1). ISSN: 1586-0140 (2005)
13. Thompson-Colon, J.A., Huber, M., Liddle, J.W.: SAE Paper No. 910404 (1991)
14. Monk, M.W., Sullivan, L.K..: SAE Paper No. 861887 (1986)
15. McCullough, D.W., Pakulsky, R.R., Liddle, J.W.: SAE Paper No. 920336 (1992)
16. Rossio, R.O., Vecchio, M.T., Abramczyk, J.E.: SAE Paper No. 930433 (1993)
17. Liu, Q., Subhash, G.: A phenomenological constitutive model for foams under large deformations. Polym. Eng Sci. **44**, 463–473 (2004)
18. Rusch, K.C.: Load-compression behavior of flexible foams. J. Appl. Polym. Sci. **13**, 2297 (1969)
19. Meinecke, E.A., Schwaber, D.M.: Energy absorption in polymeric foams. J. Appl. Polym. Sci. **14**, 2239 (1970)
20. Nagy, A., Ko, W.L., Lindholm, U.S.: Mechanical behavior of foamed materials under dynamic compression. J. Cell. Plastics **10**, 127 (1974)
21. Sherwood, J.A., Frost, C.C.: Constitutive modeling and simulation of energy absorbing polyurethane foam under impact loading. Polym. Eng Sci. **32**, 1138 (1992)
22. Chou, C.C., Zhao, Y., Chai, L., Co, J., Lim, G.G.: SAE Paper No. 952733 (1996)
23. Faruque, O., Liu, N., Chou, C.C.: SAE Paper No. 971076 (1997)
24. Zhang, J., Kikuchi, N., Li, V., Yee, A., Nuscholtz, G.: Constitutive modeling of polymeric foam material subjected to dynamic crash loading. Int. J. Impact Eng. **21**, 369 (1998)
25. Roff, W.F., Scott, J.R.: Fibres, Films, Plastics and Rubbers – A Handbook of Common Polymers. Butterworths, London (1971)

Chapter 5
The Effect of an Interphase on Micro-Crack Behaviour in Polymer Composites

P. Hutař, L. Náhlík, Z. Majer, and Z. Knésl

Abstract The present chapter is focused mainly on polypropylene based composites filled by mineral fillers, but some results can be generalized for other particle composites. The stiffness and toughness of the composite is modelled using a three-phase continuum consisting of the polymer matrix, mineral particles and an interphase between them. It is shown that the effect of the interphase on the macroscopic characteristics of the composite is decisive. Generally, the addition of the mineral filler to the polymer matrix leads to ebrittlement of the composite. The computational methodology presented quantifies the effect of the microstructure properties and morphology on the macroscopic material response. It is shown that properties of the interphase control both the stiffness and embrittlement of the particulate composite. Primarily, the interaction of micro-cracks with coated particles is studied. It is concluded that in some cases of the microscopic particles, size and specific interphase properties the addition of mineral fillers can lead to a good balance between fracture toughness and stiffness. Linear elastic fracture mechanics is used for calculations.

Keywords Particulate polymer composite · Mineral filler · Embrittlement · Multiscale modelling

5.1 Introduction

Applications of polymers in praxis are limited by low values of their mechanical properties as, for example, low stiffness and low strength. To extend their use in praxis, inorganic particular fillers are frequently added in order to improve polymeric particulate composites (PPC). These are then frequently used in many engineering applications and are of great importance due to the possibility of modifying mechanical properties and reducing the price/volume ratio of the resulting material [1, 2]. Polymeric particulate composites usually used are composed from

---

P. Hutař (✉)
Institute of Physics of Materials, Academy of Sciences, Žižkova 22, 616 62 Brno, Czech Republic
e-mail: hutar@ipm.cz

J. Murín et al. (eds.), *Computational Modelling and Advanced Simulations*,
Computational Methods in Applied Sciences 24, DOI 10.1007/978-94-007-0317-9_5,

soft polymer matrix and rigid mineral fillers. It is known that the addition of rigid particles to the soft matrix causes generally higher stiffness of the PPC but leads simultaneously to an embrittlement of the resulting composite with lower toughness in comparison with a native polymer matrix. A survey of this development with regard to mechanical properties as a function of volume fraction, particle size and particle/matrix interface was undertaken recently e.g. in [3]. However, where fracture toughness is concerned, the behaviour is more complex and in some particular cases an increase of fracture toughness was found in experiments. In these cases, it is necessary to fulfill certain requirements [4]:

(I) The particle should be of small size (less then 5 μm)
(II) The aspect ratio should be close to 1
(III) The particles must deboned prior to the yield strain of the matrix
(IV) The particles should be dispersed homogenously in the matrix – aggregation should be avoided

The addition of a second component to the polymer matrix modifies all the characteristics of the polymer and to achieve its optimum application potential the microstructure morphology and properties of single phases of the composite have to be considered. In this sense an important effect is connected to the creation of a third phase existing between particle and matrix called an interphase (sometimes the term mesophase is also used, see [5]).

The interphase is formed as a consequence of different chemical and mechanical properties of the matrix and particles and/or special chemical treatment of the particles during composite processing [6]. This region controls the adhesion between particle and matrix, acts as a crucial element transferring the applied load from matrix to particles and can play an important role in damage evolution in the composite [7, 8]. Its properties influence the overall mechanical behaviour of the composite and have to be considered as a special phase with material properties different from those of matrix and particles, see Fig. 5.1.

Although some analytical and semi-analytical models based on homogenization techniques have been developed to evaluate the effective material properties of the

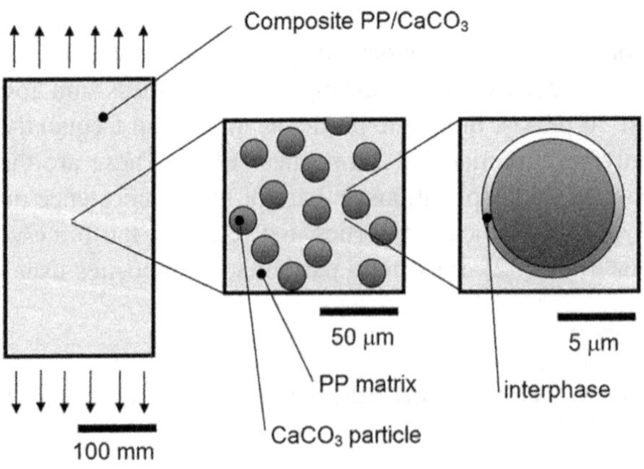

Fig. 5.1 Multiscale view on the structure of the particulate composite PP/$CaCO_3$

particle reinforced composites, they are often reduced to specific cases. Numerical models seem to be a well suited approach to describe the behaviour of these materials, because there is no restriction on the geometry, on the material properties and on the number of phases, see [9]. Moreover, numerical modelling of the composite as a three phase continuum makes it possible to predict its overall behaviour as a function of the composition properties of single phases. It reduces the number of corresponding experiments and can provide hints for improving the application potential of the composite.

In the following the influence of the interface between particles and matrix on the overall stiffness and toughness of the composite is numerically modelled and studied. As a first approximation, we assume that the interphase is a well defined area between matrix and particle, which can be described by its thickness and particular material properties. The estimation of interphase properties is problematic. Many methods have been used and sometimes contradictory results have been obtained and published [1]. The thickness of the interphase and its material properties depends mainly on matrix and particle chemical composition and on the particle chemical treatment during composite processing. The effect of the particle morphology and interphase properties on overall mechanical properties of the composite is primarily studied in the following.

Specifically, the aim of the article is to estimate the effect of the interphase between polypropylene matrix filled by rigid particles $CaCO_3$. The estimations of the interphase thickness for PP-$CaCO_3$ system varies from 0.012 to 0.16 μm [1]. In the papers [1, 6], the interphase thickness is correlated with the work of adhesion and for uncoated particles is estimated as 0.117 μm. This value will be considered in the following. First, the overall elastic stiffness of the composite containing periodically dispersed coated particles is analysed. In the second part, toughening of the composite induced by interaction of micro-cracks with coated particles is considered. It is shown that the fracture toughness of the composite can be increased due to shielding of rigid particles by a softer interphase.

5.2 Estimation of the Composite Stiffness

In order to predict the overall mechanical properties of the PP composite an approach of homogenization within a mesoscopic level has been performed. For parametric study of the properties of the three-phase composite, the finite element method and unit cell models provide an effective tool [10, 11]. By assuming the spherical particles to be packed in simple-cubic symmetry array, as in Fig. 5.2, only one-eighth of the particle embedded in the cube is needed for analysis. The geometry of the representative unit cell (representative volume element-RVE) for the three phase material model is shown in Fig. 5.2.

The material properties characterizing the composite corresponding to calcium carbonate ($CaCO_3$) – filled polypropylene (PP) at room temperature are used. The Young modulus of the particles and value of Poisson's ratio are $E_p = 72$ GPa, $\nu_p = 0.29$. The corresponding parameters of the net polymer matrix (PP material) are $E_m = 1.9$ GPa, $\nu_m = 0.29$.

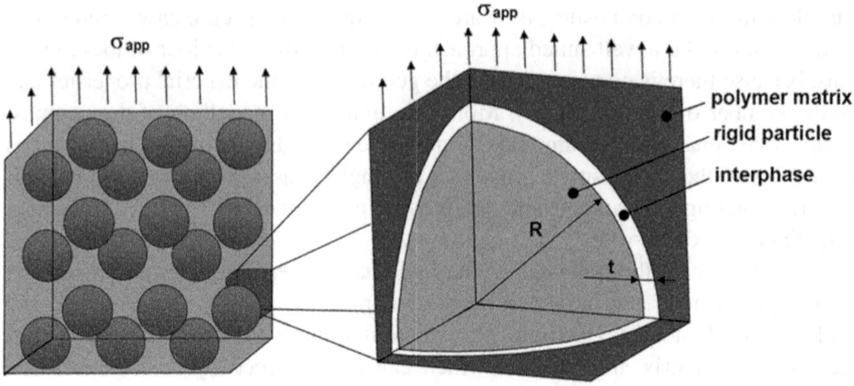

Fig. 5.2 Model of the representative unit cell used for calculations, R is a particle radius and t the interphase thickness

Firstly, the exactness of the used numerical model is verified. To this end a perfect bonding is assumed between the matrix and the particles and the composite is modelled as a two-phase continuum, i.e. without the interphase. Results of the finite element calculations corresponding to the present model answer well to estimation of the stiffness for a two-phase composite based on semi-empirical equations see [12].

The elastic properties of a three-phase composite made up of spherical coated particles and distributed in a continuum matrix have also been studied in the literature, see e.g. [3, 13]. Again, the conclusions of the papers correspond with the present results.

5.2.1 Influence of Space Distribution

Particulate-filled composites behave isotropically. A simple numerical model developed and used here to compute the stiffness of the three-phase particulate composite is based on the assumption of periodically distributed particles dispersed in a continuous matrix and particle packing arrangements in the form of simple cubic lattice is assumed. In reality, particles are randomly distributed in the matrix. With the aim to estimate the influence of random particle distribution on the overall stiffness of the composite the structure with irregular particle distribution has been modelled, see [12]. For calculations the interphase thickness $t = 0.117\mu\text{m}$ has been used.

Results show that the values of effective stiffness for random particle distribution are close to those obtained for simple cubic arrangement with the same volume fraction. Consequently, in the all following cases, the spatial distribution of particles is supposed to be simple cube symmetric in the three dimensional space.

5.2.2 Influence of Aspect Ratio

Most results in the literature concerning the evaluation of effective elastic properties of composite containing periodically dispersed particles are related to spherical

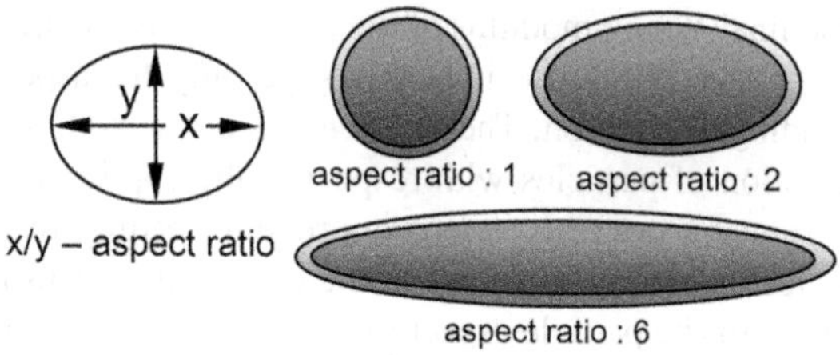

Fig. 5.3 Different particle shapes used for numerical calculations

particles. However, in praxis, the conditions of the spherical shape of particles are frequently not exactly fulfilled. With the aim to investigate the influence of the particle shape on overall elastic stiffness of the composite, the effect of the aspect ratio on the mechanical response of the composite was investigated in the following. Three different particle types with aspect ratio from 1 to 6 were used for modelling, see Fig. 5.3.

Again a particulate composite including an interphase is here modelled as a three-phase continuum using a representative volume element with corresponding particle shapes. The interphase thickness $t = 0.117\mu m$ has been again used for calculations. Further, Young modulus of the interphase has been taken from 50 to 1,900 MPa (1,900 MPa is the value close to the Young modulus of the matrix) and the effect of the interphase on the mechanical response of the composite was estimated for given particle shapes, see Fig. 5.4.

In the case of the $CaCO_3$ particles, used in our research, the typical aspect ratio of the particles is smaller than 2. In this case, the effect of the particle aspect ratio on

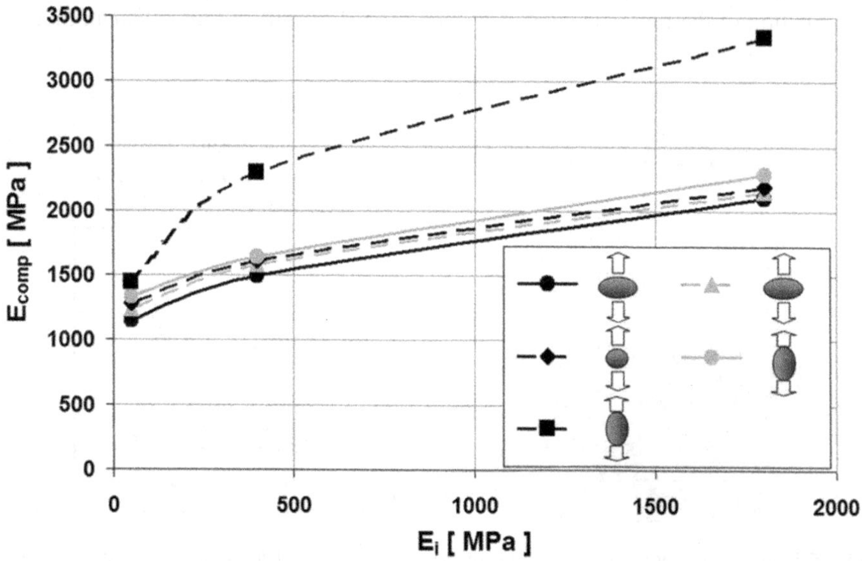

Fig. 5.4 Dependence of the overall Young modulus of the composite on the Young modulus of the interphase for different aspect ratio of the particles and different orientation of the external loading. The particle volume fraction $V_p = 10\%$ and particle size 1 μm is considered

the final Young modulus of the composite is smaller than 5%, and can be neglected. A greater difference is here visible only for aspect ratio 6 and for the particular loading orientation. The difference is connected, at least partially, to the special orientation of particles with respect to the applied load. In this case the FEM model assumes a perfectly oriented structure leading to anisotropy of mechanical properties. In reality, particles in the composite are usually randomly oriented and the effect of the particle aspect ratio is here overestimated.

5.3 Micro-Crack Behaviour in Particulate Composite

At the present time there is no general consensus on the mechanism of toughening in the studied composite. The toughening of the composite based on the interaction of micro-cracks with particles is studied here.

The basic idea of the suggested toughening mechanism consists in the fact that the fracture toughness of the composite can be increased due to a change of interaction between coated particles and micro-cracks propagating in the matrix. To this aim micro-crack behaviour in a three-phase composite with homogenously distributed coated particles is numerically simulated in the following.

A two-dimensional model with plane strain conditions was assumed for the computation. As only particles located close to the crack tip significantly influence micro-crack behaviour a unit cell with four particles can be regarded as representative for the composite structure.

The geometry and parameters of the model are schematically shown in Fig. 5.5. Due to the high stress concentration around the micro-crack tip and enormously different elastic material properties of each component, a strongly non-homogenous distribution of the finite elements has to be used. Typically, a finite element model

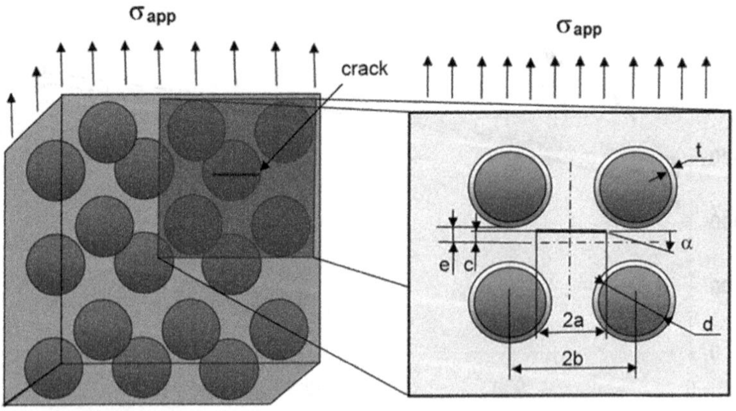

Fig. 5.5 Two dimensional model of the micro-crack in particulate composite used for FEM computations

contains approximately 50,000 two-dimensional isoparametric elements concentrated mainly in the crack tip vicinity and close to the material interfaces. The interaction of a micro-crack with coated particles is controlled by the stress state around its crack tip. Due to the non-homogeneity of the stress field in the particulate composite, mixed mode loading conditions are induced at the crack tip and the stress field is characterized by corresponding values of the stress intensity factors K_I and K_{II}. Generally, a crack propagates in a direction leading to zero values of K_{II}. Several criteria for determination of the crack propagation direction exist in the literature, see e.g. [14]. Due to numerous similarities and small differences between the individual criteria, the maximum tangential stress (MTS) criterion [15] was used. The direction of the crack propagation (α – see Fig. 5.5) can be expressed by the following equation:

$$\alpha = \arccos\left(\frac{3K_{II}^2 + K_I\sqrt{K_I^2 + 8K_{II}^2}}{K_I^2 + 9K_{II}^2}\right), \tag{5.1}$$

The behaviour of the micro-crack (the length of the crack $2a$ is comparable with the distance between particles $2b$) is studied for particle size 1 μm and 25% of volume fraction, see Fig. 5.6 for results. Note that negative values of α mean that the micro-crack is attracted to the particle, see Fig. 5.6.

A special case $E_i = E_m$ means uncoated particles with perfect adhesion (i.e. zero interphase thickness) between the particle and the matrix. In this case the angle of further crack propagation α is always positive, see Fig. 5.6 ($E_i = 1{,}900$ MPa), meaning that the micro-crack tends to avoid rigid particles and grows preferentially in the net matrix, see Fig. 5.7a) In the case of coated particles with interphase Young

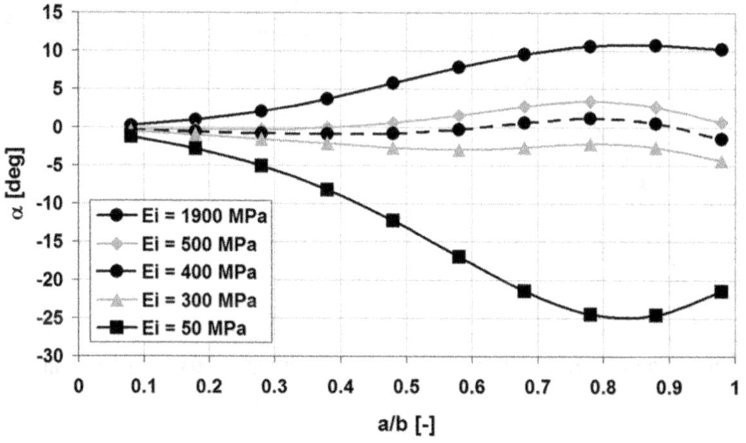

Fig. 5.6 Dependence of the crack propagation direction α and ratio *a/b* for different values of Young modulus of the interphase. Results correspond to particle size 1 μm and $V_p = 25\%$

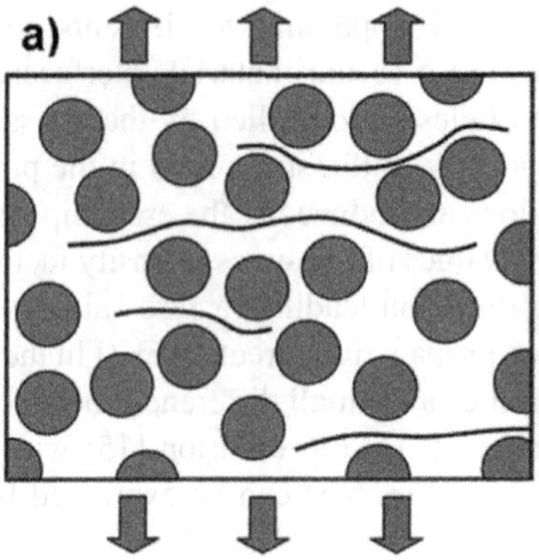

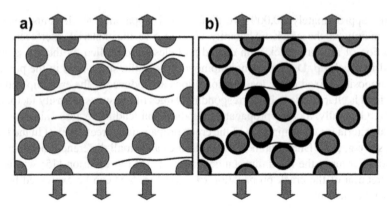

Fig. 5.7 (**a**) Micro-crack propagation in the case of rigid particles without an interphase and (**b**) micro-crack propagation in the case of coated particles

modulus $E_i < 1,900$ MPa, the angle of further crack propagation decreases by a drop of E_i.

Typically, for $E_i < 400$ MPa the calculated angles of deflection α are negative and stiff particles are shielded by a softer interphase. This means that the micro-crack tends to be attracted by the particle. Consequently, the damage mechanism of the composite is different from the previous case. The change of α as a function of the interphase modulus E_i for one particular *a/b* ratio is presented in Fig. 5.8.

In the case of small values of the Young modulus (compared with those of the particles and the matrix) of the interphase, $E_i < 200$ MPa, stiff particles are strongly shielded by the interphase and the micro-cracks path ends in a collision with particles, see Fig. 5.7b).

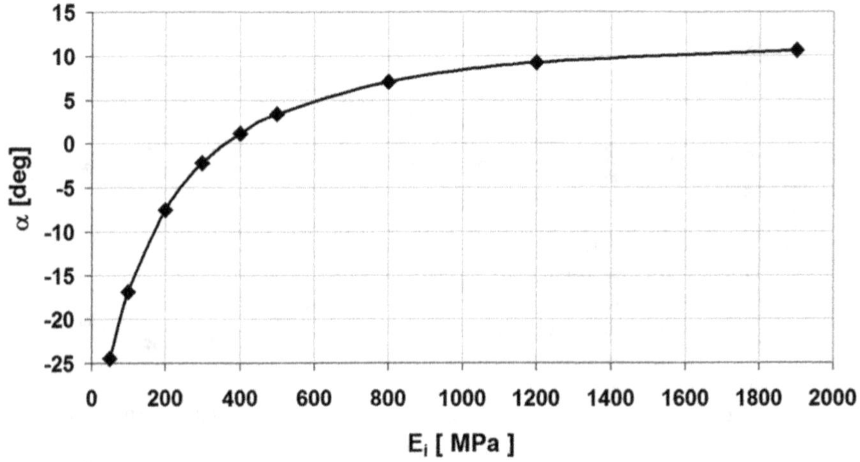

Fig. 5.8 Dependence of the crack propagation direction α on Young modulus of the interphase. Results correspond to particle size 1 μm, $V_p = 25\%$ and $a/b = 0.78$

When the crack tip touches the interphase, the interphase is damaged due to the high stress gradient close to the crack tip and the particle and the matrix can be fully debonded. Consequently, the crack is blunted, the stress singular field is changed to a regular one and the crack can remain arrested near the particle. The blunting of micro-cracks in connection with debonding can contribute to an increase in the fracture toughness of the composite.

Another factor influencing the micro-crack path in matrix is constraint. The lower constraint level stabilizes crack propagation direction and the high constraint level helps towards direction change, e.g. [16]. Generally, the level of the constraint characterized by the biaxiality factor can help to describe crack behaviour more accurately [14]. The T-stress value characterizes the constraint level and can be used to improve a classical one-parameter linear elastic fracture mechanics description of the singular crack tip stress field in the variety of the crack configurations, outer geometries and loading conditions [16]. Usually, the biaxiality factor B is used as the expression to characterize the constraint level [17]:

$$B = \frac{T\sqrt{\pi a}}{K_I}, \tag{5.2}$$

where K_I is the value of the corresponding stress intensity factor and a is a crack length. The level of constraint was investigated here for both coated and un-coated particles in the PP matrix. To extract the elastic T-stress values the direct method was used [18].

The results presented in Fig. 5.9 show that the existence of softer interphase between particles and the matrix increase the constraint level for a crack in the

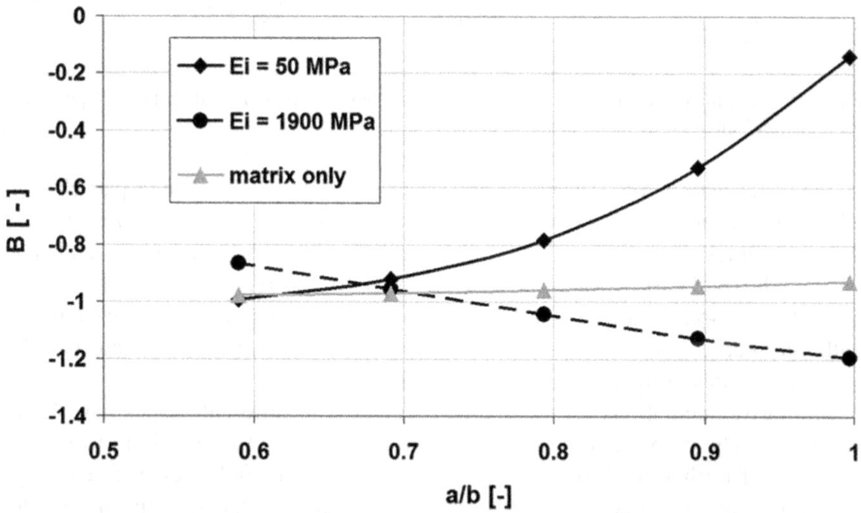

Fig. 5.9 Dependence of biaxiality B on ratio a/b for coated particle, uncoated particle and pure matrix

vicinity of the particle. The change of the crack propagation direction to the coated particle is then easier and the final configuration corresponding to the collision of the micro-crack with the coated particle, followed by its blunting is more probable.

5.3.1 Influence of Particles Morphology

The $CaCO_3$ particles usually have a shape with aspect ratio between 1 and 2 but some irregular shapes and/or sharp particles can be found. The effect of the different particle morphology and interphase imperfections on crack behaviour at the micro-level is studied in the following chapter.

The material properties of the matrix and particles corresponding to those presented in previous chapters are used. Interphase Young modulus $E_i = 400$ MPa was assumed for all computations. For this value of E_i the interphase starts to be important for the overall mechanical response of the proposed polymer composite.

Only some typical results of the extended parametric study are presented here; for more details see [19]. The basic cases studied are shown on Fig. 5.10, namely a spherical particle (aspect ratio 1), an elliptical particle (aspect ratio 2), an irregular particle (aspect ratio 2), and a partially coated irregular particle. For the purposes of calculation a simplified 2D model of a micro-crack interacting with a single particle has been used, see Fig. 5.10.

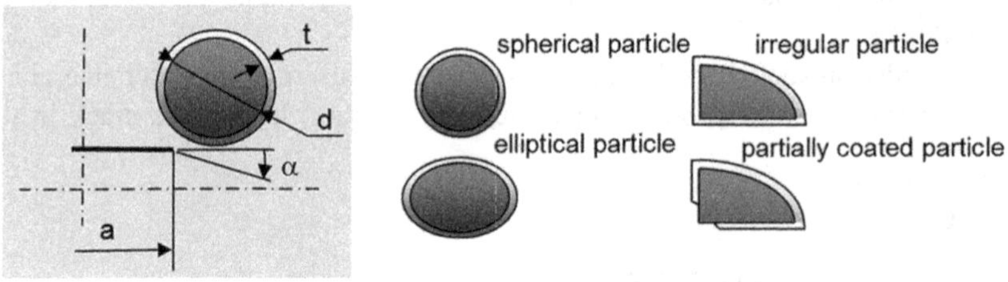

Fig. 5.10 Simplified model of the micro-crack interacted with particle and particle shapes used for numerical simulations

First, the interaction of a micro-cracks growing close to uncoated particles of elliptical and irregular shape was studied. The results show that in all cases, where the micro-crack tip is situated closer to the particle, the micro-crack has a tendency to avoid the rigid particle and propagate freely in the matrix. Similarly to the results presented for spherical particles, a coincidence of elliptical and/or sharp particles with micro-cracks is in this case sporadic. Therefore the toughness depends mainly on the properties of the matrix itself, see [19].

Similarly, calculations for the coated particles with different shapes have been performed. The corresponding results are shown in Fig. 5.11. Prevailing values of the angle of further crack propagation α are negative; this indicates that generally the micro-crack is attracted by the particles. Therefore, independently of particle shapes, the final configuration corresponds to a crack with its tip situated at the interphase between matrix and particles. Differences between crack behaviour close

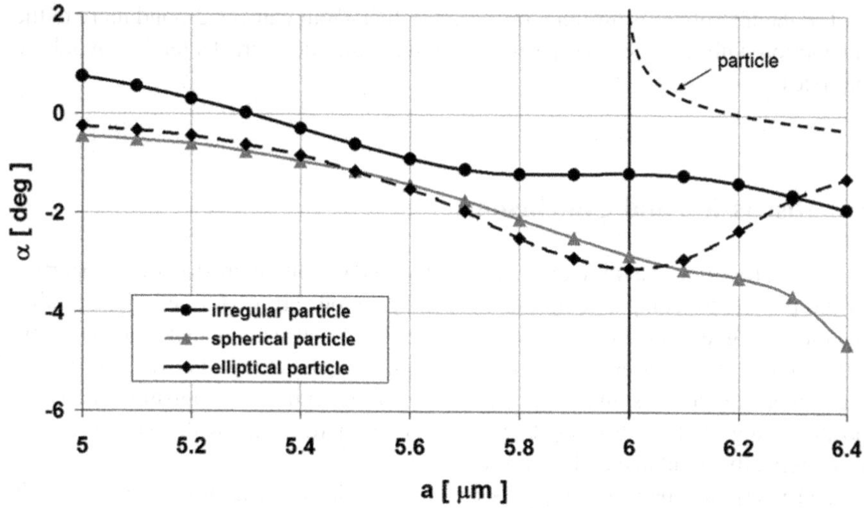

Fig. 5.11 Dependence of the crack propagation direction α on the crack length for different perfectly coated particle shapes. The particle position is indicated by a *dash line*

to the different particles are not very important and probably occur mainly due to the geometrical non-similarity of the cases studied.

Finally, the interaction between the micro-crack and fully and partially coated irregular (sharp) particle has been analysed, see Fig. 5.12. Results indicate that the imperfection of the interphase can significantly alter micro-crack behaviour. When

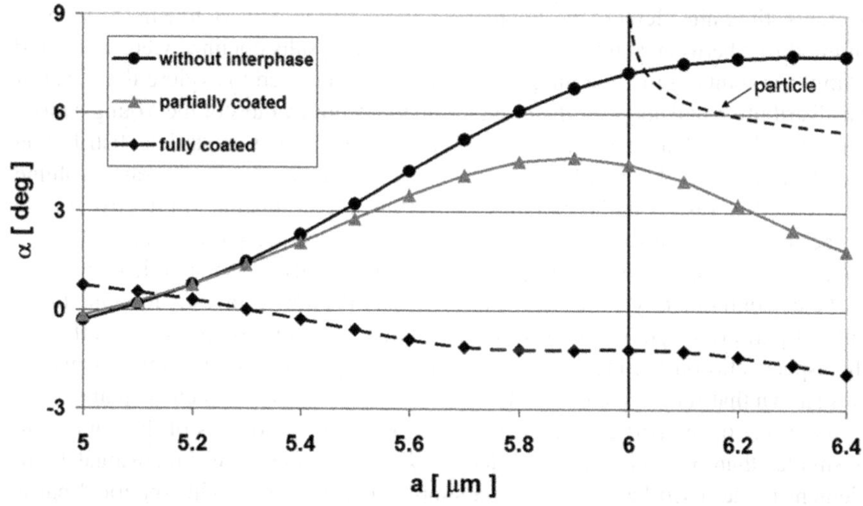

Fig. 5.12 Dependence of the crack propagation direction α on the crack length for irregular particles with and without perfect interphase. The particle position is indicated by a *dash line*

part of the interphase is missing, micro-crack behaviour can correspond more to the interaction with the uncoated particles. In this case micro-cracks tend to avoid the particles.

5.4 Discussion and Conclusions

In this contribution the finite element simulations relating microstructure properties of the particulate composite and its overall behaviour are performed. Specifically, the model suggested results in relation between geometrical and material characteristics of the single phases (i.e. particles, interphase and matrix) and overall values of the composite stiffness and toughness. A new toughening mechanism of the composite is suggested. To this aim the interaction of micro-cracks with coated particles is numerically simulated and analysed.

The interphase between particles and matrix plays a determining role here. It acts as a crucial bridge transferring the applied load from the continuous matrix to the discrete particles and its properties directly influence the overall mechanical behaviour of the composite. The numerical calculations presented focus on both the stiffness and toughness of the composite aspects. The effect of the interphase properties and particle shape is primarily considered. The composite is modelled as a three phase continuum and the finite element method is used for numerical simulations. Since the polymer exhibits minimal yielding, linear elastic fracture mechanics is used for calculation.

First a simple numerical model was introduced to compute the overall Young's modulus of a three phase composite made up of regularly distributed spherical coated particles. Due to the large elastic mismatch between particles and matrix stiffness, the finite element mesh which was used had to be very fine in the vicinity of interfaces between single composite phases. The results obtained were compared with those found in the literature. The model was then used to explore the effect of the distribution and shape of the particles on the overall values of the Young's modulus. It is shown that the values of effective stiffness for random particle distribution are close to those obtained for a simple cubic arrangement with the same volume fraction. Most results in the literature concerning the evaluation of effective elastic properties of a composite containing periodically dispersed particles are related to spherical particles. However, in practice, the conditions of the spherical shape of particles are frequently not exactly fulfilled. With the aim of investigating the influence of particle shape on the overall elastic stiffness of the composite, the effect of the aspect ratio on the mechanical response of the composite was investigated. It was shown that for a typical aspect ratio of $CaCO_3$ particles (which is smaller than 2) the effect of the particles shape on the final Young's modulus of the composite is smaller than 5% and can be neglected. Therefore, the three dimensional finite element model used here, assuming a simple cubic structure with spherical particles, can be considered appropriate for estimation of polymer particulate composite stiffness.

Generally, the presence of rigid particles in the polymer matrix leads to higher stiffness of the composite. The existence of the interphase between matrix and particles can significantly change stress distribution inside the composite structure and, consequently its global behaviour. The predicted overall macro Young's modulus tends to increase with the increasing volume density of stiff particles and decreases with reduction of the interphase Young's modulus see Fig. 5.4. The addition of stiff reinforcement calcium carbonate particles to the polypropylene matrix improves monotonic global properties such as Young's modulus and strength. For smaller values of the interphase Young's modulus $E_i < 500$ MPa the stiff particles are shielded by a soft interphase and the Young's modulus of the composite is then comparable to or even less than for the net matrix.

Although much work has been done to improve the fracture toughness of the particle-filled composites, there is still no general consensus on the mechanism of toughening. In this contribution specific attention is paid to the influence of the interface between matrix and particles on the behaviour of micro-cracks. Despite the practical meaning, the problem cannot be considered as solved. The influence of particle shape and properties of the interphase on the interaction was analyzed in details. All results are obtained for one size of the particle 1 μm, which was found to be realistic from the experimental point of view (see e.g. [20]).

In the case of perfect adhesion between particles and matrix (zero thickness of interphase) the micro-crack avoids regions with rigid particles and grows preferentially in the net matrix only, see Fig. 5.7. This trend is changed due to the existence of an interphase. If the Young's modulus of the interphase is less than those of particles the effect of the rigid particles is partially shielded by a softer interphase. The final effect on micro-crack behaviour is negligible in the case of interphase Young's modulus $E_i > 800$ MPa, see Fig. 5.8. For smaller values of the interphase Young's modulus ($E_i < 400$ MPa) stiff particles are completely shielded by the softer interface and the micro-crack is attracted by the particles. A higher constraint level for coated particles in comparison with uncoated can also contribute to this interaction between micro-crack and particle. As a result, the crack coincides with the coated particle and the crack tip touches the interphase. Due to high stresses, the bonds between the particle and matrix are damaged and the particle is fully debonded. As a result, the crack is blunted, the stress singular field is changed to a regular one and the crack stays arrested near the particle, see Fig. 5.13. The blunting of micro-cracks in connection with debonding contributes to an increase in the fracture toughness of the composite.

The effect of different particle shapes on the mechanism was also studied. Results show that differences between crack behaviour close to particles of different shapes are not significant and their influence on toughening is not decisive. On the other hand, the imperfections of the interphase can change crack interaction and consequently decrease the contribution to the toughening process.

The computational methodology presented here provides a powerful tool to transfer the effects of particular microstructural properties to a macroscopic material response. Consequently, numerical calculations can help design polypropylene based composites with specifically requested material properties. The results of this

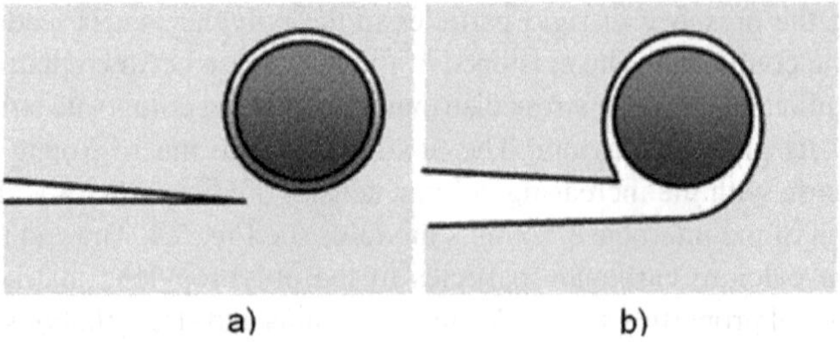

Fig. 5.13 (**a**) Micro crack approaching particle covered by interphase, (**b**) The particle is fully debonded and the micro-crack is blunted

contribution can facilitate the finding of the proper micro-structural properties of the composite leading to an optimum relation between decrease in stiffness and increase in toughness caused by the existence of coated $CaCO_3$ particles distributed in the PP matrix.

Acknowledgements This research was supported by grants 106/07/1284 and 106/08/1409 of the Czech Science Foundation and grant KJB 200410803 of the Grant Agency of Academy of Sciences of the Czech Republic.

References

1. Pukánszky, B.: Interfaces and interphases in multicomponent materials: past, present, future. Eur. Polym. J. **41**, 645–662 (2005)
2. Elias, H.G.: An Introduction to Plastics, 2nd edn. Wiley-VCH GmbH&Co., Weinheim (2003)
3. Fu, S.-Y., Feng, X.-Q., Lauke, B., Mai, Y.-W.: Effects of particle size, particle/matrix interface adhesion and particle loading on mechanical properties of particulate–polymer composites. Composites B **39**, 933–961 (2008)
4. Zuiderduin, W.C.J., Westzaan, C., Huétink, J., Gaymans, R.J.: Toughening of polypropylene with calcium carbonate particles. Polymer **44**, 261–275 (2003)
5. Theocaris, P.S.: The Mesophase Concept in Composites. Springer, Berlin (1987)
6. Moczó, J., Fekete, E., Pukánszky, B.: Acid-base interactions and interphase formation in particulate-filled polymers. J. Adhes. **78**, 861–876 (2002)
7. Kim, G.M., Michler, G.H.: Micromechanical deformation processes in toughened and particle filled semicrystalline polymers: 2. Model representation for micromechanical deformation processes. Polymer **39**, 5699–5703 (1998)
8. Jančář, J., Dibenedetto, A.T., Dianselmo, A.: Effect of adhesion on fracture toughness of calcium carbonate – filled polypropylene. Polym. Eng. Sci. **9**, 559–563 (1993)
9. Kari, S., Berger, H., Rodriguez-Ramos, R., Gabbert, U.: Computational evaluation of effective material properties of composites reinforced by randomly distributed spherical particles. Compos. Struct. **77**, 223–231 (2007)
10. Tsui, C.P., Tang, C.Y., Lee, T.C.: Finite element analysis of polymer composite filled by interphase coated particles. J. Mater. Process. Technol. **117**, 105–110 (2001)
11. Wu, Y., Dong, Z.: Three dimensional finite element analysis of composites with coated spherical inclusions. Mater. Sci. Eng. A **203**, 314–323 (1995)
12. Majer, Z.: Fracture-mechanics model of particulate composite. PhD.Thesis, Brno University of Technology, Brno (2009)
13. Lorca, J., Elices, M., Termonia, Z.: Elastic properties of sphere-reinforced composites with a mesophase. Acta Mater. **48**, 4589–4597 (2000)

14. Anderson, T.L.: Fracture Mechanics – Fundamentals and Applications. CRC Press, Boca Raton, FL (1995)
15. Erdogan, F., Sih, G.C.: On the crack extension in plates under plane loading and transverse shear. J Basic Eng. **85**, 519–527 (1963)
16. Seitl, S., Knésl, Z.: Two parameter fracture mechanics: fatigue crack behavior under mixed mode conditions. Eng. Fract. Mech. **75**, 857–865 (2008)
17. Leevers, P.S., Radon, J.C.: Inherent stress biaxiality in various fracture specimen geometries. Int. J. Fract. **19**, 311–325 (1982)
18. Owen, D.R., Fawkes, A.J.: Engineering Fracture Mechanics: Numerical Method and Applications. Pineridge Press, Swansea (1983)
19. Majer, Z., Hutař, P., Náhlík, L., Knésl, Z., Nezbedová, E.: Crack behaviour in polymeric composites filled by rigid particles. Proceedings of ICF12, Ottawa (2009)
20. Nezbedová, E., Knésl, Z., Hutař, P., Majer, Z., Veselý, P.: Modelling of fracture behaviour of PP particulate composites on micro-level. Proceedings of SAMPE meeting, Paris (2008)

Chapter 6
Temperature Fields in Short Fibre Composites

Vladimír Kompiš, Z. Murčinková, and M. Očkay

Abstract The paper deals with computational models for temperature fields in short fibre composites reinforced by very large aspect ratio fibres. Method of Continuous Source Functions (MCSF) is used for simulation of temperature fields. MCSF is a boundary meshless method reducing the problem considerably. In MCSF the continuity conditions of temperature fields, expressed in collocation (discrete) points of the fibre-matrix interface, are used to compute intensities of continuous source functions of heat sources and heat dipoles located along the fibre axis. Because of large gradients in the intensities of source functions, the NURBS were chosen to define the distribution. Number of unknown intensities is usually much smaller than the number of collocation points and the system of resulting equations is solved in the Least Square (LS) sense. Examples of computational simulations for single fibre, two fibres and regularly distributed fibres in a matrix are presented.

Keywords Meshless method · Fiber · Large aspect ratio · Heat conduction

6.1 Introduction

Short Fibre/Tube Composites (SFC) are often defined to be materials of future with excellent electro-thermo-mechanical (ETM) properties. Understanding the behaviour of such composite materials is essential for structural design. Our goal is to understand the principles of short fibre thermal behaviour in micro-scale. To achieve mentioned goal, we have to make accurate computational simulation of interactions: matrix-fibre, fibre-fibre and fibre-boundary/macrostructure. The mentioned interactions have influence on global composite behaviour.

V. Kompiš (✉)
Armed Forces Academy of General Milan Rastislav Štefánik, Demänová 393,
031 19 Liptovský Mikuláš, Slovakia
e-mail: vladimir.kompis@aos.sk

J. Murín et al. (eds.), *Computational Modelling and Advanced Simulations*,
Computational Methods in Applied Sciences 24, DOI 10.1007/978-94-007-0317-9_6,

Typical properties of short fibre composites are following:

- large aspect ratio of fibres. The word "short" means relatively short fibre. The aspect ratio of fibres (tubes) with diameter 20 nm and length 1 mm is 1:50,000,
- large gradients in all ETM fields along the fibres and in the matrix,
- large number of reinforcing elements/fibres in small volume (thousands or millions of fibres contained in a micro-dimension),
- different properties in different directions by unidirectional reinforcement,
- large difference between stiffness, strength, heat and electric conductivity of matrix materials and fibre materials.

The local and directional changes of ETM properties enable novel principles for designing despite the fact that such changes are difficult to simulate. Using the SFC, we can change the mechanical properties of structure without changing the shape or dimensions of structure.

In the simulations, the domain-type, the boundary-type, the mesh-free methodologies, or some combinations of different type methodologies can be used. In generally, the most effective method is the method which enables to obtain good accuracy with smallest computational effort.

The present methods mostly used to simulate the temperature fields in solids are Finite Element Method (FEM) and other volume discretization methods, Boundary Element Method (BEM), Meshless and Mesh-Reducing Methods (MRM).

The main FEM and BEM disadvantage is very fine meshing to keep gradient and aspect ratio either in volume or in surface. BEM defines the problem by corresponding fundamental solution and leads to singular integral equations and full matrix after discretization. Because of large gradients and large aspect ratio of fibre surface, many boundary elements are necessary to solve the problem by required accuracy. Similar properties have also present formulations using meshless and mesh-reducing methods.

The Trefftz type formulation (T-FEM) provides the considerable reduction of number of elements by use of large finite elements. In such formulation, the shape functions satisfy the governing equations in corresponding domain/subdomain [1, 2]. However, they are not so efficient for complex interaction problems with more than one fibre.

Large reduction of the problem enables the Fast Multi-pole (FM) BEM [3] by which the kernel functions of boundary integrals are substituted by corresponding truncated Taylor series expansion and resulting discrete dipoles and moments are substituted for the continuum in far field interaction. However, classical BEM formulation describes the near field interaction and thus, the method is not as efficient as it is in the case of inhomogeneous materials when the aspect ratio of inclusions is small.

In this paper the MCSF, which was first applied to linear elasticity [4], is documented for simulation of stationary problems of heat conduction. This method determines the field distributions in short-fibrous composite.

6.2 Method of Continuous Source Functions – MCSF

The basic equation is derived from the boundary integral equation approach introducing a source function. In case of structural analysis, the source function is a unit force; in case of thermal analysis it is a unit heat source. The source functions produce fields of displacement, stress, temperature, heat flow in the body (matrix). The derivative of source function in corresponding direction is called a dipole and it is composed of two collinear forces acting in the opposite directions or a heat sink and a heat source acting at the same point.

The MCSF is a boundary meshless method reducing the problem considerably. The continuity conditions of temperature fields, expressed in collocation (discrete) points of the fibre-matrix interface, are used to compute intensities of 1D continuous source functions, i.e. of heat sources and heat dipoles located along the fibre axis. Because of large gradients in the intensities of source functions, the NURBS (Non-Uniform Rational B-Splines) were chosen to define the distribution. Number of unknown intensities is usually much smaller than the number of collocation points and the system of resulting equations is solved in the Least Square (LS) sense. Examples of computational simulations for single fibre in a matrix and regularly distributed fibres in a matrix are presented in examples below.

In presented approach, the fibres are assumed to be ideal conductors. All fields are split into homogeneous (matrix without fibres) and non-homogeneous part. Such assumption is not accurate, and so, temperature in a fibre is taken to be constant (Fig. 6.1). However, it is quite good approximation, if the fibres are not too long and the heat conductivity of fibres is much greater than that of the matrix. These facts allow us to considerably simplify the mathematical model of the heat conduction in SFC.

MCSF does not need any mesh. Continuity of temperature fields between the matrix and fibres is satisfied in discrete nodes – collocation points, only. Domain of the solution is the matrix, and the internal parts of fibres are supposed to be outside of this domain. The source functions are continuously distributed along the

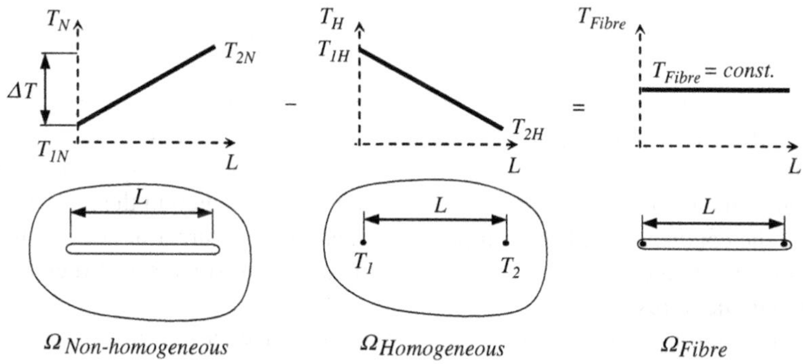

Fig. 6.1 Homogeneous and non-homogeneous part of thermal analysis

fibre axis. They are defined by shape function and their intensities. The intensity of shape function is modelled by 1D quadratic NURBS along the axis in order to satisfy boundary conditions (continuity of temperature between corresponding fibre and matrix).

6.2.1 Source Functions

The source functions for thermal analysis are heat sources and heat dipoles. Heat source is scalar quantity. Heat dipole is a heat source and heat sink acting at the same point by approaching each other in some coordinate direction. Mathematically, it is a derivative of heat source in corresponding direction (i.e. the heat dipole is a vector). Heat sources and heat dipoles act in source points. They can be distributed discretely in source points or continuously. 1D distribution along fibre axis of the source functions is used in our models. This is crucial for reduction of this kind of problem.

Temperature field by a unit heat source acting in arbitrary point of infinite domain is a fundamental solution and it is given by

$$T = \frac{1}{4\pi\, r} \tag{6.1}$$

where r is distance of the field point t and source point s, where the heat source is acting.

Temperature field by a unit heat dipole in x_i direction is

$$T = \frac{1}{4\pi}\left(\frac{1}{r}\right)_{,i} = -\frac{1}{4\pi}\frac{1}{r^2}r_{,i} = -\frac{1}{4\pi}\frac{x_i}{r^3} \tag{6.2}$$

The index after comma denotes partial derivative in corresponding coordinate direction.

Heat flow in x_i direction is given by

$$q_i = -k\frac{\partial T}{\partial x_i} = -kT_{,i} = k\frac{1}{4\pi}\frac{x_i}{r^3} \tag{6.3}$$

where k is coefficient of heat conductivity of the matrix. The material of a fibre is considered to be ideal conductor, i.e. its coefficient of conductivity is many orders better than that of the matrix. Usually the heat conductivity of fibres is much larger than that of the matrix. If the fibres are unidirectionally oriented in the matrix, they improve the conductivity of the composite material in fibre direction, i.e. the material can be good conductor in the fibre direction and good isolator in perpendicular direction to the fibres.

Heat flow through a surface with the normal n is defined as

$$q_n = q_i n_i \tag{6.4}$$

The total contribution of the source functions to the temperature and heat flow in corresponding collocation point is given by integrals along all source function paths (fibre axes). Because of finite thickness of the fibre the integrals are quasi-singular with corresponding weak or strong singularities. Of course the closest points will influence the fields more than the far points.

The integration can be made analytically [4] or numerically. The analytical integration is possible for straight fibres, but the integral kernels are more complicated if the fibres are curved and numerical integration is necessary. After numerical experiments on behaviour of source functions in composite material and about their influence to physical fields following rules were chosen for generation of collocation points and numerical integration:

- Collocation points are necessary to be chosen denser at the ends of fibres and by the points nearest to the ends of closest fibres.
- Because of very large differences in denominator (distance between source and collocation points) the whole integration path is split into integration elements. In our models the smallest element (closest to the collocation point) have to be equal to the fibre diameter and the others are about as large as the distance of its closest point from the collocation point. In this way same number of Gauss integration points can be used for all elements with about equal error from all integration elements in the model and five Gauss points rule is used in our computations giving good accuracy.

6.2.2 MCSF Model

If it is assumed that the thermal conductivity of fibres is by several orders larger than that of matrix material then the temperature of the fibre can be considered to be constant. Heat sources and dipoles 1D-continuously distributed along the fibre axis are used to simulate the interactions of fibres with the matrix and with other fibres (Fig. 6.2).

Collocation points are located on the interface fibre-matrix. The continuity of temperature is satisfied in the collocation points.

6.2.3 Shape Functions

The distribution of source functions is 1D. The 1D elements use quadratic NURBS shape functions [5–7]. The NURBS curve is defined by its order, set of weighted control points and knot vector. NURBS curve takes form (6.5), where fractional part can be denoted as rational basis function.

$$C(u) = \sum_{i=1}^{k} \frac{N_{i,n} w_i}{\sum_{j=1}^{k} N_{j,n} w_j} \mathbf{P}_i \qquad (6.5)$$

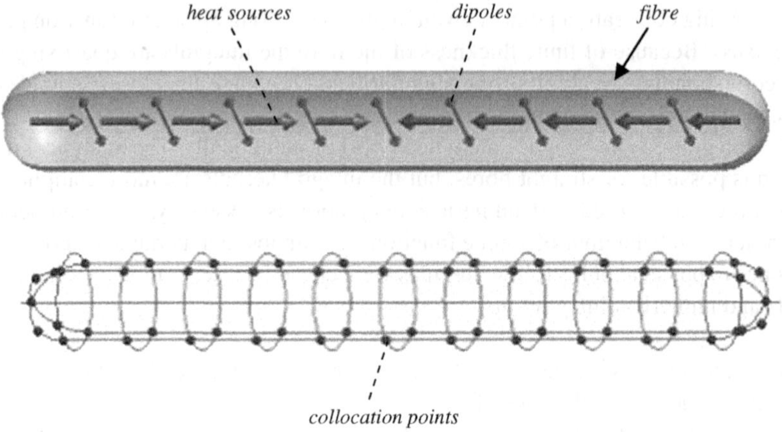

Fig. 6.2 Distribution of source functions and collocation points

where k is number of control points, P_i, u is parameter (in our case it is linear dimension in fibre direction), $N_{i,n}$ is the basis function of i-th control point and degree n (quadratic NURBS are used in our model, i.e. $n = 3$) and w_i are weights corresponding to the control point.

The denominator is a normalizing factor equal to one, if all weights are taken equal to one. This can be seen from the partition of unity property of basis functions. The (6.5) can be rewritten as:

$$C(u) = \sum_{i=1}^{k} R_{i,n}\mathbf{P}_i \tag{6.6}$$

where the fraction in (6.5) is denoted as $R_{i,n}$. $R_{i,n}$ are the rational basis functions.

The order of a NURBS curve defines the number of nearby points that influence any given point on the curve. The curve is represented mathematically by a polynomial of degree one less than order of the curve. Hence, second-order curve (which are represented by linear polynomials) are called linear curves, third-order curves are called quadratic. The number of control points must be greater than or equal to the order of the curve.

The positions of set of control points determine the shape of the curve (Fig. 6.4). The change by moving one control point allows making localized changes without affecting the overall shape of the curve. Each control point influences the part of the curve nearest to it but little or no effect on parts of the curve that are farther away. Each point of the curve is computed by taking a weighted sum of a number of control points. The weight of each point varies according to the governing parameter. For a curve of degree n, the weight of any control point is only nonzero in $n + 1$ intervals of the parameter space. At the boundaries of intervals, the basis functions go smoothly to zero.

The knot vector is a sequence of parameter values that determines where and how the control points affect the NURBS curve. The points demarcating the intervals are known as knot points. The ordered list of knot is knot vector. The number of knots is always equal to number of control points plus curve degree plus one. The knot vector divides the parametric space in the intervals. Values in the knot vector should be in non-descending order. The knot vector for basis functions in the Fig. 6.3 is (1.5, 1.5, 1.5, 2, 3, 4, 4.5, 5.5, 5.5, 5.5). To control exact placement of the endpoints of curve, we use multiple identical knots. Notice, that basis function $N_{4,3}$ associated with control point B_4 have larger value as $N_{1,3}$, $N_{2,3}$, $N_{3,3}$ and $N_{5,3}$ and NURBS curve would be pulled towards control point B_4. At $u = 1.5$, all the basis functions except $N_{0,3}$ have value 0, so control point B_0 is the only one to effect the curve and thus the curve coincides with that control point. Each control point has its own basis function.

Basis functions $N_{i,n}$ are computed as:

$$N_{i,n} = f_{i,n}N_{i,n-1} + g_{i+1,n}N_{i+1,n-1} \tag{6.7}$$

We can write the functions f and g as:

$$f_{i,n}(u) = \frac{u - k_i}{k_{i+n} - k_i} \text{ and } g_{i,n}(u) = \frac{k_{i+n} - u}{k_{i+n} - k_i} \tag{6.8}$$

The sum of basis functions for particular value of the parameter is unity. It is partition of unity property of basis function as we can see in the Fig. 6.3.

Figure 6.4 illustrates influence of $x(u)$ and $y(u)$ on smoothness of approximated curve. The smoothness is decreased or increased by shifting the coordinate, $x(u)$, and the function, $y(u)$, values, respectively. Different values are distinguished by different colours. Such shifting of values can be caused e.g. by errors, position of ends of integration elements, heat source intensity values, etc.

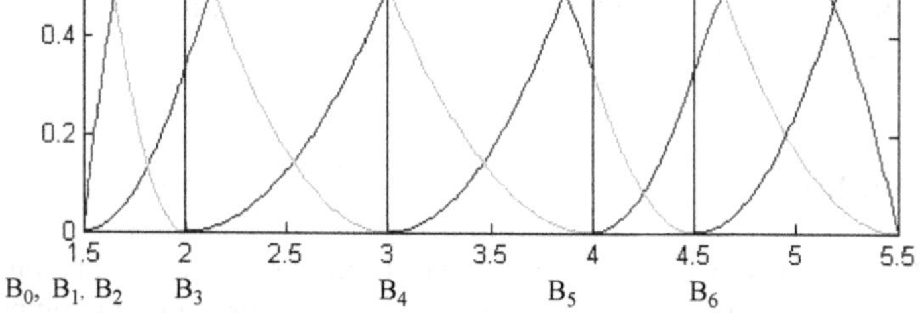

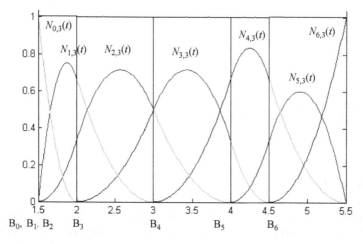

Fig. 6.3 NonUniform basis functions for a curve with multiple identical knots at the beginning and end

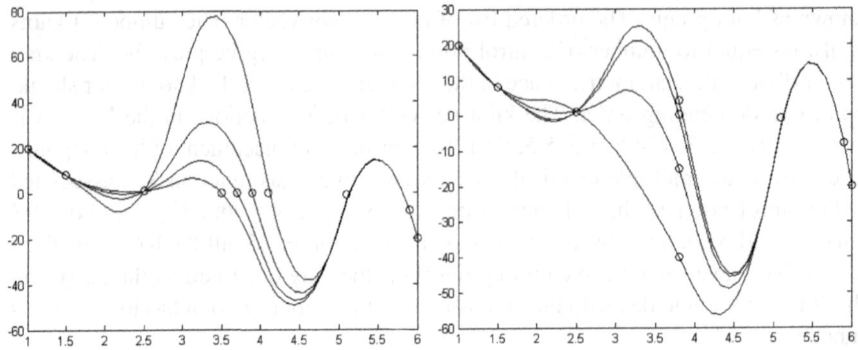

Fig. 6.4 NURBS smoothness control influenced by coordinate, *left*, and by the function values, *right*, in one point

Fig. 6.5 Modification of NURBS 3D curve through the control points (*circles*)

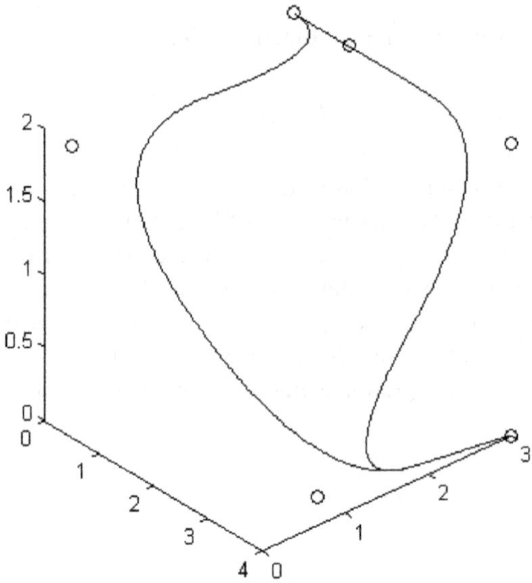

Figure 6.5 documents definition of 3D curve by control points and corresponding NURBS shape functions. This can be conveniently used for definition random curved fibres in a matrix.

6.3 Examples

Three different kinds of problems are shown. In the first example, one only fibre is situated in the matrix. Naturally, the results of single fibre do not comprise the influence of other fibres. In the second problem, the patch of regularly distributed fibres, the interaction of the fibres is studied. The interaction just of two fibres is considered by computations in the third problem. It will serve for simulation of irregularly distributed fibres in a matrix in a model for parallel computations.

6.3.1 Single Fibre

The first two examples are computational results for temperature fields in a material containing single fibre in a matrix. Length of the fibre is 100 times larger than its radius in the Figs. 6.6 and 6.7 and 1,000 times larger than the radius in the Fig. 6.8.

Figure 6.6 shows heat source intensity along the single fibre (z-direction is direction of fibre axis). After integration of intensity of heat sources, we obtain heat flow along axis in a single fibre in the axis direction (Fig. 6.7). The maximum heat flow is in the middle of the fibre.

Figure 6.8 shows the largest heat flow in the ends of fibre through the fibre surface, i.e. in perpendicular direction to the fibre axis. In these parts are also the largest temperature gradients.

Distribution of different quantities along fibres is similar for fibres with different aspect ratios, only the gradients of the functions are different in each case.

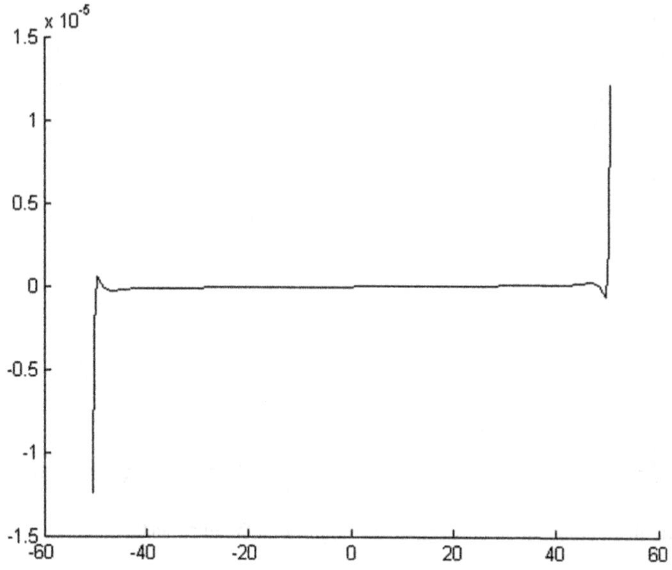

Fig. 6.6 Heat source intensity along single fibre ($L = 100R$)

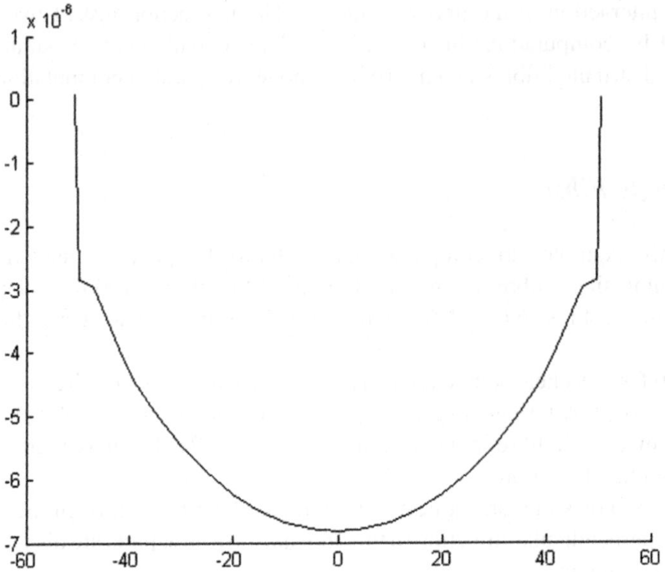

Fig. 6.7 Heat flow along single fibre ($L = 100$R) in axis direction

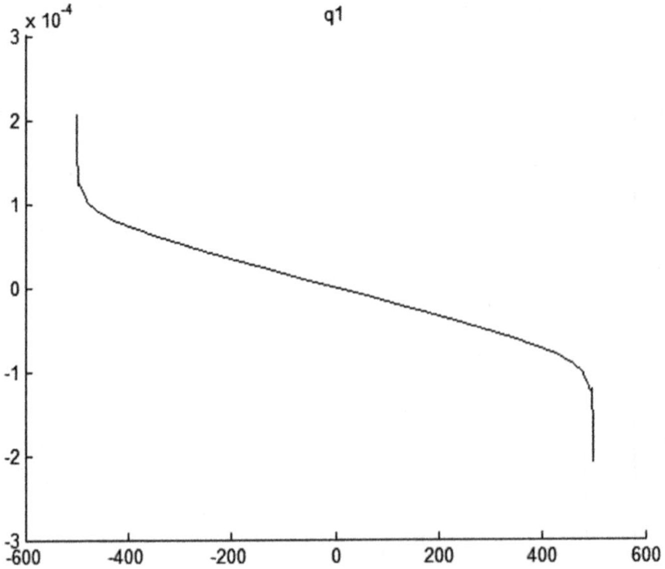

Fig. 6.8 Heat flow along single fibre ($L = 1000$R) through the fibre surface

6.3.2 Patch of Fibres

The models can involve not only one fibre but also patch of fibres which are straight and distributed regularly in the matrix. The patches of fibres in the next example (Fig. 6.9) consist of overlapping and non-overlapping rows of fibres. One can see the strong interaction between fibres in case of overlapping fibres. This is very different from the case of non-overlapping rows of fibres.

Figures 6.10 and 6.11 illustrate heat source intensity and heat flow in fibre with and without overlap. The aspect ratio of fibres is 1:500 (fibre diameter: fibre length, R is radius and L is length). Dashed and continuous curves are for non-overlapping fibres with smaller and larger gap in fibre direction. One can see that the fibres with larger gap in their axes direction with overlapping transmit more heat than those with smaller gap. It is because of larger difference of temperatures in the points connecting the ends of fibres.

Comparing the cases with or without overlap, the fibres with overlap transmit much more heat and the ends of fibres interact very strongly with the nearest fibres. Notice large gradients in heat source intensities on the ends of fibres and also in the points closest to the ends of neighbour fibres. These parts require very fine distribution of control points for definition of shape functions in order to keep good accuracy and numerical stability of the models.

The similar behaviour of patches of fibres is documented also in the Figs. 6.12 and 6.13, which present results of fibres with and without overlap of fibres with aspect ratio 1:50 (shorter fibres than those in previous case). Different distribution of fibres and corresponding line styles are defined in the figure. The smaller gap with overlapping, this time in perpendicular direction to fibre axis, results in higher percentage of fibres in the volume and larger heat flow and thus better conductivity in

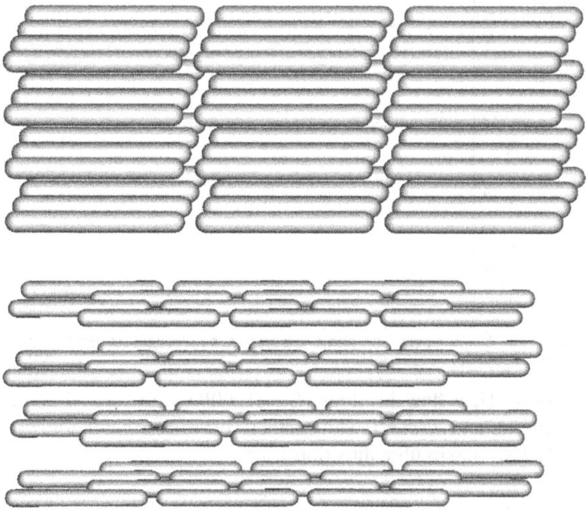

Fig. 6.9 Patch of non-overlapping (*upper part*) and overlapping (*lower part* of the figure) rows of fibres

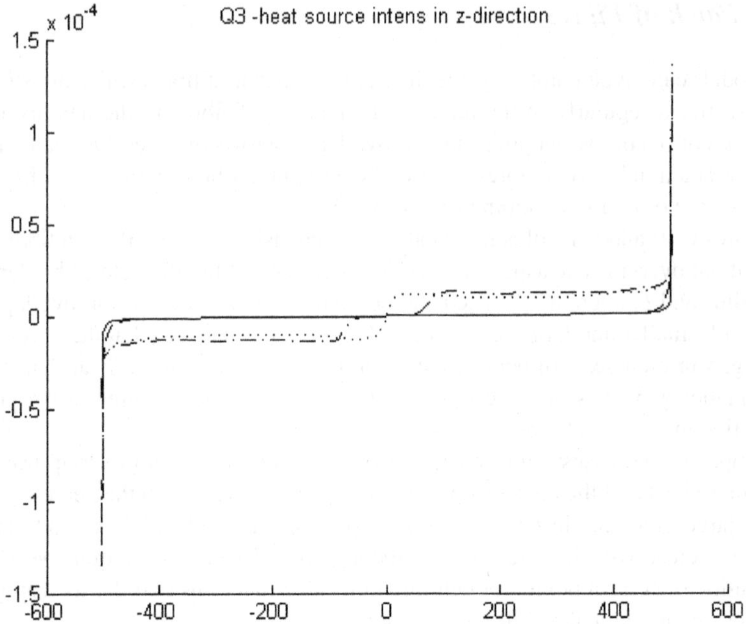

Fig. 6.10 Heat source intensity in fibres ($L = 1000$R) without overlap (*continuous*, *dashed*), with smaller 16R (*dotted*) and larger 160R (*dot-dashed*) gap in fibre direction

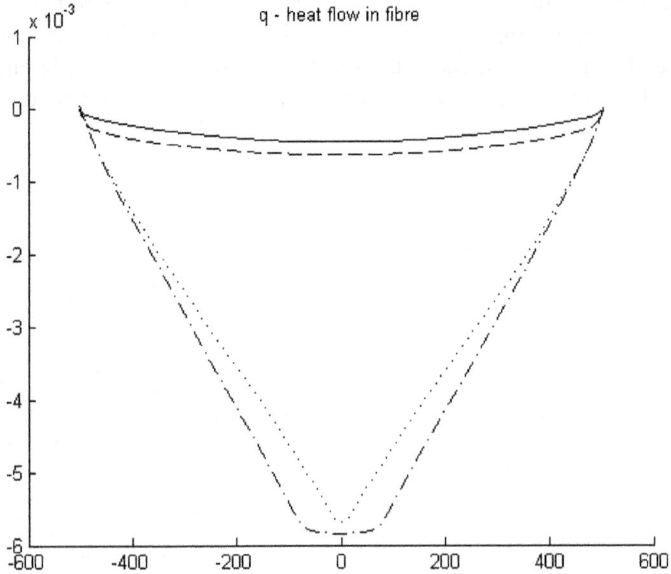

Fig. 6.11 Heat flow in fibres ($L = 1000$R) with (*dotted* and *dot-dashed*) and without overlap (*continuous* and *dashed*), with smaller 16R (*dashed* and *dotted*) and larger 160R (*continuous* and *dot-dashed*) gap in fibre direction

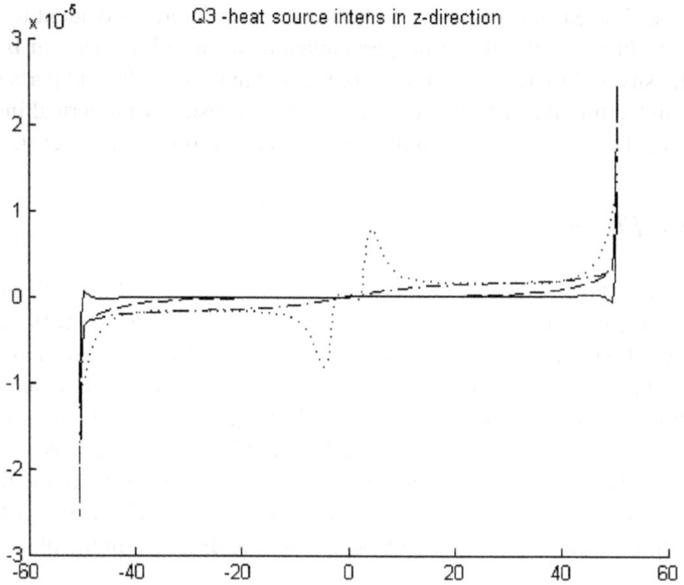

Fig. 6.12 Heat source intensity in fibres ($L = 100R$) with (*dotted* and *dot-dashed*) and without overlap (*continuous* and *dashed*), with smaller 2R (*continuous* and *dotted*) and larger 16R (*dashed* and *dot-dashed*) gap in perpendicular direction

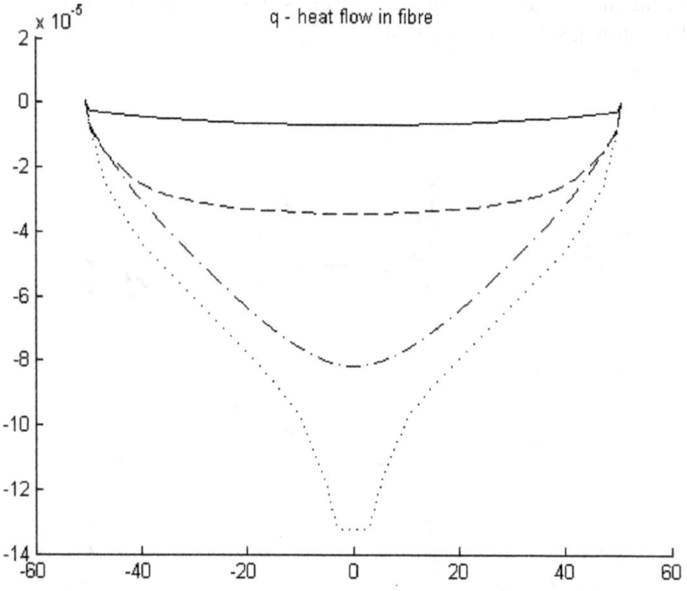

Fig. 6.13 Heat flow in fibre direction ($L = 100R$) with (*dotted* and *dot-dashed*) and without overlap (*continuous* and *dashed*), with smaller 2R (*continous* and *dotted*) and larger 16R (*dashed* and *dot-dashed*) gap in perpendicular direction

the fibre direction. Stronger interaction between ends of fibres is evident by smaller gap between fibres in the direction perpendicular to the fibres axes in both heat source intensity and in heat flow in the fibre. The waviness in the end parts (continuous) and in the middle part (dotted) in Fig. 6.12 is caused by numerical instability of the model. The accuracy is not influenced much by this effect, however.

6.3.3 Two Fibres in a Matrix

Next examples show results of heat flow in the fibres' axes and through sides perpendicular to fibres' axes in Figs. 6.14, 6.15, 6.16, 6.17, 6.18 and 6.19. The first three figures show the case for fibres of equal length and with the overlap equal to 50R in the fibres' axes direction, the last three figures contain similar results but for different fibres with length equal to 100R and 50 R, respectively. The Figs. 6.15 and 6.18 give the sum of the flow through the fibres, which is obtained as sum of the flows through the opposite sides of the fibres. Recall that the normal to the fibre surface has opposite sign in the points on opposite sides of the fibre and this function gives the total flow in direction perpendicular to fibre direction in the plane of both fibres.

The computational models for simulation of the interaction of fibres have to satisfy the energy balance in the fibres, which is not kept automatically. For a pair of fibres this condition is identical to setting the flow in longitudinal direction of corresponding fibre equal to zero in both ends and it is achieved by solving the problem for following boundary conditions:

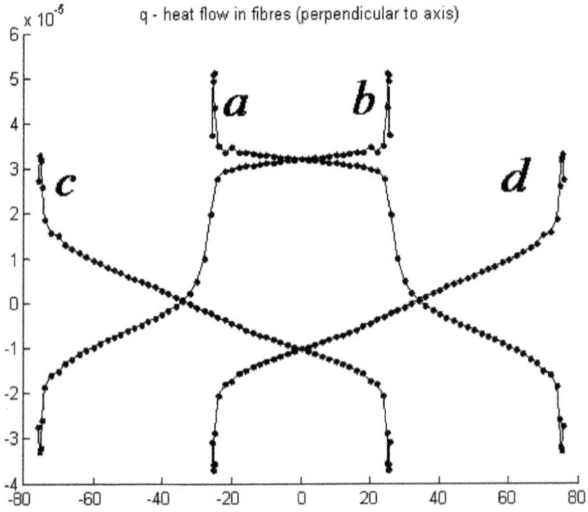

Fig. 6.14 Heat flow in direction perpendicular to fibre axes ($L = 100\text{R}$) in sides closer (a and b) and farther to neighbour fibre (c and d)

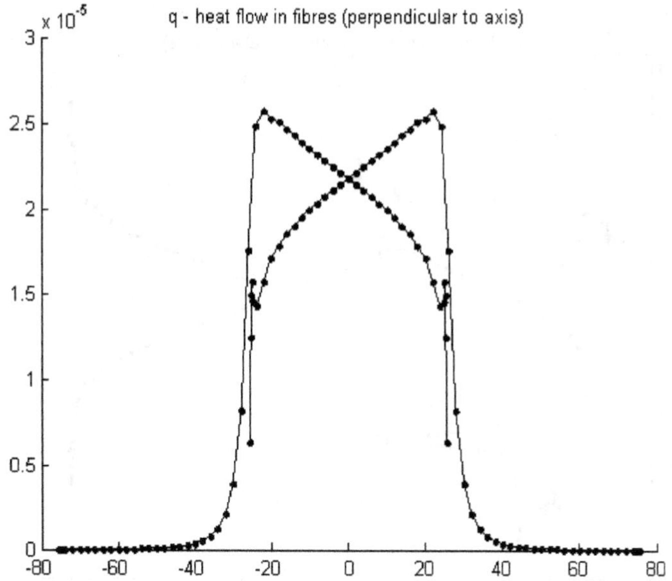

Fig. 6.15 Heat flow through the fibres in direction perpendicular to fibre axes ($L = 100\text{R}$)

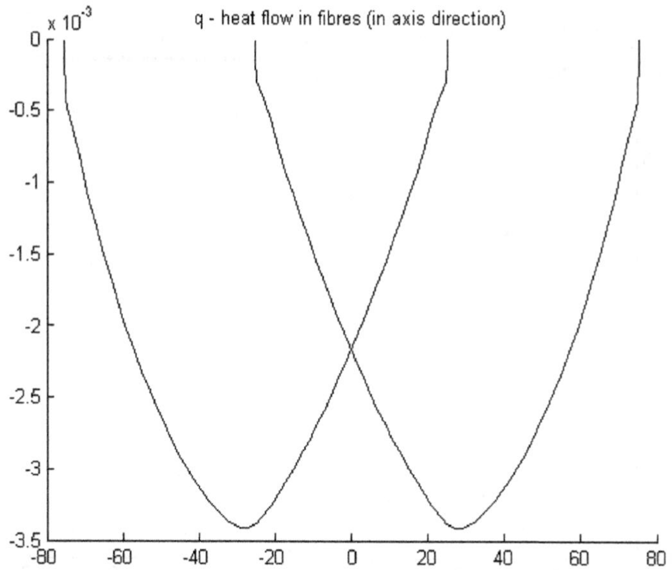

Fig. 6.16 Heat flow in direction of fibre axes ($L = 100\text{R}$)

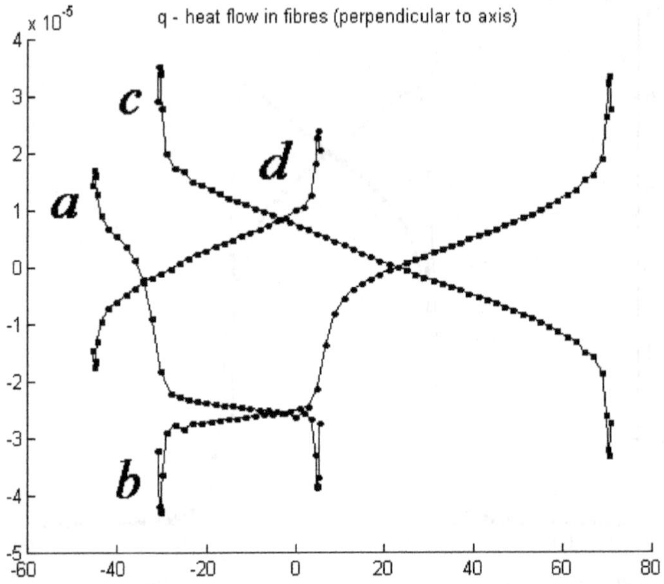

Fig. 6.17 Heat flow in direction perpendicular to fibre axes ($L = 100R$ and 50R) in sides closer (*a* and *b*) and farther to neighbour fibre (*c* and *d*)

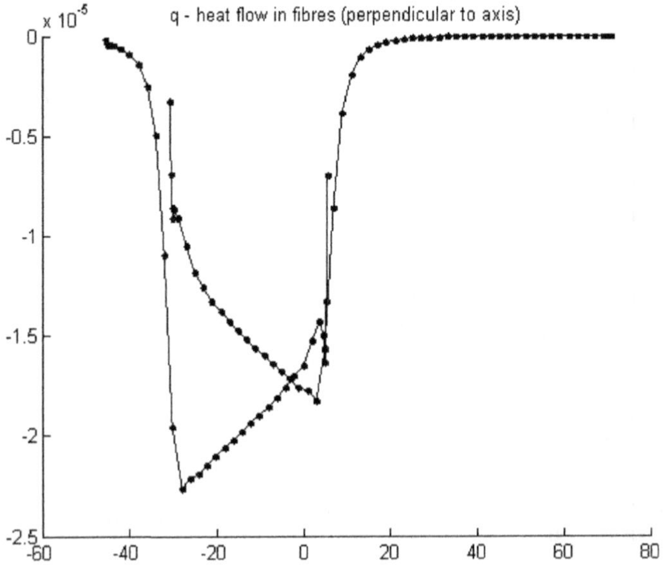

Fig. 6.18 Heat flow through the fibres in direction perpendicular to fibre axes ($L = 100R$ and 50R)

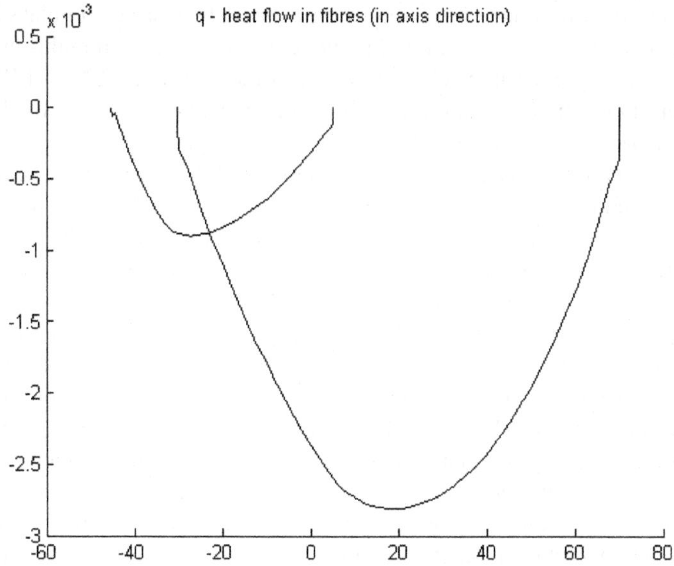

Fig. 6.19 Heat flow in direction of fibre axes ($L = 100$R and 50R)

According to the Fig. 6.1 we prescribe the temperature as given for the non-homogeneous case for each fibre. The problem is solved for another two r.h.s. The first one, the temperature is set equal to one for one fibre and zero for the other fibre. The second r.h.s. is defined conversely to the previous for the fibres. The condition of the longitudinal flow in both ends of both fibres gives the resulting flows and at the same time one obtains the temperature change for both fibres in consequence of the interaction.

In the examples above, the temperature gradient in fibres' axes was prescribed to 10^{-6} (non-dimensional units are used here) and the temperature increased by 1.220×10^{-5} in the first fibre and decreased by the same value in the second one for equal fibres. For unequal fibres, the temperature increase by 1.998×10^{-5} in the longer fibre and decreased by 0.188×10^{-5} in the shorter fibre.

Typically, problem of interaction of two fibres in a matrix leads to solution of 300 $\times$ 200 or more equations. If a Control Volume Element (CVE) for homogenization contains 100 fibres, there will be 100 $\times$ 50 interactive pairs of fibres. To compute complete interactions for homogenization, it is effective to make the parallel algorithm for the problem. The temperature changes in all fibres and heat flow are obtained as the sum of all interactions.

6.4 Conclusions

The MCSF is a boundary meshless method using collocation technique. It provides good accuracy even for large aspect ratio of fibres. Continuous heat source and dipole models enable to simulate both near and far fields in materials reinforced by fibres and they introduce large reduction of models. Correct simulation of the far fields is necessary for evaluation of homogenization process in multi-level modelling. Similar accuracy can be obtained by other methods like BEM or FEM when the number of elements is by several orders larger than in presented method (see [1] for the comparison of the methods).

The presented model assumes ideal conductivity of the fibre. As the heat flow in fibre can exceed the largest heat flow in the matrix by several orders, if the aspect ratio of fibres is large, the assumption of ideal conductivity can be incorrect, however, iterative corrections can improve the accuracy in very few steps (usually not more than four steps are necessary to obtain final solution).

Correct simulation of the interaction of fibres is important for evaluation of composite properties. The high gradients in the fields require some experience with choose of control points for definition of the continuous source functions along the fibre axis. The examples show that short fibres can contribute to change considerably properties of composite material and that MCSF is a method which enables to simulate large gradient fields in such composite very effectively.

Our next aims are: Further reduction of models (iterative solutions for more general problems), parallel computational models for complex microstructure containg many fibres and multilevel models (homogenization) and modelling of imperfect straight and curved fibres in order to be able to simulate more complicated microstructure of the composite materials.

Acknowledgments Support of the DSSI, grant agencies APVV (grant No. APVT-20-035404) and RTO-NATO (grant No. 001-AVT-SVK) and VEGA (grant No. 1/0140/08) for this research is gratefully acknowledged.

References

1. Kompiš, V., Murčinková, Z., Grzhibovskis, R., Rjasanow, S., Qin, Q.-H. Computational simulation methods for composites reinforced by short fibres. In: Yuan, Y., Cui, J., Mang, H.A. (eds.) Computational Structural Engineering. Springer, Shanghai, pp. 63–70 (2009)
2. A Guide to PROCISION 3.5.: Precision Analysis Inc. Mississauga (1999)
3. Liu, Y.L., Nishimura, N., Otani, Y., Takahashi, T., Chen, X.L., Munakata, H.: A fast boundary element method for the analysis of fibre-reinforced composites based on a rigid-inclusion model. ASME J. Appl. Mech. **72**, 115–128 (2005)
4. Kompiš, V., Štiavnický, M., Kompiš, M., Murčinková, Z., Qin, Q.-H.: Method of continuous source functions for modeling of matrix reinforced by finite fibres. In: Kompiš, V. (ed.) Composites with Micro- and Nano-Structure. Springer, Dordrecht, pp. 27–46 (2008)
5. Rogers, D.R.: An Introduction to NURBS. Morgan Kaufmann Publishers, London (2001)
6. Inglesias, A.: Computer-Aided Design and Computer Graphics: B-splines and NURBS Curves and Surfaces. University of Calabria, Cosenza (2001)
7. Reali, A.: An Isogeometric Analysis Approach for the Study of Structural Vibrations. University of Pavia, Pavia (2004)

Chapter 7
Simulation of Distributed Detection of Ammonia Gas

O. Sýkora, J. Aubrecht, R. Klepáček, and L. Kalvoda

Abstract A basic theoretical model of a distributed fiber optic system for detection and location of ammonia leaks is proposed and employed in computer simulations of principal system characteristics. The simulated system consists of an optical time-domain reflectometer (OTDR) interrogating a sensing optical fiber composed of silica core and polymer cladding doped with a reagent. The simulated processes include gas diffusion into the cladding, conversion of the reagent in presence of the analyte, time- and space-evolution of polarizability and absorption coefficient, propagation of the primary light pulse along the fiber, interaction of the pulse with the cladding and generation of a backscattered light signal registered afterwards by the OTDR detector. Juxtaposition of the simulated results and the demands laid on a real system results in several recommendations for further development and optimization of such sensing systems.

Keywords Fiber optic sensors · OTDR method · Ammonia sensor

7.1 Introduction

Clean air, free of harmful adulterants, is one of the vital conditions necessary for our biosphere to flourish. Even restricting ourselves on involuntary events only, the risks associated with leaks of potentially harmful gaseous substances are becoming more and more real with the all-encompassing industrial development. If such accident is the case, the resulting injury typically strongly depends on the concentration level of the noxious agent and the exposition time. Thus, among the measures applied to reduce the hazards (on the first place, of course, focused on increase of the intrinsic safety of the technology in question itself), one of the priorities lies in installation of a warning system providing an early pollutant detection, characterization of the event's parameters and fast dissemination of the warning.

L. Kalvoda (✉)
Department of Solid State Engineering, Faculty of Nuclear Science and Physical Engineering, Czech Technical University in Prague, Prague, Czech Republic
e-mail: ladislav.kalvoda@fjfi.cvut.cz

J. Murín et al. (eds.), *Computational Modelling and Advanced Simulations*, Computational Methods in Applied Sciences 24, DOI 10.1007/978-94-007-0317-9_7,

It is the truth that exploration of majority of potentially harmful or even lethal gases proceeds within industrial facilities. But, there are some exceptions, such as carbon monoxide, propane-butane or ammonia, where the risk of a massive leakage can also occur in public areas. Focusing in further on ammonia, it is fairly extensively used in large-scale cooling systems employed, e.g., in dairies, breweries, abattoirs, hypermarkets, logistic centers, refrigerator vehicles and vessels, ice hockey halls and ice rinks. Large storage tanks with liquid ammonia can be found in chemical plants utilizing Haber-Bosh reaction and using ammonia to produce nitrogen fertilizers. A massive and permanent release of ammonia also occurs in farming facilities housing animals or keeping poultry. Especially in the later case, any long-lasting malfunction of the venting system providing the necessary air circulation can lead to disaster.

In all the mentioned cases, any reliable and fast detection and location of ammonia leaks is of a considerable public importance due to the malignant nature of ammonia gas, especially at higher concentrations. According to review [1], the concentration level of ammonia in air safe for human perception is about 50 ppm, but even lower concentrations can be harmful for the respiratory system if a long-lasting exposition occurs. In such case, the allowed concentration is set to 20 ppm. Medium concentrations (500–1,000 ppm) can lead to a serious attack on the respiratory system; high concentrations (5000–10,000 ppm) can be lethal.

The ammonia detection systems used at present are based on signal networks of conductometric sensing heads. The limited gas selectivity and lack of long-term stability of the sensing elements is a serious problem of this solution [1]. From the economical point of view, the concept suffers a rapid rise of purchase costs and electric power consumption when the number of installed heads is increasing. Numerous schemes have been proposed and tested to overcome the drawbacks [1, 2]. One of the promising concepts is to employ an optical fiber as the sensing element with the cladding layer sensitized by a suitable reagent selective to ammonia gas. The two most extensively tested working principles used in this regard utilize light absorption and fluorescence analysis. Interaction of a reagent with ammonia leads to changes of the original spectral distribution of optical transmittance (absorption-based sensors) or to changes of stimulated light emission (fluorescence-based sensors). The later optical properties can be then effectively interrogated by some of the recent optical reflectometric methods; the optical time domain reflectometry (OTDR) employed in our model represents likely the most common variant.

Initial experimental results obtained in tests of the real sensing system of such type [2] convicted us that definition of a comprehensive theoretical model and its utilization in simulations of basic sensor functions could provide a very useful tool for further development and optimization of the system. The following sections of this article are devoted to introduction of the model, description of the applied simulation strategies and discussion of the results and their implications for construction of real systems.

7.2 Proposed Model of the Sensing System

7.2.1 General Features

The system to be analyzed consists of two main parts, namely (i) the sensing optical fiber working as a chemo-optical, absorption-based transducer and (ii) an interrogating OTDR unit (Fig. 7.1).

The fiber is of a step-index, multimode type with the core radius a (cm), the overall radius b (cm), the core refractive index n_0, the cladding refractive index n_1 and the numerical aperture $NA = (n_0^2 - n_1^2)^{1/2}$. The cladding is supposed to be sensitized by an ammonia-sensitive reagent R possessing a particle density $N^R(\text{cm}^{-3})$, molecular extinction coefficient $\varepsilon^R(\text{cm}^{-1})$ and molecular polarizability $\delta^R(\text{cm}^3)$. The intended sensing process includes two steps: (i) diffusion of ammonia (the analyte) into the cladding and (ii) conversion of the reagent in presence of the analyte. The conversion is supposed to be a reversible, concentration-controlled composition/decomposition reaction of organo-metallic reagent $R = (\text{L}_n - \text{Me})^{p+}(\text{A}^-)_p$ (L – organic ligand, n – number of ligands in the reagent molecule, m – number of ligands in ammonia complex, Me – the central metal ion, p – the oxidation state of Me, A – a selected univalent counter-anion):

$$(\text{L}_n - \text{Me})^{p+} + (\text{A}^-)_p + m(\text{MH}_3) \leftrightarrow ((\text{NH}_3)_m - \text{Me})^{p+} + (\text{A}^-)_p + nL \quad (7.1)$$

The OTDR unit launches a short rectangular monochromatic light pulse into the fiber. The pulse is characterized by its irradiance $I(z = 0) = I_0(\text{W/cm}^2)$, duration τ (s), spatial width $w = \tau c/n_0$(cm) and wavelength $\lambda(\text{cm}^{-1})$. For sake of simplicity, the modal structure of electro-magnetic (EM) waves transported within the fiber core (each mode generally characterized by the pair (l,m) of the azimuth index l and radial index m [3]) is approximated by the first mode $(l, m) = (0, 1)$. Moreover, the power density distribution across the core is supposed to be uniform, with the value $I^{\text{core}}(z)$ independent of the fiber radius $r = (x^2 + y^2)^{1/2}$ and the azimuth angle $\varphi = arctan(y/x)$. The parameter $\gamma_{(0,1)} = \gamma(\text{cm}^{-1})$ characterizing the decay of the radial evanescent component of the power density within the fiber cladding is then obtained as solution to the characteristic equation [3]

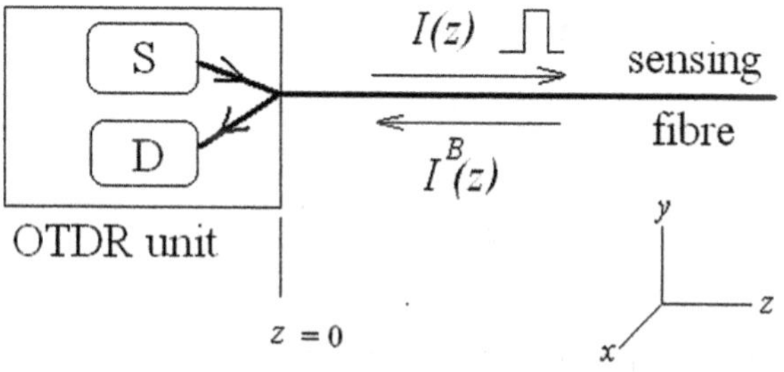

Fig. 7.1 Schema of the simulated sensing system and definition of the reference co-ordinate system. S – the pulsing laser diode, D – the light detector, I – the light intensity propagating along the $+z$ direction, I^B – the intensity propagating along the $-z$ direction

$$X\frac{J_1(X)}{J_0(X)} = Y\frac{K_1(Y)}{K_0(X)} \tag{7.2}$$

The parameters X and Y can be expressed as

$$X = a\left\{\left(\frac{2\pi NA}{\lambda}\right)^2 - \gamma^2\right\}^{1/2}, \; Y = \gamma a \tag{7.3}$$

The symbols J_l and K_l represent the Bessel functions of the first kind and the modified Bessel functions of the second kind, respectively. We have obtained $\gamma = 1.26\ \mu m^{-1}$ for the above specified fiber.

As the primary pulse propagates along the $+z$ direction, it is experiencing several principal interactions: absorption and Rayleigh scattering caused by reagent molecules and Rayleigh scattering occurring on the materials composing the core and the cladding matrix. All the processes lead to reduction of the primary pulse intensity. The progressive interference of spherical waves created at individual scattering centers is, at the same time, giving rise to two planar waves propagating along the $+z$ and $-z$ directions. The first wave is omitted since its intensity is negligible compared to the intensity of the primary pulse. Time evolution of the intensity of the later wave ($I^B(t)$) is registered at the OTDR unit detector. The temporal co-ordinate is converted to the spatial position using $z = ct/(2n_0)$. The usual $I^B(t)$ course (Fig. 7.2) contains two Fresnel reflections originating at both ends of the tested

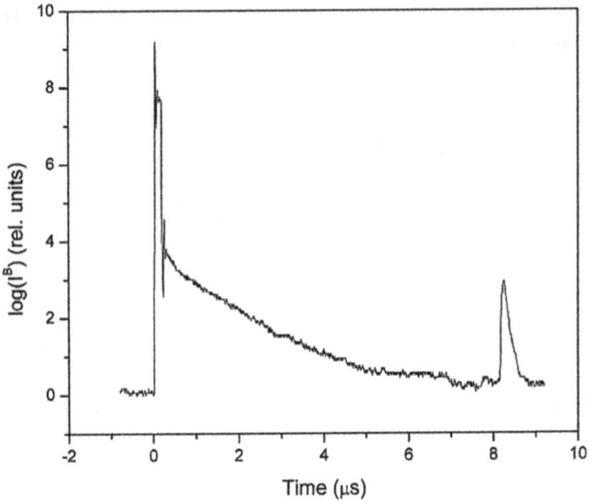

Fig. 7.2 Typical I^B-course obtained with a multimode plastic-clad silica fiber (length 834 m, $b/a = 120\,\mu m/100\,\mu m$); OTDR unit Photodyne 5500 XFA ($\tau = 100$ ns), $\lambda = 850$ nm, $I_0 = 30\mu W$

fiber and an intermediate part providing us with information about optical properties along the tested fiber length. Our simulations are focused on this intermediate part of the course.

7.2.2 Chemo-Optical Transducer

Our chemo-optical transducer is represented by a multimode fiber comprising a silica core ($a = 100\,\mu\text{m}$, $n_0 = 1.465$) and a polysiloxane cladding ($b = 120\,\mu\text{m}$, $n_1 = 1.455$). Numerical aperture of the fiber is $NA = 0.171$ and the critical angle $\theta_c = 6.7°$. Thus, the fiber can be considered as weekly guiding and our "first-mode" approximation is acceptable.

The cladding layer is supposed to be doped with a real organo-metallic reagent R [2, 4]: L = 5-(4′-dimethylaminophenylimino)quinolin-8-one, Me = Cu, A = Br, $p = 2$, $n = 2$ and $m = 4$. The molecular weight of the ligand (M^L) and the reagent (M^R) is 277.38 and 773.5 g/mol, respectively. Reagent molecules are supposed to be distributed within the cladding polymer with the mass concentration $u^R(r,z,t)$:

$$u^R(r,z,t) = (M^R/N_A)N^R(r,z,t) \tag{7.4}$$

Here N_A is the Avogadro number. Let's now expose the fiber to ammonia within the interval of its length $z_1 \le z \le z_2$. Obviously, the mass concentration of ammonia in the cladding volume ($u^A(r,z,t),\ r \le b,\ t > 0$) will start to grow according to the second Fick's law

$$\frac{\partial u^A}{\partial t} = \frac{\partial}{\partial r}\left\{D\left(\frac{\partial u^A}{\partial r}\right)\right\} - qu^R u^A \tag{7.5}$$

For the sake of simplicity, we suppose that (i) the coefficient of ammonia solubility in the cladding is equal to unity, i.e. the concentration $u^A(r = b - \Delta r) = u^A(r = b + \Delta r)$ for any $\Delta r > 0$, and (ii) that the diffusion coefficient of ammonia gas D in the cladding does not change with r.

For we did not find any suitable experimental data related to diffusion of ammonia in polysiloxanes, we had to adopt the following approximation. In [5], the value $D = 2 \times 10^{-10}\ \text{cm}^2/\text{s}$ was obtained for molecular oxygen diffusing through a Sylgard 184 cross-linked matrix.[1] Taking in account the fact that NH_3 molecule is slightly smaller in size then O_2 and neglecting possible Columbic and polarization effects, we suppose that the D-value for ammonia can be roughly estimated as $D \cong 10^{-9}\ \text{cm}^2/\text{s}$.

The second term on the right side of (7.5) quantifies the trapping process accompanying the chemical reaction of ammonia with reagent. The parameter q ($\text{cm}^3\text{g}^{-1}\text{s}^{-1}$) can be understood as a speed constant related to a first order reaction

[1] Sylgard (Dow Corning) is a commercial thermo-cross-linkable fiber cladding material based on polysiloxane formulation.

between ammonia and reagent: A + R $\leftrightarrow$ AR adopted as the approximation of the exact reaction course (7.1). Apparently, the radial concentration distribution of reagent $u^R(r, z, t)$ will also change during the exposition:

$$\frac{\partial u^R}{\partial t} = -qu^A u^R \tag{7.6}$$

The boundary conditions of the processes described by (7.5) and (7.6) are given as

$$u^A(r < b, t = 0) = 0 \tag{7.7a}$$

$$u^A(r = b, t \geq 0) = u_0^A = const \tag{7.7b}$$

$$u^R(a \leq r \leq b, t = 0) = u_0^R = const. \tag{7.7c}$$

The differential equation (7.5) and (7.6) together with the boundary conditions (7.7) can be transformed into the following difference equations by setting $t = i\,\Delta t,\ r = a + j\Delta r;\ \Delta r = a + j(b - a)/100;\ i, j \in \mathrm{N}_0,\ j = 0, ..100.$[2]

$$u_j^{A,i+1} = u_j^{A,i} + \frac{\Delta t D}{(\Delta r)^2}(u_{j+1}^{A,i} - 2u_j^{A,i} + u_{j-1}^{A,i}) - q\Delta t u_j^{R,i} u_j^{A,i} \tag{7.8}$$

$$u_j^{R,i+1} = (1 - q\Delta t u_j^{A,i}) u_j^{R,i} \tag{7.9}$$

$$u_j^{A,i=0} = 0, j = 0, ..100 \tag{7.10a}$$

$$u_{j=0}^{A,i} = u_0^A, i \geq 0 \tag{7.10b}$$

$$u_j^{R,i=0} = u_0^R, j = 0, ..100 \tag{7.10c}$$

The derived explicit difference schema is numerically stable if the temporal (Δt) and spatial (Δr) step satisfy the stability condition

$$D\frac{\Delta t}{(\Delta r)^2} < \frac{1}{2} \tag{7.11}$$

Equations (7.8), (7.9), (7.10) and (7.11) allow for calculation of the sought radial and temporal evolution of the ammonia and reagent concentrations using a simple iteration schema. Starting at $t = 0$ ($i = 0$), the concentration profiles of ammonia and reagent are set according to (7.10) and the time step Δt calculated from (7.11).

[2] Further on, we suppose that the radial co-ordinate is discretized into 100 intervals.

Then, the vector $u_j^{A,i=1}$ is obtained from (7.8) and using (7.9) we get the new $u_j^{R,i=1}$. The later value enters Relation (7.8) and the process is repeated.

Knowledge of the radial profile of the reagent concentration is required for the consecutive evaluation of the interactions between the interrogating light pulse and the fiber matter. Before we can do that, we need to quantify several important material constants: the extinction coefficient and the molecular polarizability of the reagent and the loss coefficient (α^M) and the average molecular polarizability (δ^M) of the core and the cladding matrix. For sake of simplicity, we suppose in further that the molecular polarizability is identical for both matrices.[3]

The molecular extinction coefficient of the reagent R (ε^R) was determined from the VIS-NIR optical absorption spectrum of the reagent dissolved in ethanol[4]: $\varepsilon^R(\lambda = 850\text{ nm}) = 1.5 \times 10^{-17}\text{ cm}^{-1}$. The molecular polarizability of the reagent (δ^R) was obtained from its electric susceptibility (calculated with aid of Kramers-Kronig relations from the optical absorption spectrum) using Claussius-Mossotti model: $\delta^R(\lambda = 850\text{ nm}) = 3.3 \times 10^{-23}\text{ cm}^3$. The polarizability of the core and cladding matrices was approximated by the polarizability of SiO_2 molecule $\delta^M = 2.86 \times 10^{-24}\text{ cm}^3$ [6]. The average concentration of scattering centers (N^M) can be then calculated from the mass density of quartz ($\rho^{SiO_2} \cong 2.2\text{ g/cm}^3$) and its molecular weight (M^{SiO_2}) as $N^M = N_A \rho^{SiO_2}/M^{SiO_2} \cong 2.21 \times 10^{22}\text{ cm}^{-3}$. The intrinsic loss of the primary pulse intensity due to the optical absorption and Rayleigh scattering occurring in the fiber matrix is characterized by a mean loss coefficient $\alpha^M = 6.9 \times 10^{-6}\text{ cm}^{-1}$ corresponding to the total loss 3 dB/km of the original fiber.

7.2.3 Light Intensity Propagating Along the Fiber

The total light power $I(z)$ (W) at the position z is expressed as

$$I(z) = I_1(z) + I_2(z) \tag{7.12}$$

$$I_1(z) = \pi a^2 I^{\text{core}}(z) \tag{7.13}$$

$$I_2(z) = 2\pi I^{\text{core}}(z) \int_a^b \exp\{-\gamma(r-a)\} r dr \tag{7.14}$$

The subscript 1 and 2 relates to the fiber core and cladding, respectively. In (7.14), the square of the Bessel functions $K_1(\gamma r)$ (giving the correct radial decay of the evanescent electric field in the cladding) is approximated by a simple exponential

[3]Both fused silica and polysiloxane resin matrix contains extended siloxane network, though differing in its density and topology.

[4]Maximum of the absorption band is observed at $\lambda_{max} = 740$ nm.

function. The integral can be then solved analytically and the mean core irradiance ($I^{core}(z)$) written as

$$I^{\text{core}}(z) = \frac{I_1(z) + I_2(z)}{\pi} \left\{ a^2 + \frac{2}{\gamma^2} (1 - \exp(-\gamma(b-a))(\gamma(b-a)+1)) \right\}^{-1} \tag{7.15}$$

The light intensity transported between two points z_1 and z_2 ($z_1 < z_2$) is damped:

$$I_1(z_2) = I_1(z_1) \exp\left(-\alpha^M w\right) \tag{7.16}$$

$$I_2(z_2) = 2\pi I^{core}(z_1)\, exp\,(-\alpha^M w)\, exp\left[-w \int_a^b exp\{-\gamma(r-a)\}\alpha^R(r, z_1 \leftrightarrow z_2) r dr\right] \tag{7.17}$$

Radial distribution of the linear absorption coefficient $\alpha^R(r,z)$ depends on the reagent concentration profile averaged over the distance $z_1 \leftrightarrow z_2$:

$$\alpha^R(r, z_1 \leftrightarrow z_2) = \frac{N_A \ln(10)\varepsilon^R}{M^R} \frac{1}{z_2 - z_1} \int_{z1}^{z2} u^R(r,z) dz \tag{7.18}$$

The practical simulation procedure of the intensity profile of the light propagating along the fiber is based on discretization of the axial co-ordinate (z). With the step size of the grid chosen as $\Delta z = w$ and the number of steps fixed to 100, the total length of our simulated fiber amounts 410 m. At every nexus, the components I_1 and I_2 of the intensity coming from the preceding node are determined using (7.16) ÷ (7.18) and the total core intensity is obtained from (7.15). The new components I_1' and I_2' are then calculated from (7.13) and (7.14) and used in evaluation of the intensity coming to the next nexus etc.

The re-distribution of the intensity between the core and the cladding ($I_1 \to I_1'$, $I_2 \to I_2'$) reconstructs the necessary intensity balance distorted primarily by the reagent absorption within the cladding layer. This schema can be also understood as a discrete application of the conditions of continuity binding together the corresponding EM field amplitudes and the radial intensities at the core/cladding boundary.

Intensity of a back-scattered wave $I^B(z_1)$ arisen due to Rayleigh scattering at the position z_2 and approaching the point z_1 is given as

$$I_1^B(z_1) = I_1(z_2) R_1 \tag{7.19}$$

$$I_2(z_2) = 2\pi I^{core}(z_1)\, R_2\, exp\,(-\alpha^M w)\, exp\left[-w \int_a^b exp\{-\gamma(r-a)\}\alpha^R(r, z_1 \leftrightarrow z_2) r dr\right] \tag{7.20}$$

$$I^B(z_1) = I_1^B(z_1) + I_2^B(z_1) \tag{7.21}$$

The Rayleigh scattering coefficients R_1 and R_2 in (7.19) and (7.20) can be expressed as

$$R_1 = \left(\pi a^2 w N^M\right) \left(S^B \frac{NA^2}{2n_0^2}\right) \left(\frac{2\sqrt{3}\pi^{5/2}\delta^M}{\lambda^2 d}\right)^2 \tag{7.22}$$

$$R_2 = \pi w \left\{ (b^2 - a^2) N^M \left(2\sqrt{3}\pi^{5/2} \frac{\delta^M}{\lambda^2 d}\right)^2 + \exp(\gamma a) \int_a^b \exp(-\gamma r)\, \delta^R(r) r dr \right\} \tag{7.23}$$

The left-most term on the right side of (7.22) provides the total of scattering centers within one "spatial cell" of the width w. The second term gives the portion of the scattered intensity re-bound into the fiber core and propagating along the $-z$ direction. The parameter S^B quantifies the backward-scattered part of the total scattered intensity; we set $S^B \equiv 1/2$. The third term is an integral efficiency of a single Rayleigh scattering source "observed" from the distance d [7]. The later parameter is approximated by the path between a scattering point 1 lying on the axis z and an "observation" point 2 on the core/cladding boundary "hit" by a "representative" ray propagating from the point 1 to the point 2 at the angle $\theta = \theta_c/2$ with respect to the axis z. Using the current fiber parameters n_0, n_1 and a, we get $d = 1.7$ mm.

The similar interpretation as for R_1 also holds for the term R_2. The radial distribution $\delta^R(r)$ in (7.23) quantifies the total efficiency of Rayleigh scattering on reagent molecules within the cladding averaged over the fiber length interval $z1 \leftrightarrow z2$:

$$\delta^R(r) = \frac{2N_A}{M^R} \left(\frac{2\sqrt{3}\pi^{5/2}\delta^R}{\lambda^2 d}\right)^2 \frac{1}{z_2 - z_1} \int_{z1}^{z2} u^R(r, z) dz \tag{7.24}$$

Integrals in (7.17), (7.20) and (7.23) can be solved numerically using a trapezoidal integration schema. Transport of the back-scattered light back to the detector is again governed by Relations (7.15), (7.16) and (7.17); the necessary re-distribution of the intensities I_1^B and I_2^B has to be again applied analogically to the model used for the light intensity propagating in the $+z$ direction.

7.3 Results and Discussion

This section gives overview of basic results obtained by application of the derived model. At first, the radial concentration profiles of ammonia and reagent within the cladding were calculated for two total exposition periods $t_e = 10^2$ and 10^4 s and the ammonia concentrations $u_0^A = 500$, 1,000, 5,000 and 10,000 ppm (Fig. 7.3).

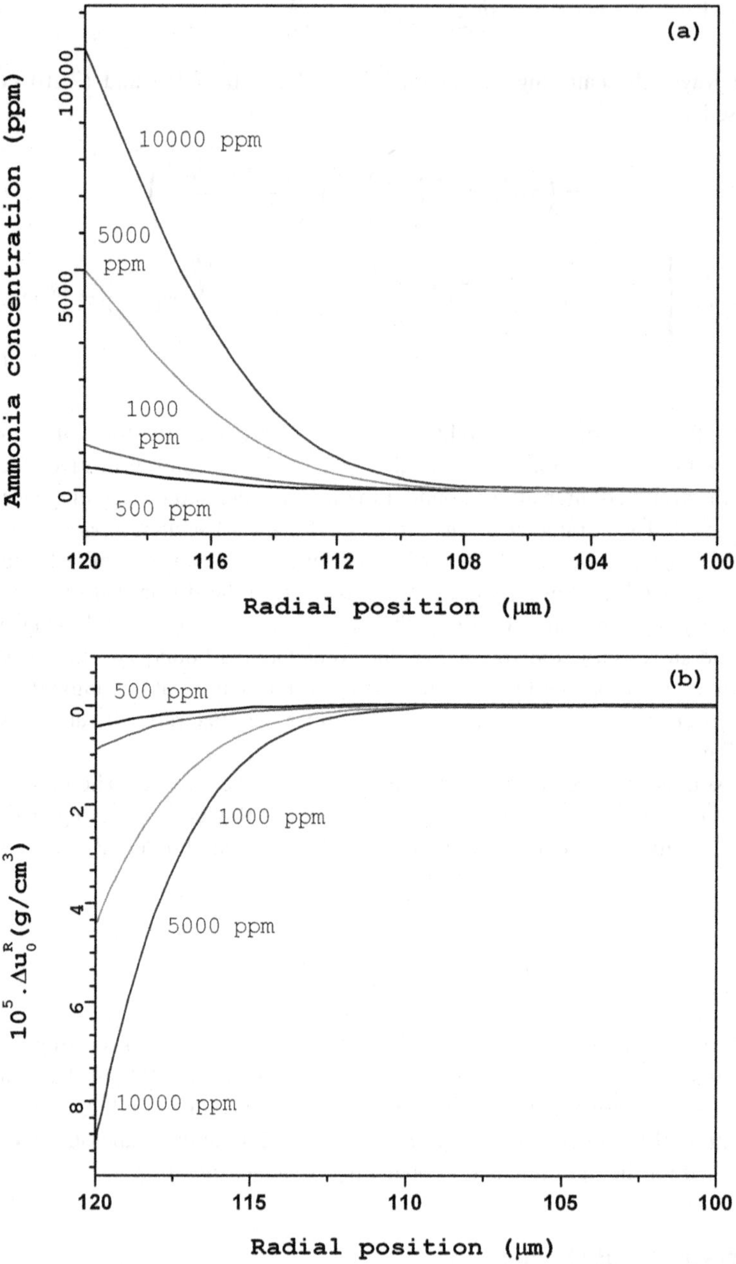

Fig. 7.3 Simulated concentration profiles of ammonia (**a**, **c**) and reagent (**b**, **d**) after exposition to the marked concentration of ammonia; $t_e = 10^2$ s (**a**, **b**), $t_e = 10^4$s (**c**, **d**). Other parameters: $D = 10^{-9}\text{cm}^2/\text{s}$, $N_0^R = 1$ mmol/l, $q = 0.5$ $(\text{cm}^3\text{g}^{-1}\text{s}^{-1})$

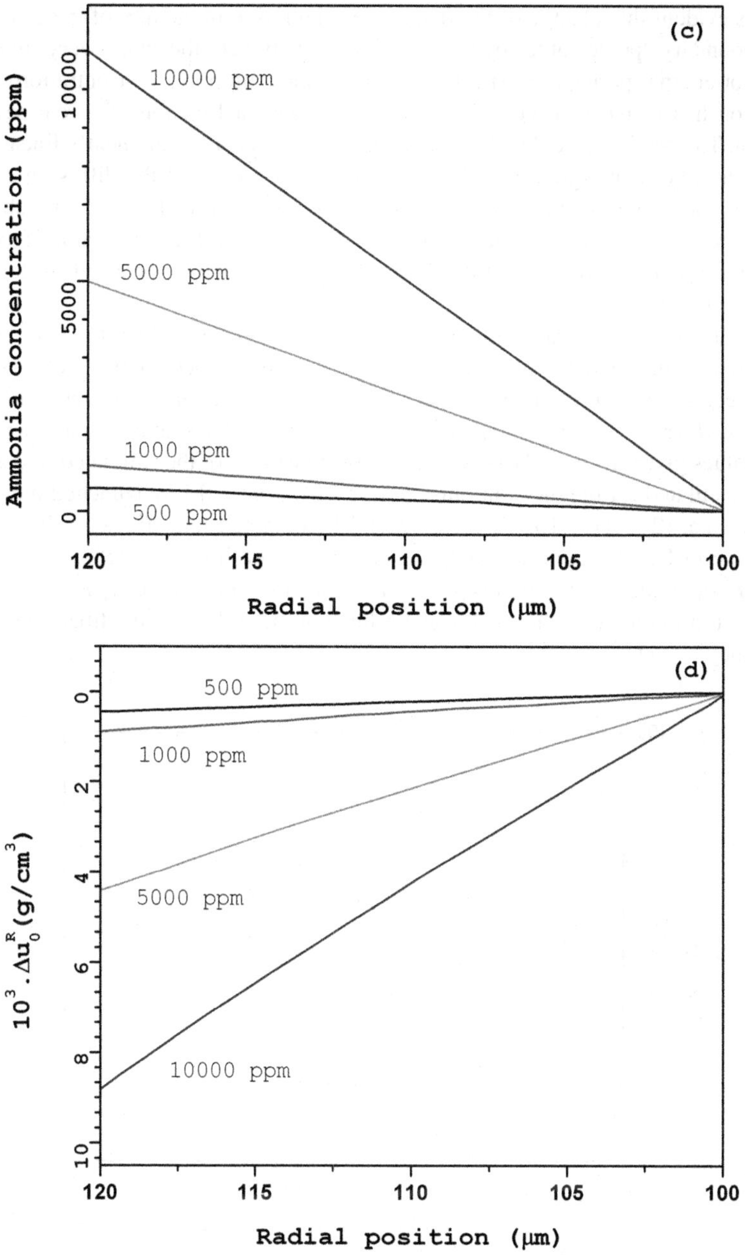

Fig. 7.3 (continued)

It is evident that change of the reagent concentration in vicinity of the cladding/core boundary "perceivable" by the evanescent EM field of the core is very small for the shorter time periods, irrespective of the applied ammonia concentration. Such situation has to have, of course, an adverse effect on function of a real sensing system. The profiles calculated for the exposition period 10^4 s are nearly linear. The fact is a direct consequence of the assumed independence of the diffusion coefficient on the radial position: $D = const \neq f(r)$. For its crucial impact on behavior of any real sensor system, the preposition about purely Fickian character of the diffusion process will be carefully evaluated and optionally corrected in the next model versions.[5]

Character of the radial profile of the reagent concentration along the fiber length then governs the character of the courses of back-scattered light intensity observed at the detector position. For convenience, the intensity slopes are shown as difference curves ($\Delta I^B(z) = I^B(z, t = i\Delta t) - I^B(z, t = 0)$) expressed in spatial domain, with the ΔI^B values given in 10^{-12} W (pW) units. Absolute order of the simulated intensities $I^B(z)$ calculated for the incident pulse power $I(0) = I_0 = 0.1$ W launched into fiber amounts ca. $(2 \div 4) \times 10^{-5}$ W (Fig. 7.4). The obtained ratio $I^B/I_0 \approx 10^{-4}$ is of the same order of magnitude as those observed experimentally (cf. Fig. 7.2, the value I^B/I_0 being in this case proportional to the ratio of the intensity $I^B(t \cong 0.3\mu s)$ and the top intensity of the 1st Fresnel reflection multiplied by the fiber connector efficiency ~ 0.1).

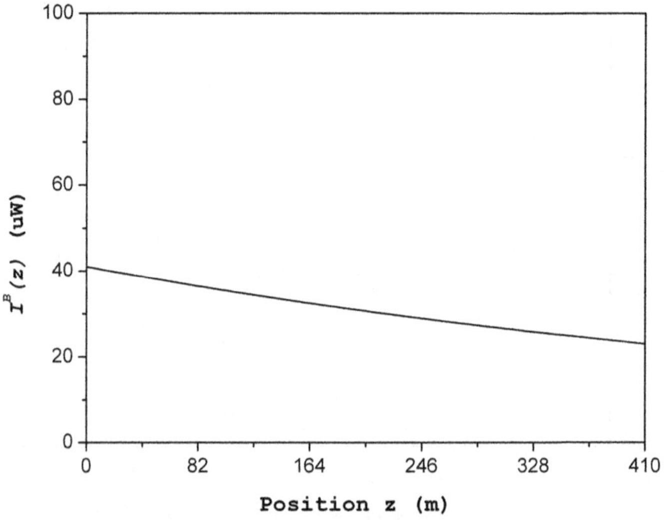

Fig. 7.4 Simulated course of the back-scattered light intensity $I^B(z)$ approaching the detector. The pulse power launched into the fiber is $I_0 = 100$ mW

[5]The ammonia accumulated within the cladding during the exposition is to be expected as the main source of possible discrepancies.

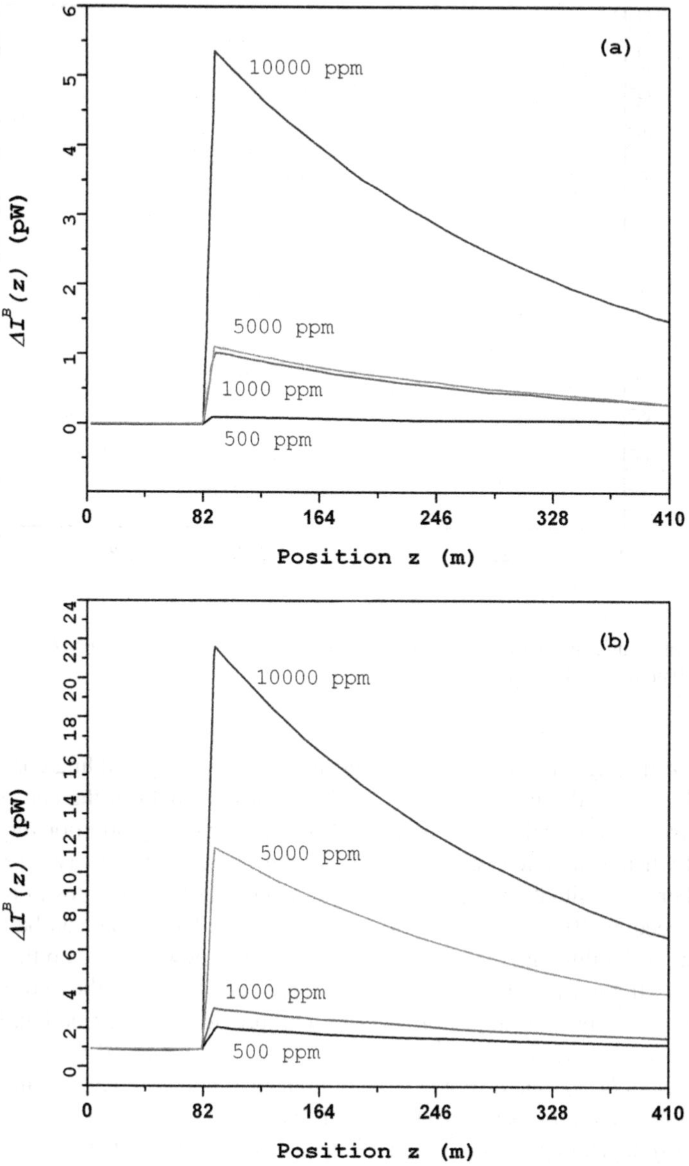

Fig. 7.5 Differential intensity $\Delta I^B(z)$ as function of ammonia concentration for two exposition periods $t_e = 10^3$ s (**a**) and 10^4 s (**b**); $I_0 = 100$ mW, $D = 10^{-9}$ cm^2/s, $N_0^R = 1$ mmol/l, $q = 0.5\,(\text{cm}^3\text{g}^{-1}\text{s}^{-1})$

Juxtaposition of the signals obtained for two different recording periods $\Delta t = 10^3$ and 10^4 s (Fig. 7.5) suggests that the applied registration time can strongly influence on the obtained concentration values (cf. the curves 1,000 and 5,000 ppm in Fig. 7.5a, b). The result follows from the combined action of a slow gradual

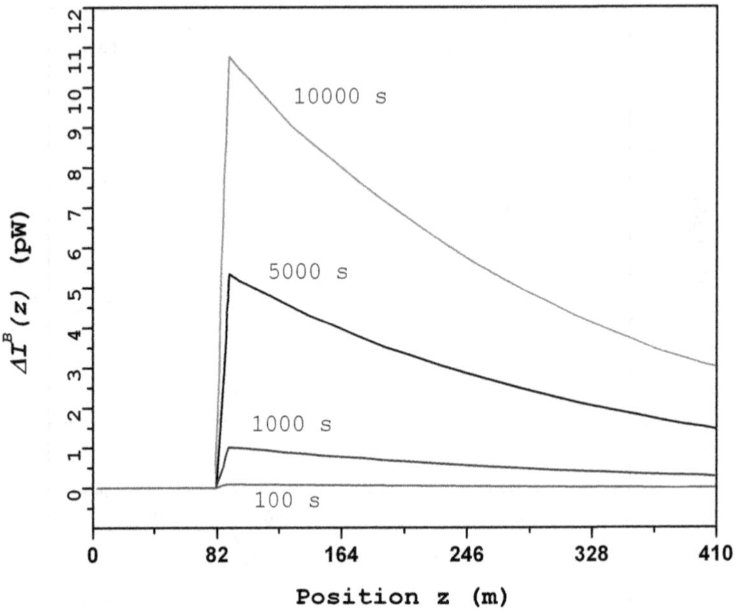

Fig. 7.6 Differential intensity $\Delta I^B(z)$ as function of the exposition period t_e; $I_0 = 100$ mW, $u_0^A = 5{,}000$ ppm, $D = 10^{-9}$ cm^2/s, $N_0^R = 1$ mmol/l, $q = 0.5\,(\mathrm{cm^3 g^{-1} s^{-1}})$

advance of the region of reagent decomposition towards the cladding/core boundary (cf. Fig. 7.3b, d) and a fast decay of the evanescent field of the core with the distance from the core/cladding boundary. The concentration proportionality is only observed when a saturated, steady state is reached (Fig. 7.5b). In case of our simulated fiber, such situation apparently occurs after a very long exposition/reading period exceeding 10^4 s (cf. Figs. 7.3c, d and 7.6). The later value can be theoretically reduced by thinning the cladding layer; in praxis however, feasibility of such approach is limited by the need to preserve protection of the silica core against embrittlement (caused by attack of water vapors) provided by the cladding layer.

Non-linear evolutions of $\Delta I^B(z)$ courses can be also obtained by variations of the parameters D, N_0^R and q (Fig. 7.7). From practical point of view, the most interesting result is the remarkable sensitivity of the response signal to variation of the parameter q depending on the design of the sensing reaction[6]. On the other hand, it is generally difficult to increase real values of the other two parameters: value of the diffusion coefficient D is limited by the protective function of the cladding mentioned above and by the micro-structural features of the cladding polymer. The maximum reagent concentration ${u_0}^R$ is then restricted by the utmost practically acceptable optical loss of the sensing fiber.

---

[6] However, sensing reversibility issues have to be carefully considered in this case.

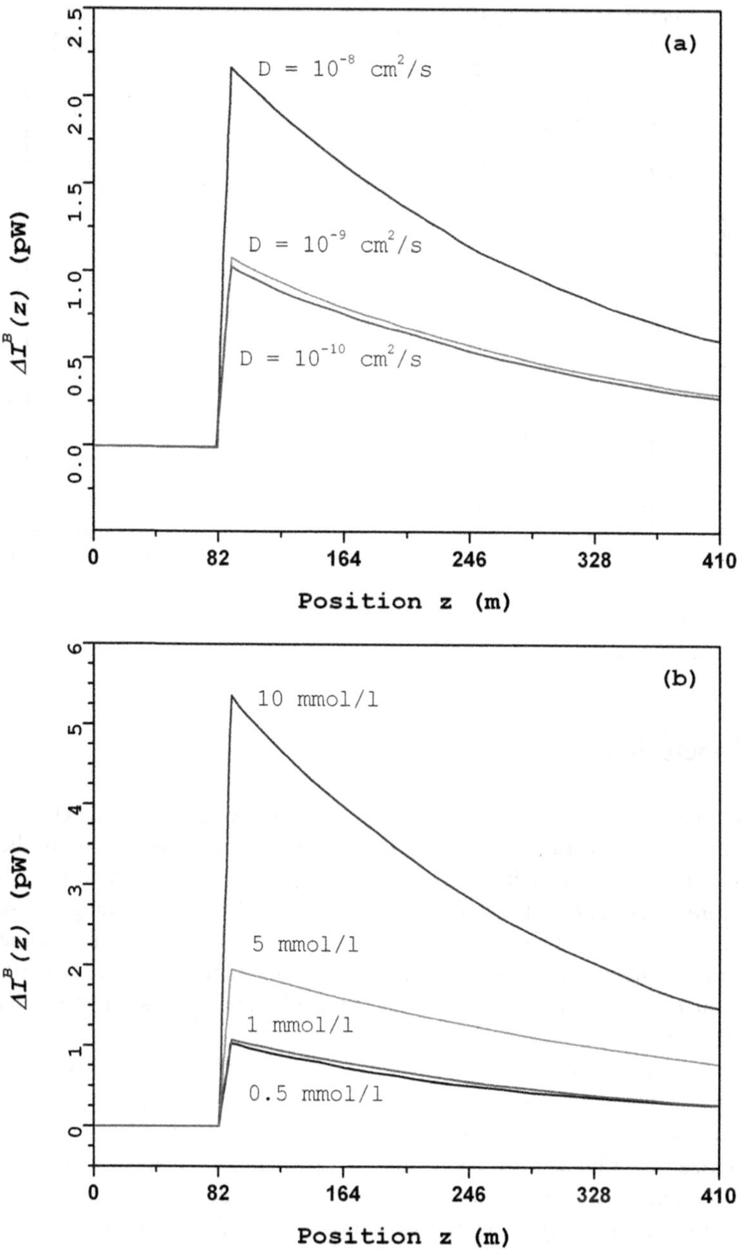

Fig. 7.7 Differential intensity $\Delta I^B(z)$ as a function of D (**a**), N_0^R (**b**) and q (**c**); $I_0 = 100$ mW, $u_0^A = 1,000$ ppm, $t_e = 10^3$ s. When not varied, $D = 10^{-9}$ cm^2/s, $N^R = 1$ mmol/l and $q = 0.5\,(\text{cm}^3\text{g}^{-1}\text{s}^{-1})$

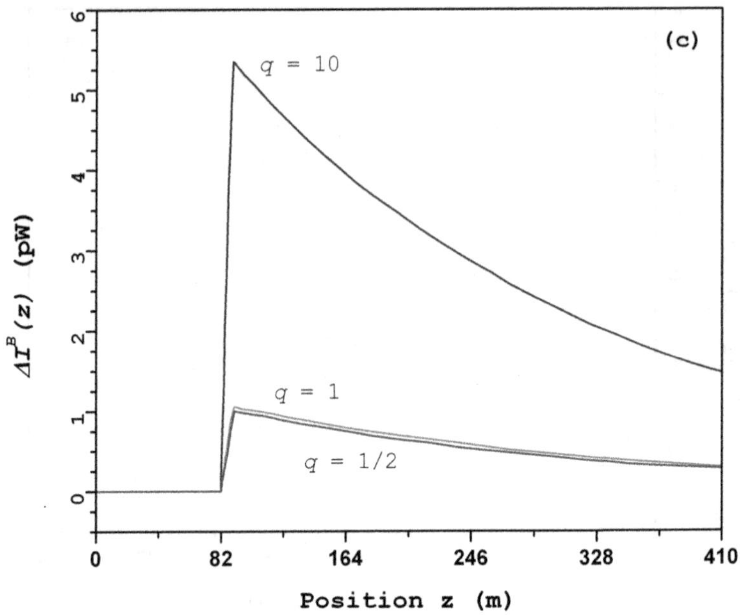

Fig. 7.7 (continued)

7.4 Conclusions

The simulation results focused on variations of some crucial parameters of the model provided us with a deeper insight into the mechanisms governing the system and offered several useful ideas for further optimization of such sensing fibers. Clearly, not every parameter can be easily varied within a broad range of values. For instance, the model suggests that it is very desirable to keep the thickness of the fiber cladding as thin as possible and, in the same time, maximize the diffusion coefficient D. But, such approach is not applicable from the point of mechanical robustness of the fiber.

Still, the achieved results show several likely passable, but also divertive directions towards practically applicable distributed sensors based on sensitized optical fibers: (i) application of reagents featuring chemical reaction scheme to ammonia characterized by a high speed coefficient, (ii) preparation of fibers with an irregular distribution of reagent maximized on the internal side of the cladding, (iii) use of a long-wavelength light source together with a small core radius and a small numerical aperture of the sensing fiber, thus maximizing the evanescent interaction between the core and the cladding, and (iv) optional application of a reagent showing stimulated fluorescence in presence of the analyte (chemo-luminescence), thus strongly increasing the registered signal resolution.

The basic adequacy of the model and its parameters is documented by the calculated back-scattered light intensities being of the same order of magnitude as the experimentally observed values. Nevertheless, some of the prepositions made with aim to keep the model simple, such as constancy of the ammonia diffusion coefficient D must be further analyzed and validated. Another important extension of the model which should be of concern in future is introduction of noise characteristics of the construction parts constituting a real sensing system (light source, light detector) with aim to evaluate resolution parameters of the system.

Acknowledgements The research has been supported by the Ministry of Education, Youth and Sports of the Czech Republic, grants MSM6840770040 and MSM6840770021.

References

1. Timmer, B., Olthuis, W., van den Berg, A.: Ammonia sensors and their applications – a review. Sens. Act. **B107**, 666–677 (2005)
2. Kalvoda, L., Aubrecht, J., Klepáček, R.: Fiber optic detection of ammonia gas. Acta Polytech. **46**, 41–46 (2006)
3. Saleh, B.A., Teich, C.M.: Fundamentals of Photonics. Wiley, New York, NY (1991)
4. Greenwood, N.N., Earnshaw, A.: Chemistry of Elements. Pergamon Press, Oxford (1984)
5. Chowdhury, S., Dodson, J., Bhethanabotla, VR., Sen, R.: Measurement of diffusion and permeation coefficients of oxygen in polymers based on luminescence quenching. In: 2007 AIChE Annual Meeting Conference Proceedings, AIChE, Salt Lake City, 8 pages (2007). ISBN 978-08169-1022-9
6. Lasaga, A.C., Cygan, R.T.: Electronic and ionic polarizabilities of silicon minerals. Am. Miner. **67**, 328–334 (1982)
7. Ingle, J.D., Crouch, S.R. Spectrochemical Analysis. Prentice-Hall, New Jersey (1988)

Chapter 8
Exact Solution of Bending Free Vibration Problem of the FGM Beams with Effect of Axial Force

Justín Murín, Mehdi Aminbaghai, and Vladimír Kutiš

Abstract In this contribution a fourth-order differential equation of the functionally graded material (FGM) beam deflection with longitudinal variation of the effective material properties has been derived where the second order beam theory has been applied for establishing the equilibrium- and kinematics beam equations. Not only the shear forces deformation effect and the effect of consistent mass distribution and mass moment of inertia but also the effect of large axial force has been taken into account. Numerical experiments will be done concerning the calculation of the eigenfrequencies and corresponded eigenmodes of chosen one-layer beams and multilayered FGM sandwich beams. Effect of the axial forces on the free vibration has been studied and evaluated. The solution results will be compared with those obtained by using a very fine mesh of 2D plane elements of the FEM software ANSYS.

Keywords FGM beams · Free vibration · Large axial force · Homogenization of material properties

8.1 Introduction

For free vibration analysis of the beam structures that are built of composite or functionally graded materials (FGMs) a very fine mesh of the classical solid elements can be used, but preparation of input data is very time consuming and the solution accuracy depends strongly on the mesh fineness (especially for beams with spatially varying stiffness and mass density). The spatial variation of the effective longitudinal elasticity modulus and mass density can be caused by longitudinal variation of both material properties and volume fraction of the FGM constituents. To avoid

J. Murín (✉)
Department of Mechanics, Faculty of Electrical Engineering and Information Technology, Slovak University of Technology, Bratislava, Slovakia
e-mail: justin.murin@stuba.sk

J. Murín et al. (eds.), *Computational Modelling and Advanced Simulations*, Computational Methods in Applied Sciences 24, DOI 10.1007/978-94-007-0317-9_8,

such difficulties new beam finite elements have been developed or exact solutions of vibration problems have been searched for. For example in [1] the dynamic analysis of 3-D composite beam element of constant stiffness restrained at their edges by the most general boundary conditions and subjected in arbitrarily distributed dynamic loading is presented.

In the contribution [2] we deal with deriving a fourth-order differential equation of the beams deflection with longitudinal variation of material properties. The linear beam theory has been used for establishing the equilibrium- and kinematics beam equations. The shear forces deformation effect and the effect of consistent mass distribution and mass moment of inertia have been taken into account too. Homogenization of the varying material properties of the one layer beam and the multilayered sandwich beam was done by extended mixture rules and laminate theory. Obtained solution results have shown a significant influence of the shear forces and the effect of consistent mass distribution and mass moment of inertia on the beam free vibration.

In this paper, which is a continuation of [2], a fourth-order differential equation of the beams deflection with longitudinal variation of the effective material properties has been derived where the second order beam theory has been used for establishing the equilibrium- and kinematics beam equations. Not only the shear forces deformation effect and the effect of consistent mass distribution and mass moment of inertia but also the effect of large axial force has been taken into account. The variation of the effective elasticity modulus and the mass density has been considered by variation of both the volume fractions and material properties of the FGM constituents. Homogenization of the varying material properties of the one layer beam and the multilayered sandwich beam is done by the extended mixture rules and laminate theory [2].

Numerical experiments will be conducted concerning the calculation of the eigenfrequencies and corresponded eigenmodes of chosen one layer beams and multilayered FGM sandwich beams. The effect of axial forces on the free vibration has been studied and evaluated. The calculated results will be compared with those obtained by using the 2D plane elements of the FEM program ANSYS [3].

8.2 Second Order Beam Theory Differential Equation for the FGM-Beam Deflection

Variations of the homogenized beam properties and the loading are shown in Fig. 8.1.

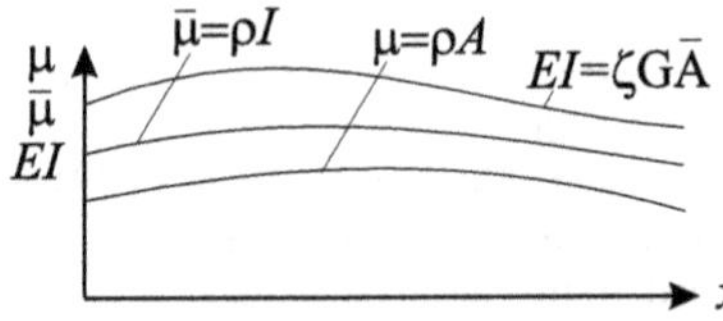

Fig. 8.1 Variation of the beam parameters

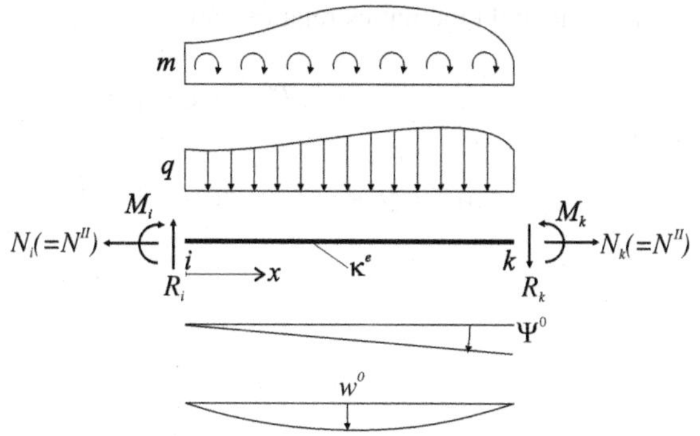

Fig. 8.2 The loads and nodal internal forces

$\mu = \rho A$ is the mass distribution, $\bar{\mu} = \rho I$ is the mass moment of inertia distribution, $\rho = \sum_{j=0}^{p_\rho} \rho_j a_j$ is the homogenized varying mass density, A is the cross-sectional area, I is the area moment of inertia, $EI = \sum_{j=0}^{p_E} EI_j\, a_j$ is the varying bending stiffness caused by varying elasticity modulus E, $G\bar{A} = \sum_{j=0}^{p_G} G\bar{A}_j\, a_j$ is the varying shear stiffness with varying shear elasticity modulus G, and $\zeta = \dfrac{EI}{G\bar{A}} = const.$, and finally $a_j = \dfrac{x^j}{j!}$ is the polynomial function. The same longitudinal polynomial variation of the homogenized shear modulus G has been assumed as it is by the homogenized elasticity modulus E.

In Fig. 8.2 the distributed loads and the nodal internal forces are shown.

m is the distributed bending moment $- m = \sum_{j=0}^{p_m} m_j\, a_j$, q is the distributed transversal force $- q = \sum_{j=0}^{p_q} q_j\, a_j$, κ^e is the bending curvature, M is the bending moment, N is the axial force (a constant axial force $N_i = N_k = N^{II}$ has been assumed) and R is the transversal force, w^0 and ψ^0 are the imperfections of the second order beam theory.

Figure 8.3 shows the beam nodal point displacements.

φ is the angle of cross-section rotation, w is the deflection, and u is the longitudinal displacement.

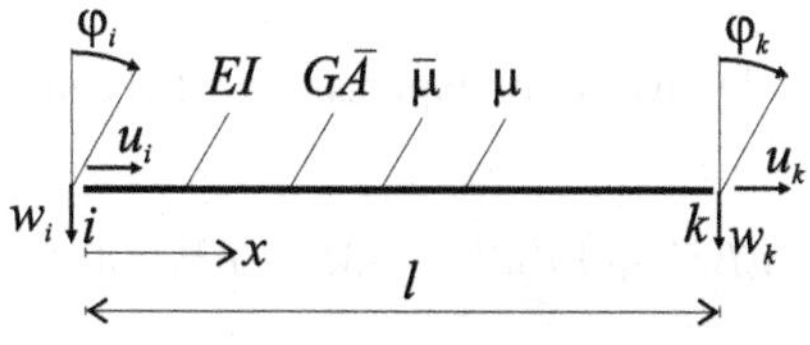

Fig. 8.3 Nodal displacements

The static and kinematics relations are

$$R' = -q - \mu\omega^2 w \tag{8.1}$$

$$M' = Q + m + \bar{\mu}\omega^2\varphi = Q + \sum_{0}^{p_m} m_j a_j + \bar{\mu}\omega^2\varphi \tag{8.2}$$

$$\varphi' = -\frac{M}{EI} - \kappa^e \tag{8.3}$$

$$w' = \varphi + \frac{Q}{G\bar{A}} = \varphi + \zeta\frac{Q}{EI} \tag{8.4}$$

where ω is the circular and $f = \dfrac{\omega}{2\pi}$ is the linear frequency, respectively, and Q is the shear force.

The dependence between the transversal R and the shear force Q is:

$$Q = R - N^{II}w' - N^{II}w^{\nu\prime} \tag{8.5}$$

and $w^\nu = a_1 w_1^\nu + a_2 w_2^\nu$, with $w_1^\nu = \psi^0 + 4\dfrac{w^0}{l}$, and $w_2^\nu = -8\dfrac{w^0}{l^2}$

Then $w^{\nu\prime} = a_0 w_1^\nu + a_1 w_2^\nu$, and $w^{\nu\prime\prime} = a_0 w_2^\nu$.

After derivation of (8.5) we get:

$$Q' = R' - N^{II}w'' - N^{II}w^{\nu\prime\prime} \tag{8.6}$$

$$Q'' = R'' - N^{II}w''' - N^{II}w^{\nu\prime\prime\prime} \tag{8.7}$$

and the derivative of (8.4) gives:

$$w'' = \varphi' + \zeta\frac{Q'EI - EI'Q}{EI^2} = \varphi' + \zeta\frac{Q'}{EI} - \zeta\frac{EI'Q}{EI^2} \tag{8.8}$$

Formula (8.3) will be applied now in (8.8) and then we get:

$$w'' = -\frac{M}{EI} - \kappa^e + \zeta\frac{Q'}{EI} - \zeta\frac{EI'Q}{EI^2} \tag{8.9}$$

and after some manipulation it reads

$$EIw'' = -M - EI\kappa^e + \zeta Q' - \zeta\frac{EI'Q}{EI} \tag{8.10}$$

Derivative of expression (8.10) can be expressed as:

$$EIw''' + EI'w'' = -M' - EI'\kappa^e + \zeta Q'' - \zeta\left(\frac{EI''Q}{EI} + \frac{EI'Q'}{EI} - \frac{EI'^2Q}{EI^2}\right) \tag{8.11}$$

After inserting expression (8.2) into (8.11) we get

$$EIw''' + EI'w'' = \\ =-Q\left(1+\frac{\zeta EI''}{EI}-\frac{\zeta EI'^2}{EI^2}\right)-m-\bar{\mu}\omega^2\varphi-EI'\kappa^e+\zeta Q''-\zeta\frac{EI'}{EI}Q' \tag{8.12}$$

After applying (8.4) in (8.12) we get:

$$EIw''' + EI'w'' + \bar{\mu}\omega^2 w' = \\ =\left(-1-\frac{\zeta EI''}{EI}+\zeta\frac{\bar{\mu}\omega^2}{EI}+\frac{\zeta EI'^2}{EI^2}\right)Q-\zeta\left(\frac{EI'}{EI}\right)Q'+\zeta Q''-m-EI'\kappa^e \tag{8.13}$$

After some manipulation we get:

$$\begin{aligned} &EI^3w''' + EI^2EI'w'' + \bar{\mu}\omega^2 EI^2 w' = \\ &\left(-EI^2 - EI\zeta EI'' + \zeta EI\bar{\mu}\omega^2 + \zeta EI'^2\right)Q- \\ &-\zeta EIEI'Q' + \zeta EI^2 Q'' - EI^2 m - EI^2 EI'\kappa^e \end{aligned} \tag{8.14}$$

Now we insert (8.5), (8.6) and (8.7) into expression (8.14):

$$\begin{aligned} &\left(m + R - N^{II}\left(w_1^v + xw_2^v\right) - \left(-\bar{\mu}\omega^2 + N^{II}\right)w' + EI'\left(\kappa^e + w''\right)\right)EI^2 + \\ &+\zeta EI'^2\left(-R + N^{II}w_1^v + N^{II}xw_2^v + N^{II}w'\right) + \\ &+\zeta\left(-\left(R - N^{II}\left(w_1^v + xw_2^v\right) - N^{II}w'\right)\left(\omega^2\bar{\mu} - EI''\right)\right. \\ &-EI'\left(\mu\omega^2 w + q + N^{II}w_2^v + N^{II}w''\right)\left.\right)EI \\ &+\left(\zeta\left(q' + w\omega^2\mu' + w'\mu\omega^2 + N^{II}w'''\right)\right)EI^2 + w'''EI^3 = 0 \end{aligned} \tag{8.15}$$

The transversal force R is derived from (8.15):

$$\begin{aligned} R = &\frac{1}{-\zeta EI\bar{\mu}\omega^2 + EI^2 - \zeta EI'^2 + \zeta EIEI''} \\ &\left(\begin{aligned} &-w'''EI^3 - mEI^2 + N^{II}\left(w_1^v + xw_2^v\right)EI^2 \\ &+\left(-\bar{\mu}\omega^2 + N^{II}\right)w'EI^2 - EI'\left(\kappa^e + w''\right)EI^2 \\ &+\zeta\left(-\mu w'\omega^2 - w\mu'\omega^2 - q' - N^{II}w'''\right)EI^2 \\ &-\zeta N^{II}\left(w_1^v + xw_2^v\right)\left(\omega^2\bar{\mu} - EI''\right)EI \\ &-\zeta N^{II}w'\left(\omega^2\bar{\mu} - EI''\right)EI \\ &+\zeta EI'\left(w\mu\omega^2 + q + N^{II}w_2^v + N^{II}w''\right)EI \\ &-\zeta N^{II}EI'^2 w_1^v - \zeta N^{II}xEI'^2 w_2^v - \zeta N^{II}EI'^2 w' \end{aligned}\right) \end{aligned} \tag{8.16}$$

The (8.16) is now derivate and inserted into the expression (8.1). So we get a differential equation of 4th order for the beams deflection with longitudinal variation of effective material properties for II – order beam theory

$$\eta_4\, w^{IV} + \eta_3\, w''' + \eta_2\, w'' + \eta_1\, w' + \eta_0\, w = K_q\, q + K_{q'}\, q' + K_{q''}\, q'' + K_m\, m + \\ +K_{m'}\, m' + K_\kappa\, \kappa^e + K_{w_1^v} w_1^v + K_{w_2^v} w_2^v \tag{8.17}$$

With the parameters and terms

$$\eta_0 = \omega^2 \begin{pmatrix} -\zeta\mu\Omega EI'^2 + \zeta EI\Gamma\mu EI' - \zeta EI\mu\Omega EI'' + \zeta EI^2\Omega\mu'' - \zeta EI^2\Gamma\mu' \\ +\mu\left(EI^2 + \zeta\left(EI'' - \omega^2\bar{\mu}\right)EI - \zeta EI'^2\right)^2 + \zeta EI\Omega EI'\mu' \end{pmatrix} \tag{8.18}$$

$$\eta_1 = \begin{pmatrix} \left(-\left(\zeta\mu + \bar{\mu}\right)\omega^2 + N^{II}\right)\Gamma EI^2 + \left(2\zeta\mu' + \bar{\mu}'\right)\Omega\omega^2 EI^2 \\ +EI\Omega EI'\left(\left(\zeta\mu + 2\bar{\mu}\right)\omega^2 - 2N^{II}\right) + EI\Omega\zeta N^{II}\left(\omega^2\bar{\mu}' - EI'''\right) \\ +\zeta N^{II} EI'\left(\Omega\left(\bar{\mu}\omega^2 + EI''\right) - \Gamma EI'\right) - EI\zeta\Gamma\Delta N^{II} \end{pmatrix} \tag{8.19}$$

$$\eta_2 = EI \begin{pmatrix} \zeta\Gamma N^{II} EI' + \Omega\left(2EI'^2 + \zeta N^{II}\left(\omega^2\bar{\mu} - 2EI''\right)\right) \\ -EI\left(\Gamma EI' + \Omega\left(-\left(\zeta\mu + \bar{\mu}\right)\omega^2 + N^{II} - EI''\right)\right) \end{pmatrix} \tag{8.20}$$

$$\eta_3 = EI\left(-\Gamma EI^2 - \left(\zeta\Gamma N^{II} - \Omega 4EI'\right)EI + \zeta\Omega N^{II} EI'\right) \tag{8.21}$$

$$\eta_4 = EI^2\Omega\left(EI + \zeta N^{II}\right) \tag{8.22}$$

$$K_{w_2^v} = -N^{II} \begin{pmatrix} \left(\zeta\Gamma\left(EI' - x\Delta\right) + \Omega\left(\zeta\left(\left(\bar{\mu} + x\bar{\mu}'\right)\omega^2 - 2EI'' - xEI'''\right) - 2xEI'\right)\right)EI \\ +\zeta xEI'\left(\Omega\left(\bar{\mu}\omega^2 + EI''\right) - \Gamma EI'\right) + (x\Gamma - \Omega)EI^2 \end{pmatrix} \tag{8.23}$$

$$K_{w_1^v} = -N^{II} \begin{pmatrix} \Gamma EI^2 - \left(\zeta\Gamma\Delta + \Omega\left(2EI' + \zeta\left(EI''' - \omega^2\bar{\mu}'\right)\right)\right)EI + \\ \zeta EI'\left(\Omega\left(\bar{\mu}\omega^2 + EI''\right) - \Gamma EI'\right) \end{pmatrix} \tag{8.24}$$

$$K_m = -EI\left(2\Omega EI' - EI\Gamma\right) \tag{8.25}$$

$$K_{m'} = -EI^2\Omega \tag{8.26}$$

$$K_q = \zeta\Omega EI'^2 - \zeta EI\Gamma EI' - \left(EI^2 + \zeta\left(EI'' - \omega^2\bar{\mu}\right)EI - \zeta EI'^2\right)^2 + \\ +\zeta EI\Omega EI'' \tag{8.27}$$

$$K_{q'} = -\zeta EI\left(\Omega EI' - EI\Gamma\right) \tag{8.28}$$

$$K_{q''} = -\zeta EI^2\Omega \tag{8.29}$$

$$K_\kappa = -EI\left(2\Omega EI'^2 + EI\left(\Omega EI'' - \Gamma EI'\right)\right) \tag{8.30}$$

with

$$\Gamma = \zeta EI'\left(\bar{\mu}\omega^2 + EI''\right) + EI\left(\zeta\bar{\mu}'\omega^2 - 2EI' - \zeta EI'''\right) \tag{8.31}$$

$$\Delta = \omega^2\bar{\mu} - EI'' \tag{8.32}$$

$$\Omega = \zeta EI\left(\omega^2\bar{\mu} - EI''\right) + \zeta EI'^2 - EI^2 \tag{8.33}$$

8.3 Calculation of the Eigenfrequencies and Eigenmodes

The above-mentioned differential equation for the beam deflection can be solved in two different ways. The first way is the new concept for solving of linear differential equations with non-constant coefficient which is based on [4]. The second solving possibility is using the software "Mathematica" [5]. Under the boundary conditions assumption the homogenous system of equations can always be established. By setting the determinant of this system of equation to zero, the eigenvalue problem has been defined. By its solution the circular eigenfrequency ω_i can be obtained. The eigenfrequency f_i can be calculated with

$$f_i = \frac{\omega_i}{2\pi} \tag{8.34}$$

whereas index i denotes the frequency number.

8.4 Numerical Experiments

8.4.1 Free Vibration Analysis of the Multilayered FGM Sandwich Beams with Effect of Large Axial Force

A 12-layer sandwich beam has been considered in Fig. 8.4. Its square cross-section is constant with height $h = 0.01$ m and width $b = 0.01$ m. The layers are built symmetrical to neutral plane of the beam, that means, the geometry and material properties of the opposite layers are the same. Other geometrical dimensions of the layers and whole beam are shown in Fig. 8.4, and these are: $h^1 = 0.004\,\text{m}$; $h^k = 0.0002\,\text{m}$ for $k \in< 2, 6 >$; $L = 0.1\,\text{m}$. Material of the layers consists of two components: NiFe – named as a matrix and denoted with index m; Tungsten – named as a fibre and denoted with index f.

Material properties of the components are constant and their values are: Tungsten (fibres) – elasticity modulus $E_f = 400$ GPa, mass density $\rho_f = 19{,}300\,\text{kgm}^{-3}$; NiFe (matrix) – elasticity modulus $E_m = 255$ GPa, mass density $\rho_m = 9200\,\text{kgm}^{-3}$ [6]. The cross-sectional area $A = 0.0001\,\text{m}^2$; the moment of inertia $I = \dfrac{bh^3}{12} = \dfrac{(0.01)^4}{12}\,\text{m}^4$; and the reduced cross-sectional area $\bar{A} = 0.833 \times 10^{-4}\,\text{m}^2$.

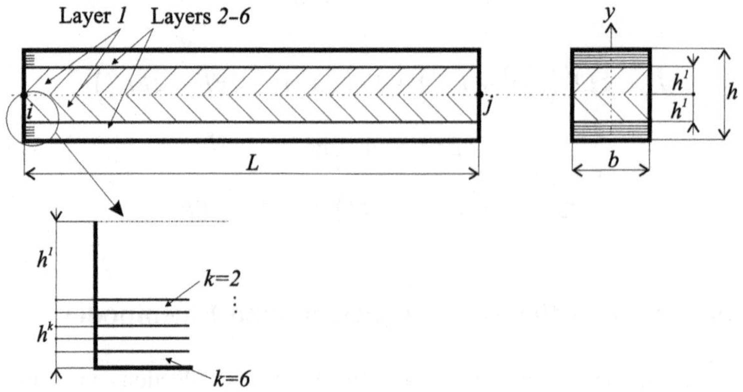

Fig. 8.4 Multilayered sandwich beam

Symmetrical layers denoted by number 1 contain the matrix (NiFe) which properties are constant over the height and width of these layers. Pairs of the symmetrical layers 2–6 were built with mixture of the both components. The volume fraction of the components is constant over the height and width of the given layer, but it varies linearly along the layer length:

$$v_f^k(x) = v_{fi}^k\left(1 + \eta_1^k x\right) = 1 - v_m^k(x) \tag{8.35}$$

where index $k \in < 2, 6 >$ is the number of symmetrical layer, v_{fi}^k is the volume fraction of the fibres in the kth layer at node i, and η_1^k is a parameter of the variation of the fibres volume fraction. The list of these parameters is given in Table 8.1. The fibre volume fractions in these layers at node j are constant and they are equal to $v_{fj}^k = 0.3$. There $v_m^k(x)$ are the matrix volume fractions in layer k.

With the extended mixture rules [2] the effective longitudinal elasticity modulus of the layers can be calculated.

The effective elasticity modules of the homogenized sandwich (for axial and transversal loading) have been calculated [2, 7], and we have got:

$$E_L^{NH}(x) = 2.782 \times 10^{11} - 1.45 \times 10^{11}x, \text{ [Pa]},$$

$$E_L^{MH}(x) = 3.128 \times 10^{11} - 3.663 \times 10^{11}x, \text{ [Pa]}$$

Table 8.1 Parameters of the fibres volume fractions variation

Layer k	2	3	4	5	6
$v_{fi}^k(-)$	0.6	0.7	0.8	0.9	1.0
$\eta_1^k(-)$	−3/0.6	−4/0.7	−5/0.8	−6/0.9	−7

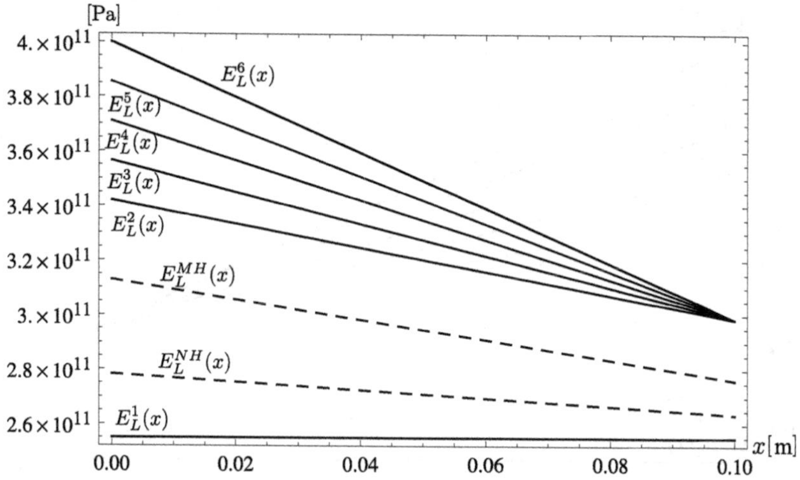

Fig. 8.5 Longitudinal distributions of the homogenized elasticity modules and the effective elasticity modules in the respective layers

The ratio $E_L^{MH}(x)/G_L^H(x) = 2.6$ has been used for the effect of shear forces assumption in this example.

Longitudinal distributions of these homogenized elasticity modules and the effective elasticity modules in respective layers are shown in Fig. 8.5. As one can see in this figure, the effective elasticity modulus for transversal loading is larger than the effective elasticity modulus for axial loading. In modal analysis of this homogenized sandwich beam the elasticity modulus for transversal loading, $E_L^{MH}(x)$ must be used.

The homogenized mass density has been calculated [2]: $\rho_L^H(x) = 10816(1 - 0.933807751x)$, $[\mathrm{kgm}^{-3}]$. Longitudinal distributions of this mass density in the homogenized beam and the effective mass densities distribution in respective layers of the sandwich beam are shown in Fig. 8.6.

Then, the effective longitudinal mass distribution is [2]:

$$\mu = \mu_L^H(x) = \rho_L^H(x)A = 10816 \times 10^{-7}\left(1 - 933807751 \times 10^{-9}x\right)\ [\mathrm{kgm}^{-1}],$$

and the effective longitudinal mass inertia moment distribution is [2]:

$$\bar{\mu} = \mu_L^H(x) = \rho_L^H(x)I = \frac{10816}{12} \times 10^{-11}\left(1 - 933807751 \times 10^{-9}x\right)[\mathrm{kgm}].$$

In order to show an influence of the layers building on the eigenfrequency, the position of the layers has been changed in such a way that layer 6 and 2, 5 and 3 have interchanged their position. Layers 1 and 3 have been stayed on their original position. New homogenized elasticity modulus is now equal to $E_L^{MH}(x) = 310.3552 - 341.272x$ [GPa]. The homogenized mass density has not been affected by the changed position of the layers, but the buckling force will change its value.

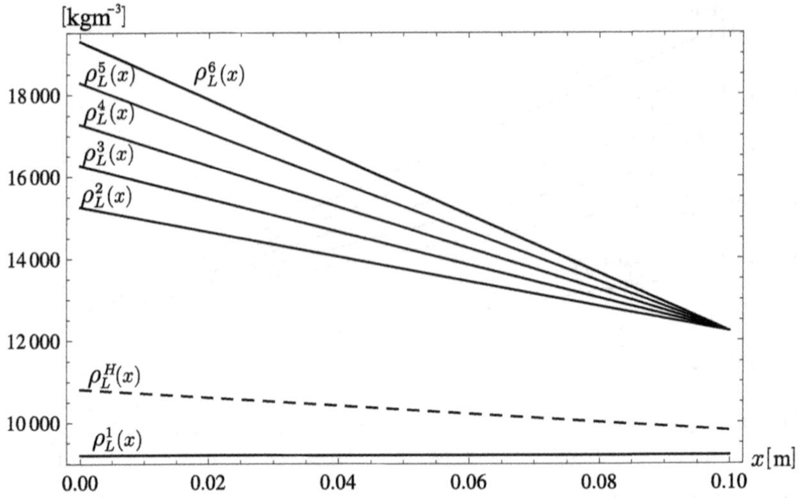

Fig. 8.6 Effective mass density distribution along the homogenized beam and the effective mass densities distribution along respective layers

8.4.1.1 Example 1 – Clamped Beam at the Left Side

The homogenized beam (Fig. 8.4) clamped at the left side has been studied by modal analysis. The first three bending eigenfrequencies and eigenmodes have been found using the differential equation (8.17) and appropriate boundary conditions. The buckling force for this beam with original layering (see Fig. 8.4) is: $N_{Ki}^{II} = -61.617\,\text{kN}$. The buckling force for this beam with interchanged layering is: $N_{Ki}^{II} = -65.463\,\text{kN}$. Effect of the axial force on the eigenfrequency has been founded in a such way that the value of axial force has been chosen as a part of the buckling force (for tension – positive axial force N^{II}, and also for compression – negative axial force N^{II}). Effect of the interchanged layering has been studied for a sandwich beam with new homogenized material properties that has been loaded by the same axial force as the original layering.

The same problem has been solved using a very fine mesh – 60,000 of 2D PLANE42 elements (with prestress effect) of the FEM program ANSYS [3] – see Fig. 8.13. With this very fine mesh, sufficient accuracy of the longitudinal and transversal variation of the elasticity modulus and the mass density has been guarantied in the real multilayer sandwich beam. The results of ANSYS as well as the results of the differential equations solution (DIFF) for compression are presented in Table 8.2. In Table 8.3 the analysis results are shown for the tensional axial force. First three rows of these tables contain eigenfrequencies of the original beam. Last three rows contain eigenfrequencies of the beam with interchanged layers. The appropriate eigenmodes (in a qualitative form, for $N^{II} = \frac{1}{2}N_{Ki}^{II} = -31\text{kN}$) are shown in Fig. 8.7. First eigenfrequency is $f_I = 885.8$ Hz for $N^{II} = 0$ in Example 1 for the original layering. First eigenfrequency $f_1 = 902.1$ Hz for $N^{II} = 0$ for the

Table 8.2 Eigenfrequencies in Example 1 for compression

N^{II} f [Hz]	ANSYS $\frac{3}{4}N^{II}_{Ki}$	DIFF $\frac{3}{4}N^{II}_{Ki}$	ANSYS $\frac{1}{2}N^{II}_{Ki}$	DIFF $\frac{1}{2}N^{II}_{Ki}$	ANSYS $\frac{1}{4}N^{II}_{Ki}$	DIFF $\frac{1}{4}N^{II}_{Ki}$
f_1	459.2	453.8	640.4	638.1	774.6	774.4
f_2	4777.3	4828.3	4893.5	4951.2	5006.2	5070.9
f_3	12899.0	13248.0	12986.0	13349.0	13017.0	13450.0
f_1	453.34	448.54	636.42	634.30	771.31	771.20
f_2	4764.3	4815.2	4880.9	4938.4	4994.1	5058.3
f_3	12873.0	13218.0	12960.0	13320.0	13017.0	13421.0

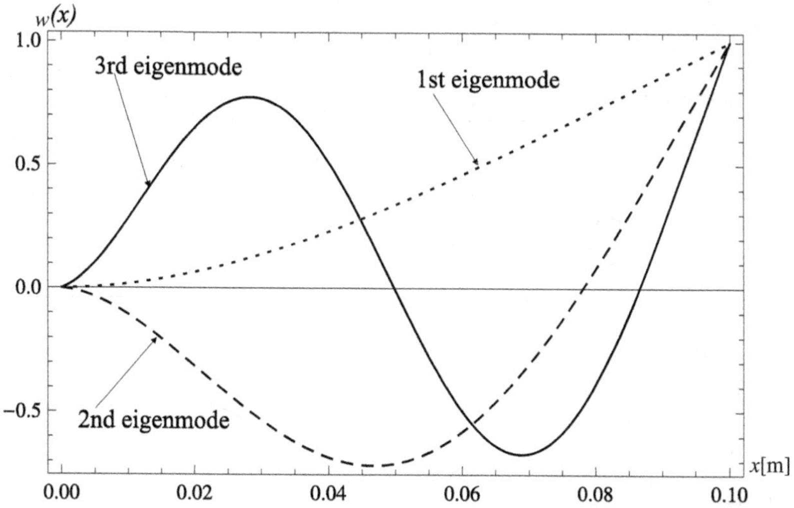

Fig. 8.7 First three eigenmodes in the Example 1 for DIFF solution (original layering)

interchanged layering. All the solution results differences (ANSYS and DIFF) for original layered beam are presented in Table 8.4.

In Fig. 8.8 the dependence of the first eigenfrequency on the axial force is shown (for DIFF solution – original layered beam). An increase of the compression axial force decreases the eigenfrequency, and an increase of the tensional axial force increases the eigenfrequency. As expected, tensional axial force makes the beam stiffener and the eigenfrequency is higher than by the compression axial force.

8.4.1.2 Example 2 – Clamped Beam at the Both Sides

Homogenized beam (Fig. 8.4) clamped at the both sides has been studied by modal analysis. The first three bending eigenfrequencies and eigenmodes have been found using the differential equation (8.17) and the appropriate boundary conditions. The buckling force of this beam with original layering is: $N^{II}_{Ki} = -877.651$ kN. The

Table 8.3 Eigenfrequencies in Example 1 for tension

N^{II} / f [Hz]	ANSYS $-\frac{3}{4}N_{Ki}^{II}$	DIFF $-\frac{3}{4}N_{Ki}^{II}$	ANSYS $-\frac{1}{2}N_{Ki}^{II}$	DIFF $-\frac{1}{2}N_{Ki}^{II}$	ANSYS $-\frac{1}{4}N_{Ki}^{II}$	DIFF $-\frac{1}{4}N_{Ki}^{II}$
f_1	1133.6	1141.3	1059.8	1065.5	977.61	981.30
f_2	5426.3	5520.1	5325.7	5411.9	5222.2	5301.1
f_3	13404.0	13846.0	13322.0	13748.0	13239.0	13649.0
f_1	1131.1	1138.9	1057.3	1063.0	974.88	978.60
f_2	5414.8	5508.1	5314.1	5399.8	5210.5	5288.9
f_3	13380.0	13817.0	13298.0	13719.0	13215.0	13620.0

Table 8.4 Solution results difference (ANSYS and DIFF) for beam with the original layering

	solution results difference (ANSYS and DIFF) (%)					
	Compression			Tension		
N^{II} / f	$\frac{3}{4}N_{Ki}^{II}$	$\frac{1}{2}N_{Ki}^{II}$	$\frac{1}{4}N_{Ki}^{II}$	$-\frac{3}{4}N_{Ki}^{II}$	$-\frac{1}{2}N_{Ki}^{II}$	$-\frac{1}{4}N_{Ki}^{II}$
f_1	1.17	0.35	0.03	0.67	0.53	0.37
f_2	1.06	1.18	1.29	1.72	1.61	1.51
f_3	2.71	2.80	3.33	3.29	3.19	3.10

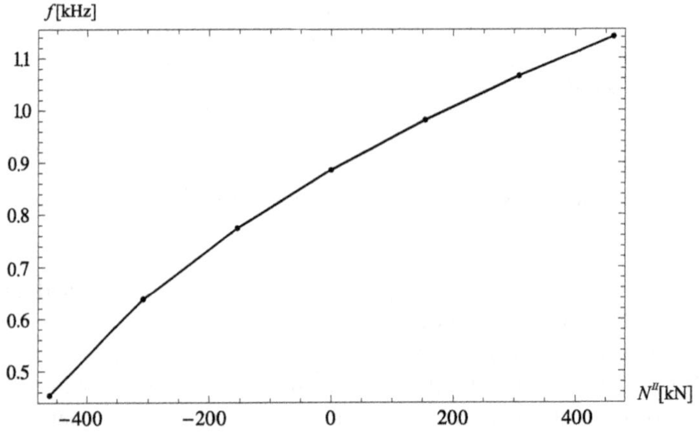

Fig. 8.8 Dependence of the first eigenfrequency on the axial force for Example 1 (original layering)

buckling force for this beam with interchanged layering is: $N_{Ki}^{II} = -976.128$ kN. Effect of the axial force on the eigenfrequency has been founded by the same way, as it was done in Example 1. The results of ANSYS as well as the results of the differential equation solution (DIFF) for compression of the original beam and this beam with the interchanged layering are presented in Table 8.5. In Table 8.6 the analysis

Table 8.5 Eigenfrequencies in Example 2 for compression

N^{II} f [Hz]	ANSYS $\frac{3}{4}N_{Ki}^{II}$	DIFF $\frac{3}{4}N_{Ki}^{II}$	ANSYS $\frac{1}{2}N_{Ki}^{II}$	DIFF $\frac{1}{2}N_{Ki}^{II}$	ANSYS $\frac{1}{4}N_{Ki}^{II}$	DIFF $\frac{1}{4}N_{Ki}^{II}$
f_1	2676.3	2626.4	3722.6	3686.6	4463.1	4484.8
f_2	10149	10222.0	11097.0	11308.0	11907.0	12367.0
f_3	19865	20607.0	20721.0	21742.0	21491.0	22820.0
f_1	2654.9	2605.9	3708.2	3672.2	4451.8	4473.0
f_2	10118.0	10188.0	11071.0	11278.0	11883.0	12262.0
f_3	19824.0	20553.0	20684.0	21691.0	21456.0	22770.0

Table 8.6 Eigenfrequencies in Example 2 for tension

N^{II} f [Hz]	ANSYS $-\frac{3}{4}N_{Ki}^{II}$	DIFF $-\frac{3}{4}N_{Ki}^{II}$	ANSYS $-\frac{1}{2}N_{Ki}^{II}$	DIFF $-\frac{1}{2}N_{Ki}^{II}$	ANSYS $-\frac{1}{4}N_{Ki}^{II}$	DIFF $-\frac{1}{4}N_{Ki}^{II}$
f_1	6272.6	6712.0	5924.5	6241.0	5520.7	5724.6
f_2	14266.0	15569.0	13776.0	14819.0	13228.0	14068.0
f_3	23908.0	26683.0	23387.0	25774.0	22815.0	24830.0
f_1	6265.8	6704.0	5917.0	6232.6	5512.4	5715.4
f_2	14249.0	15540.0	13758.0	14796.0	13209.0	14007.0
f_3	23882.0	26640.0	23359.0	25730.0	22785.0	24784.0

results are shown for the tensional axial force. First three rows of these tables contain eigenfrequencies of the original beam. Last three rows contain eigenfrequencies of the beam with interchanged layers. The appropriate eigenmodes of the beam with the original layering (in a qualitative form, for $N^{II} = \frac{1}{2}N_{Ki}^{II} = -439\text{kN}$) are presented in Fig. 8.9. All the solution results differences (ANSYS and DIFF for original layered beam) are presented in Table 8.7.

In Fig. 8.10 a dependence of the first eigenfrequency on the axial force is shown (for DIFF solution for original layered beam). The first eigenfrequency of this beam calculated with DIFF is 5147.7 Hz for $N^{II} = 0$. As can be seen from Fig. 8.10, effect of the axial force on the eigenfrequency is of the same character as it was in Example 1.

8.4.1.3 Example 3 – Beam Simply Supported at Both Sides

Homogenized beam (Fig. 8.4) simply supported at the both sides has been studied by modal analysis. The first three bending eigenfrequencies and eigenmodes have been found using the homogenous part of the differential equation (8.17) and the appropriate boundary conditions. The buckling force of this beam with original layering is: $N_{Ki}^{II} = -235.972$ kN. The buckling force for this beam with interchanged layering (see Fig. 8.4) is: $N_{Ki}^{II} = -262.4$ kN. The solution results of ANSYS as well as the solution results (for both the original and interchanged layering) of

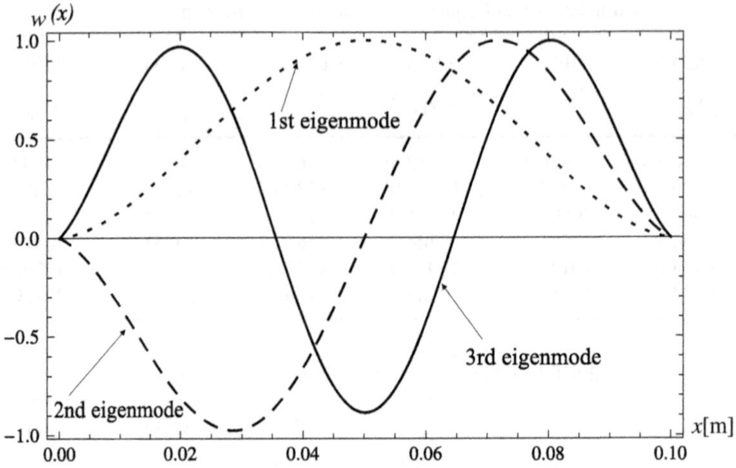

Fig. 8.9 First three eigenmodes of the original layered beam in the Example 2 for DIFF solution

Table 8.7 Solution results difference (ANSYS and DIFF) for original layering

	Solution results difference (ANSYS and DIFF) (%)					
	Compression			Tension		
N^{II} / f	$\frac{3}{4}N_{Ki}^{II}$	$\frac{1}{2}N_{Ki}^{II}$	$\frac{1}{4}N_{Ki}^{II}$	$-\frac{3}{4}N_{Ki}^{II}$	$-\frac{1}{2}N_{Ki}^{II}$	$-\frac{1}{4}N_{Ki}^{II}$
f_1	1.86	0.96	0.48	7.00	5.34	3.70
f_2	0.72	1.91	3.86	9.13	7.57	6.35
f_3	3.73	4.93	6.18	11.61	10.21	8.83

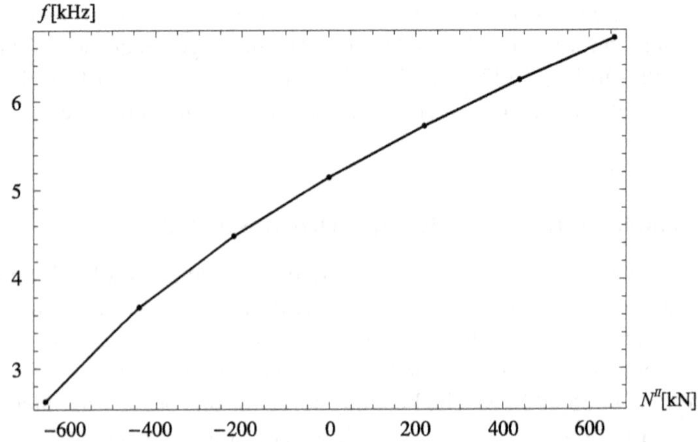

Fig. 8.10 Dependence of the first eigenfrequency on the axial force for Example 2

Table 8.8 Eigenfrequencies in Example 3 for compression

N^{II} f [Hz]	ANSYS $\frac{3}{4}N_{Ki}^{II}$	DIFF $\frac{3}{4}N_{Ki}^{II}$	ANSYS $\frac{1}{2}N_{Ki}^{II}$	DIFF $\frac{1}{2}N_{Ki}^{II}$	ANSYS $\frac{1}{4}N_{Ki}^{II}$	DIFF $\frac{1}{4}N_{Ki}^{II}$
f_1	1200.2	1188.3	1692.3	1685.0	2064.0	2063.8
f_2	8111.9	8136.4	8412.0	8473.3	8696.4	8779.3
f_3	17867	18200.9	18125	18536.7	18375	18864.9
f_1	1190.2	1181.8	1685.3	1678.3	2058.4	2058.2
f_2	8092.5	8117.3	8393.5	8453.8	8678.7	8777.3
f_3	17832.0	18162.0	18092.0	18497.0	18343.0	18826.0

Table 8.9 Eigenfrequencies in Example 3 for tension

N^{II} f [Hz]	ANSYS $-\frac{3}{4}N_{Ki}^{II}$	DIFF $-\frac{3}{4}N_{Ki}^{II}$	ANSYS $-\frac{1}{2}N_{Ki}^{II}$	DIFF $-\frac{1}{2}N_{Ki}^{II}$	ANSYS $-\frac{1}{4}N_{Ki}^{II}$	DIFF $-\frac{1}{4}N_{Ki}^{II}$
f_1	3096.2	3152.5	2879.7	2918.6	2640.9	2664.3
f_2	9705.5	9941.6	9470.2	9700.3	9224.2	9378.1
f_3	19301.0	20124.1	19080.0	19816.8	18852.0	19504.7
f_1	3092.7	3148.8	2875.8	2914.7	2636.6	2660.0
f_2	9690.5	9966.9	9454.6	9683.2	9208.0	9391.0
f_3	19271.0	20087.0	19050.0	19779.0	18821.0	19467.0

the differential equations (DIFF) for compression are presented in Table 8.8. In Table 8.9 the analysis results are presented for tensional axial force. First three rows of these tables contain eigenfrequncies of the original beam. Last three rows contain eigenfrequncies of the beam with interchanged layers. The appropriate eigenmodes of the original layered beam (in a qualitative form, for $N^{II} = \frac{1}{2}N_{Ki}^{II} = -118\,\text{kN}$) are shown in Fig. 8.11. All solution results differences (ANSYS and DIFF for original layered beam) are presented in Table 8.10.

In Fig. 8.12 a dependence of the first eigenfrequency on the axial force is shown (for DIFF solution). First eigenfrequency of this beam with the original layering calculated by DIFF is 2383.05 Hz for $N^{II} = 0$. As can be seen from Fig. 8.12, effect of the axial force is of the same character as it was in Example 1 and 2.

8.4.2 Free Vibration Analysis of One-Layer FGM Beams with Effect of the Axial Force

The composite simply supported beam (Fig. 8.13) of quadratic cross-section m^2 is made as a mixture of matrix – NiFe, and fibres – W (tungsten). The variation of tungsten volume fraction is: $v_f(x) = -400x^2 + 40x$. The material properties of tungsten (elasticity modulus and mass density) are constant and their values are: $E_m = 400\,\text{GPa}$, $\rho_f = 19300\,\text{kgm}^{-3}$. Material properties of the matrix are

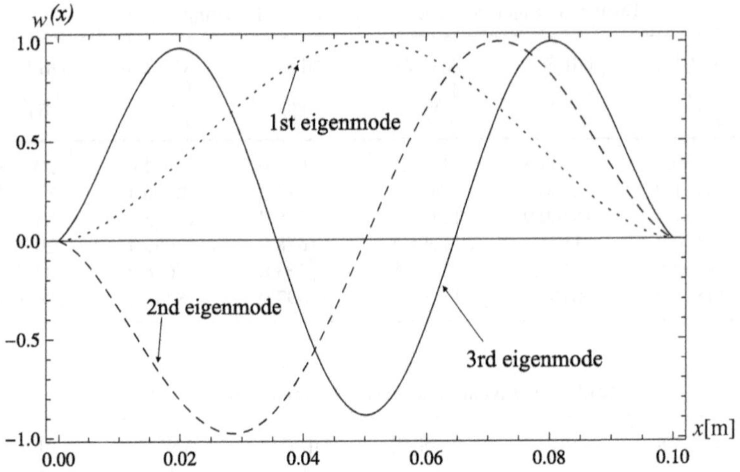

Fig. 8.11 First three eigenmodes in the Example 3 for DIFF solution for original layering

Table 8.10 Solution results difference (ANSYS and DIFF) for original layering

	Solution results difference (ANSYS and DIFF) (%)					
	Compression			Tension		
N^{II} / f	$\frac{3}{4}N_{Ki}^{II}$	$\frac{1}{2}N_{Ki}^{II}$	$\frac{1}{4}N_{Ki}^{II}$	$-\frac{3}{4}N_{Ki}^{II}$	$-\frac{1}{2}N_{Ki}^{II}$	$-\frac{1}{4}N_{Ki}^{II}$
f_1	1.00	0.43	0.01	1.81	1.35	0.88
f_2	0.30	0.73	0.95	2.43	2.42	1.66
f_3	1.87	2.27	2.66	4.26	3.86	3.46

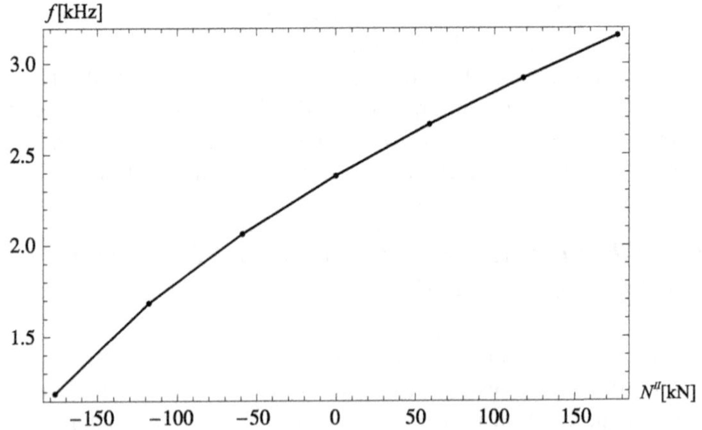

Fig. 8.12 Dependence of the first eigenfrequency on the axial force for Example 3 – original layering

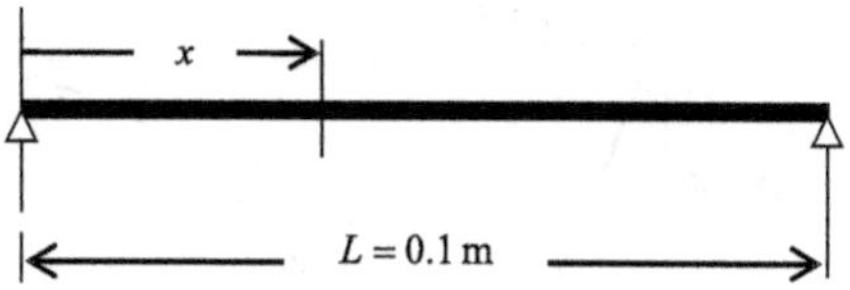

Fig. 8.13 Composite simply supported beam

constants as well: $E_m = 255$ GPa, $\rho_m = 9200\,\mathrm{kgm}^{-3}$. The cross-sectional area $A = 0.0001\,\mathrm{m}^2$; the area moment of inertia $I = \frac{A^2}{12}\,\mathrm{m}^4$; and the reduced cross-sectional area $\bar{A} = 0.833\times 10^{-4}\,\mathrm{m}^2$.

The effective longitudinal elasticity modulus $E_L(x)$ and the effective longitudinal mass density $\rho_L(x)$ of the homogenized beam are calculated using the extended mixture rules [2]: $E_L(x) = -58000x^2 + 5800x + 255$ [GPa], $\rho_L(x) = 9200 + 404000x - 4040000x^2$ $[\mathrm{kgm}]^{-3}$ (case 1). To show the effect of the variation of several material parameters the homogenized material properties have been changed to: $E_L(x) = 58000x^2 - 5800x + 400$ [GPa], $\rho_L(x) = 19300 - 404000x + 4040000x^2$ $[\mathrm{kg/m}]^3$ (case 2). Their longitudinal distributions are shown in Figs. 8.14 and 8.15. The ratio $\frac{E_L(x)}{G_L(x)} = 2.6$ has been used for the effect of shear forces assumption in this example.

The first three eigenfrequencies and eigenmodes have been found using the differential equation (8.17) and the appropriate boundary conditions. The same problem has been solved using a very fine mesh – 2800 of 2D PLANE42 elements of the FEM program ANSYS. Again, the numerical results of ANSYS and the results of the exact differential equation are presented in Tables 8.11 (for compression) and 8.12 (for tension). First three lines in these tables contain theeigenfrequnciess for case 1, the last three lines contain the eigenfrequncies for

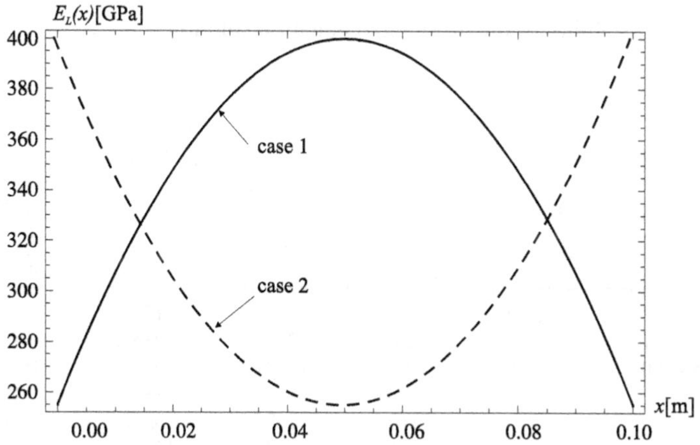

Fig. 8.14 Distribution of the effective longitudinal elasticity modulus

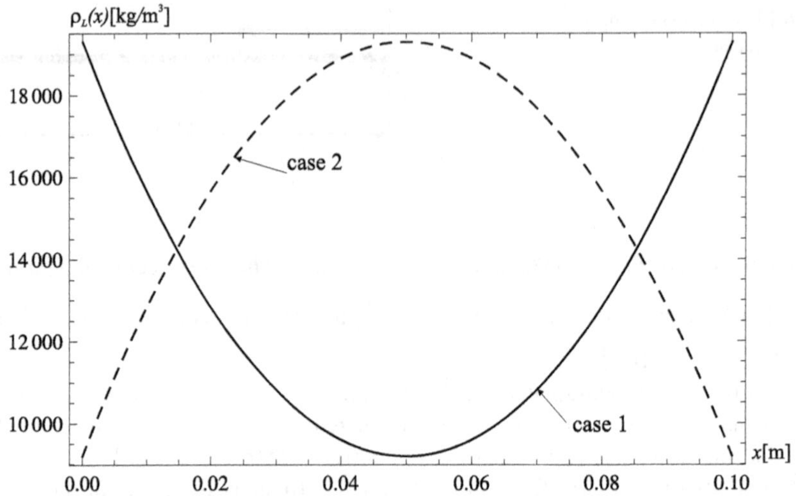

Fig. 8.15 Distribution of the effective longitudinal mass

the case 2. The critical buckling force is $N_{Ki}^{II} = -302.681$ kN in the case 1, and $N_{Ki}^{II} = -220.0$ kN in the case 2. To show an effect of different material properties distribution, the same N^{II} has been assumed in the both cases and it was calculated as a part of the buckling force from the case 1.

Any convergent eigenfrequencies has been found for the $N^{II} = \frac{3}{4}N_{Ki}^{II} = -227$ kN in the case 2 (compression) because such an axial force is larger than the appropriated buckling force. As shown in Tables 8.11 and 8.12 the eigenfrequencies are higher in the case 2 comparing to case 1 for all solutions excluding the first eigenfrequency with $N^{II} = \frac{1}{2}N_{Ki}^{II} = -151$ kN. This discrepancy could be given by numerical instability of the differential equation solution.

Table 8.11 Eigenfrequencies of the simply supported one-layer FGM-beam for compression

N^{II} f [Hz]	ANSYS $\frac{3}{4}N_{Ki}^{II}$	DIFF $\frac{3}{4}N_{Ki}^{II}$	ANSYS $\frac{1}{2}N_{Ki}^{II}$	DIFF $\frac{1}{2}N_{Ki}^{II}$	ANSYS $\frac{1}{4}N_{Ki}^{II}$	DIFF $\frac{1}{4}N_{Ki}^{II}$
f_1	1038.8	1023.9	1460.9	1449.4	1776.0	1771.8
f_2	7067.1	7035.9	7350.7	7613.6	7350.8	7648.6
f_3	15844.0	15912.5	16100.0	16231.1	16340.0	16539.4
f_1	–	–	1280.8	1267.7	1847.6	1837.2
f_2	–	–	7710.7	7676.3	8058.9	8062.2
f_3	–	–	17011.0	17072.6	17315.0	17458.8

Table 8.12 Eigenfrequencies of the simply supported one-layer FGM-beam for tension

N^{II} f [Hz]	ANSYS $-\frac{3}{4}N_{Ki}^{II}$	DIFF $-\frac{3}{4}N_{Ki}^{II}$	ANSYS $-\frac{1}{2}N_{Ki}^{II}$	DIFF $-\frac{1}{2}N_{Ki}^{II}$	ANSYS $-\frac{1}{4}N_{Ki}^{II}$	DIFF $-\frac{1}{4}N_{Ki}^{II}$
f_1	2660.0	2708.0	2474.0	2506.3	2271.6	2289.9
f_2	8545.8	8748.5	8328.7	8486.2	8104.2	8219.0
f_3	17212.0	17729.3	17005.0	17438.6	16794.0	17146.9
f_1	3188.8	3242.3	2920.5	2963.5	2618.6	2637.6
f_2	9260.1	9460.8	8984.1	9130.4	8693.2	8792.1
f_3	18417.0	18937.1	18157.0	18557.7	17888.0	18215.8

8.5 Conclusions

The fourth-order differential equation with non-constant parameters of the beam deflection with longitudinal variation of the homogenized effective material properties has been established in this paper. The 2nd order beam theory has been used for establishing the equilibrium- and kinematics beam equations. The shear forces deformation effect and the effect of consistent mass distribution and mass inertia moment have been taken into account too. Effect of large axial force has also been considered. The longitudinal variation of the effective longitudinal elasticity modulus and mass density in the homogenized beam can be caused by longitudinal and transversal symmetric variation of the elasticity modulus, the mass density and the volume fraction of the FGM constituents in the sandwich layers. Homogenization of these material properties has been done by extended mixture rules and laminate theory. Obtained results show that an increase of the negative axial force (compression) decreases the eigenfrequency, and increase of the positive axial force (tension) increases the eigenfrequency. As expected, positive axial force makes the beam stiffener and the eigenfrequency is higher than in the case of a negative axial force.

It was also shown that the eigenfrequencies of the FGM beam can be influenced by transversal and longitudinal variation of material properties and mass density significantly.

The results of the differential equation were compared with the ones obtained using a very fine mesh of the 2D plane stress elements of the FEM program ANSYS. A good agreement of the two solution results has been achieved.

Acknowledgements This research was supported by Grant Agency VEGA 1/0093/10.

References

1. Sapountzakis, E.J., Mokos, V.G.: Vibration analysis of 3-D composite beam elements including warping and shear deformation effect. J. Sound Vib. **306**(3–5), 818–834 (2007)
2. Murín, J., Aminbaghai, M., Kutiš, V.: Exact solution of the bending vibration problem of FGM beam with variation of material properties. In: Papradakis, M., Kojic, M., Papadopoulos V. (eds.) SEECCM 2009, Rhodes, Greece, 22–24 June (2009)

3. ANSYS 11 Theory Manual (2008)
4. Rubin, H.: Analytische Lösung linearer Differentialgleichungen mit veränderlichen Koeffizienten und baustatische Anwendung. Bautechnik **76**, 316–327 (1999)
5. Wolfram, S.: MATHEMATICA 5. Wolfram research, Inc., Champaign (2003)
6. Love, B.M., Batra, R.C.: Determination of effective thermomechanical parameters of mixture of two elastothermoviscoplastic constituents. Int. J. Plast. **22**, 1026–1061 (2006)
7. Altenbach, H., Altenbach, J., Kissing, W.: Mechanics of Composite Structural Elements. Engineering – Monograph (English). Springer, Berlin (2003)

Chapter 9
Wavelet Analysis of the Shear Stress in Soil Layer Caused by Dynamic Excitation

A. Borowiec

Abstract The paper deals with shear stresses analysis of fully saturated Nevada sand layer by means of Wavelet Transform. The layer modelled as two-phase Biot's porous medium, is subjected to the impulse wave. Results, obtained from program called Swandyne, were analyzed using MathCad. The Wavelet Transform was carried out to calculate proper frequency spectrum along with duration time of separate frequency. Tangential stresses were considered because of they importance in creating slip plane and liquefaction. In order to study the dynamic response of saturated soil layer as an initial-boundary problem, a FEM code was implemented. The tangential stress distribution in the layer for chosen time steps shows that the phenomena well known under static load takes place during dynamic excitation as well. The shear stresses in the reference point, are analysed using two methods, classical Fourier Transform and the modern Wavelet Analysis. Thanks to Wavelet Transform we can retrieve such important information as: the time when the dominant frequency is reached and which frequencies appears and/or disappears in time.

Keywords Soil dynamics · Porous media · Signal processing · Wavelet analysis

9.1 Introduction

Main goals of the paper are: to follow the shear stresses behavior for sand layer under dynamic loading and to use the signal processing method (Wavelet Transform here) to assess the results. From the mechanical point of view to solve problem mentioned, we need to formulate the governing set of equations choosing between wave or Lagrangian formulation. In the paper soil layer is modeled as two-phase Biot's porous medium [1] with modern contributions [2] and $\mathbf{u} - p$ simplification. Biot derived his set of equations of motion from Lagrangian formulation, the kinetic

A. Borowiec (✉)
Division of Soil – Structure Interaction, Faculty of Civil Engineering, Cracow University of Technology, ul. Warszawska 24, 31-155, Kraków, Poland
e-mail: anabo@pk.edu.pl

J. Murín et al. (eds.), *Computational Modelling and Advanced Simulations*,
Computational Methods in Applied Sciences 24, DOI 10.1007/978-94-007-0317-9_9,

energy and its dissipation. The soil here was considered as a mixture of two phases – solid skeleton and the fluid (represented by pore pressure), both phases had the separate equation of motion. Dealing with highly inelastic material, like soil, we have to chose the proper constitutive model. What is more the saturated sand subjected to dynamic loading is susceptible to the liquefaction. For that reason the Pastor-Zienkiewicz Mark III constitutive relation was chosen here. Last but not least was to decide which numerical method to implement.

9.2 Governing Set of Equations in Soil Dynamic Consolidation Theory – FEM Implementation

Liquefaction and associated shear deformations continue to cause damage during dynamics excitations. In order to study the dynamic response of saturated soil layers as an initial-boundary problem, a Finite Element Method was considered. The equations of motion for elastic skeleton saturated with a pore fluid, given by Biot [3], consist of two variables' fields: one for pore pressure p and the other for skeleton displacements $\mathbf{u}$. Introducing the Voigt notation, widely used in FEM, we can now write the overall dynamic equilibrium equation for soil-fluid mixture as:

$$\mathbf{L}^{\mathrm{T}}\sigma + \rho\mathbf{b} - \rho\ddot{\mathbf{u}} - \rho_f\dot{\mathbf{w}} = 0 \tag{9.1}$$

where ρ is a density of the soil-fluid mixture, ρ_f is fluid density, $\boldsymbol{\sigma}$ is the total stress 6-dimensional vector (according to the Voigt notation), $\mathbf{b}$ is a body force vector, $\mathbf{L}$ is the spatial derivative operator, $\mathbf{u}$ and $\mathbf{U}$ are the solid and fluid displacements respectively. The vector $\dot{\mathbf{w}}$ represents the fluid acceleration relative to the solid:

$$\dot{\mathbf{w}} = n\left(\ddot{\mathbf{U}} - \ddot{\mathbf{u}}\right) \tag{9.2}$$

The equilibrium equation related to the pore fluid can be written as:

$$-\nabla p + \rho_f\mathbf{b} - \rho_f\ddot{\mathbf{u}} - \frac{\rho_f}{n}\dot{\mathbf{w}} - \frac{\rho_f g}{k}\mathbf{w} = 0 \tag{9.3}$$

where p is pore pressure, n and k are the material porosity and permeability, and g is the gravitational acceleration. Now using (9.2) we can rewrite the (9.3) to the form:

$$-\nabla p + \rho_f\mathbf{b} - \rho_f\ddot{\mathbf{U}} - \frac{1}{\xi}\mathbf{w} = 0 \tag{9.4}$$

where $\xi = k/(\rho_f g)$ and is called the modified permeability coefficient.

Finally fluid inflow $\nabla^{\mathrm{T}}\mathbf{w}$, the storage (represented here by $\dot{\boldsymbol{\varepsilon}}$) and pore pressure changes in time $\dot{p}$, are expressed as:

$$\nabla^{\mathrm{T}}\mathbf{w} + \alpha\mathbf{m}\dot{\boldsymbol{\varepsilon}} + \frac{\dot{p}}{Q} = 0 \tag{9.5}$$

where Q is a function of bulk modulus K and α is so called Biot constant. Now a simplified numerical framework, known as $\mathbf{u} - p$ formulation is implemented [4]. Fluid inertia is generally small and we shall frequently omit it, thus terms $\rho_f \dot{\mathbf{w}}$ in (9.1) and $\rho_f \ddot{\mathbf{U}}$ in (9.4) are neglected. Finally the variable $\mathbf{w}$ is eliminated and the set of equations is reduced to a followings:

$$\mathbf{L}^{\mathbf{T}}\sigma + \rho\mathbf{b} - \rho\ddot{\mathbf{u}} = 0 \tag{9.6}$$

$$\nabla^{\mathrm{T}}\xi\left(-\nabla p + \rho_f \mathbf{b}\right) + \alpha\mathbf{m}\dot{\varepsilon} + \frac{\dot{p}}{Q} = 0 \tag{9.7}$$

in which the (9.6) and (9.7) represent the equilibrium of solid and fluid phase respectively. The total stresses $\boldsymbol{\sigma}$ are composed of effective stresses $\boldsymbol{\sigma}'$ and pore pressures, according to the Terzaghi's principle, i.e. $\boldsymbol{\sigma} = \sigma' + \alpha\mathbf{m}p$.

Introducing the FEM approach and forming the weak form of the formulation, after some algebraic operation, we can express (9.6) and (9.7) as an uncoupled set of equations:

$$\begin{bmatrix} \mathbf{M} & 0 \\ 0 & 0 \end{bmatrix} \begin{Bmatrix} \ddot{\tilde{\mathbf{u}}} \\ \ddot{\tilde{p}} \end{Bmatrix} + \begin{bmatrix} 0 & 0 \\ \mathbf{Q}^{\mathrm{T}} & \mathbf{S} \end{bmatrix} \begin{Bmatrix} \dot{\tilde{\mathbf{u}}} \\ \dot{\tilde{p}} \end{Bmatrix} + \begin{bmatrix} \mathbf{K} & -\mathbf{Q} \\ 0 & \mathbf{H} \end{bmatrix} \begin{Bmatrix} \tilde{\mathbf{u}} \\ \tilde{p} \end{Bmatrix} = \begin{Bmatrix} \mathbf{f}^{(1)} \\ \mathbf{f}^{(2)} \end{Bmatrix} \tag{9.8}$$

where $\mathbf{M}$ and $\mathbf{K}$ are the mass and stiffness matrixes respectively, $\mathbf{Q}$ is the discrete gradient operator coupling the soil and fluid phase, $\mathbf{H}$ is the permeability matrix and $\mathbf{S}$ the compressibility matrix. Vectors $\mathbf{f}^{(1)}$ and $\mathbf{f}^{(2)}$ represent the effects of body forces along with prescribed Boundary Conditions for the solid-fluid mixture and fluid phase.

Taking into account plastic behaviour of soils, we are introducing Pastor – Zienkiewicz mark III model into the consistent elasto-plastic matrix $\mathrm{D^{ep}}$ [2, 4]. The model mentioned above is the most suitable for materials susceptible to liquefaction. In order to calculate matrix $\mathrm{D^{ep}}$ and to solve set of (9.8) Newton – Raphson iterations must be carried out.

Using Newmark time integration procedure soil skeleton acceleration and pore pressure first derivative in time can be expressed as:

$$\ddot{\tilde{\mathbf{u}}}_{\mathrm{n+1}} = \ddot{\tilde{\mathbf{u}}}_{\mathrm{n}} + \Delta\ddot{\tilde{\mathbf{u}}}_{\mathrm{n}}, \quad \dot{\tilde{\mathbf{p}}}_{\mathrm{n+1}} = \dot{\tilde{\mathbf{p}}}_{\mathrm{n}} + \Delta\dot{\tilde{\mathbf{p}}}_{\mathrm{n}} \tag{9.9}$$

At the same time Taylor expansion for $\dot{\tilde{\mathbf{u}}}_{\mathrm{n+1}}$, $\tilde{\mathbf{u}}_{\mathrm{n+1}}$, $\tilde{\mathbf{p}}_{\mathrm{n+1}}$ allows us to reduce the number of variables in equation set (9.8) so that we have pore pressure first derivative according to time and soil skeleton acceleration only. With the transition Jacobian we obtain, at each time step, the following set of equations:

$$\begin{Bmatrix} \boldsymbol{\Psi}^{(1)}_{\mathrm{n+1}} \\ \boldsymbol{\Psi}^{(2)}_{\mathrm{n+1}} \end{Bmatrix}^{\mathrm{i}} + \begin{bmatrix} \mathbf{M}_{\mathrm{n+1}} + (\mathbf{K}_{\mathrm{T}})_{\mathrm{n+1}}\beta_2\Delta t^2 & -\mathbf{Q}_{\mathrm{n+1}}\bar{\beta}_1\Delta t \\ \mathbf{Q}^{\mathrm{T}}_{\mathrm{n+1}}\beta_1\Delta t & \mathbf{S}_{\mathrm{n+1}} + \mathbf{H}_{\mathrm{n+1}}\bar{\beta}_1\Delta t \end{bmatrix}^{\mathrm{i}} \begin{Bmatrix} \mathrm{d}(\Delta\ddot{\tilde{\mathbf{u}}}_{\mathrm{n}}) \\ \mathrm{d}(\Delta\dot{\tilde{\mathbf{p}}}_{\mathrm{n}}) \end{Bmatrix}^{\mathrm{i}} = 0 \tag{9.10}$$

where $\mathbf{K}_T$ is tangential stiffness matrix, β_1, β_2 are the Newmark constants. The "n" index denotes time steps and "i" – iterations resulting from non-linearity. Form numerical practice it is known that $\overline{\beta}_1 = \beta_1 = \beta_2 = 1/2$ can be applied.

Summing up for dynamic fully non-linear problem of porous medium we need to solve at each time step the system of (9.10), which additionally depends on the solution from previous time step. From other, known in the literature, formulations (full $\mathbf{u} - \mathbf{w} - p$ and $\mathbf{u} - \mathbf{u}^f$) which are establishing time integration of higher order, raising the amount of variables consequently, the $\mathbf{u} - p$ simplification is the most convenient in numerical analysis.

9.3 Constitutive Relation

Pastor – Zienkiewicz mark III constitutive relation was proposed in 1985 by Pastor Zienkiewicz and Leung for saturated soils, mostly sands, subjected to dynamic excitation, where liquefaction had to be considered. This Generalised Plasticity model is well known and calibrated. The plastic potential is defined here as:

$$g = q - M_g \cdot p \cdot \left(1 + \frac{1}{\alpha}\right)\left[1 - \left(\frac{p}{p_g}\right)^{\alpha}\right] \tag{9.11}$$

p_g is the point where the plastic potential crosses the axiator axis p and M_g represents the slope of Critical State Line (see Fig. 9.1). Yield function is describe by analogical equation:

$$f = q - M_f \cdot p \cdot \left(1 + \frac{1}{\alpha}\right)\left[1 - \left(\frac{p}{p_g}\right)^{\alpha}\right] \tag{9.12}$$

In general: $M_f \neq M_g$ and M_f/M_g ratio depends on soil relative density (D_r), here $M_f/M_g = 0{,}2$.

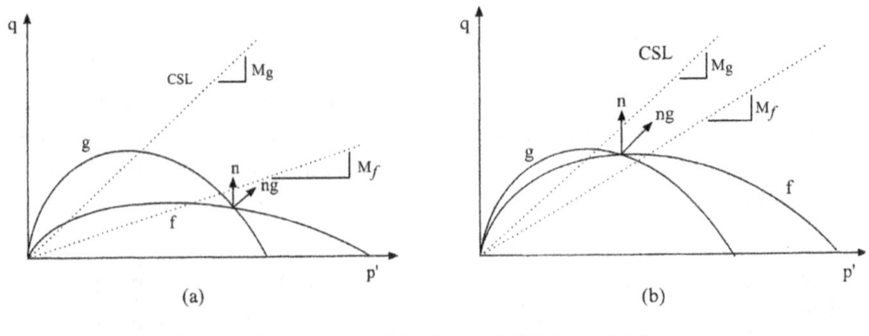

Fig. 9.1 Yield envelope *f* and plastic potential *g* for sand: (**a**) loose, (**b**) dense

9.4 Numerical Calculation

Program called DIANA – SWANDYNE II was used in the calculations. It is the Finite Element Method program for static and dynamic consolidation with Newmark scheme time integration. The formulation $\mathbf{u} - p$ means that we have two unknown fields of variables – skeleton displacements $\mathbf{u}$ and pore pressure p (scalar). Because of constitutive law for soil skeleton, we obtained fully nonlinear system of equations, which was solved by iterative Newton – Raphson procedure at each time step. Dynamic load, in the form of accelerations, was applied to the bottom of the layer.

Very sudden changes in tangential stresses, cause soil instability and pore pressure excess, leading to liquefaction. In our example the sand was treated as a fully saturated layer (11.0×17.8 m), under undrained conditions, resting on immovable base (see Fig. 9.2). Based on numerical experience of Zienkiewicz and Pastor, 4 – node quadrilateral elements were used in FEM discretisation (see Fig. 9.2). Free surface of the water was assign to the top of the layer (small dots at Fig. 9.2 denoting $p = 0$) and the left and right boundary had the same vertical displacements (so called tied nodes).

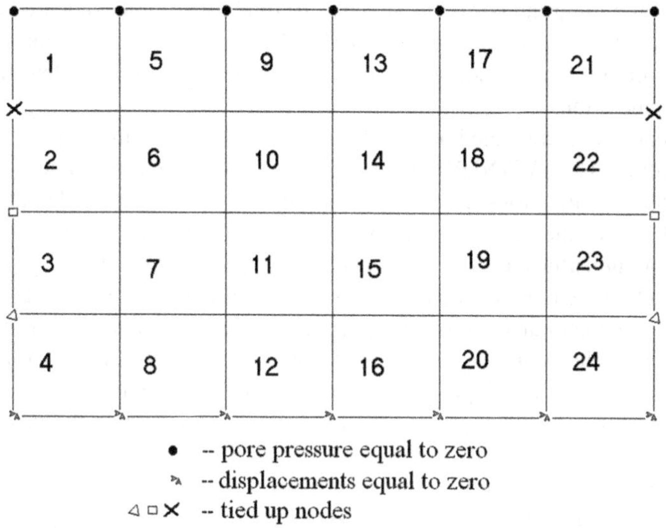

Fig. 9.2 Sand layer geometry (with BC) and geotechnical parameters used in calculation

Geotechnical parameters for Nevada sand i.e. void ratio e, permeability coefficient k and relative density D_r, as well as the constitutive model parameters, were taken from laboratory test conducted under VELACS project. It was international program carried out in the eighties, which results were accessible from web sites. VELACS stands for VErification of Liquefaction Analysis by Centrifuge Studies [5]. The geotechnical parameters were following (compare with Fig. 9.3):

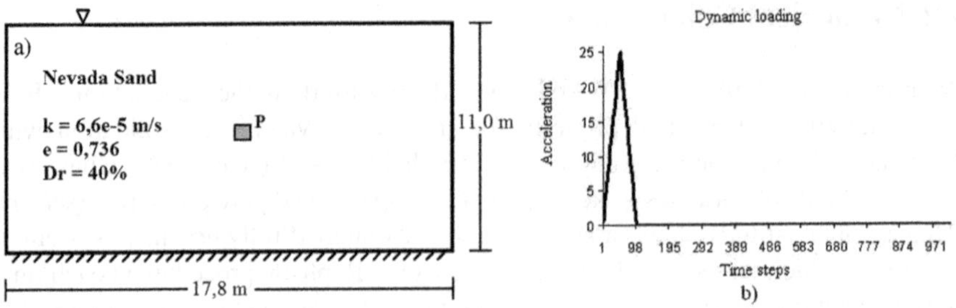

Fig. 9.3 Sand layer: (**a**) geometry and geotechnical parameters, (**b**) dynamic excitation according to the time steps

- void ratio: $e = 0.736$,
- permeability coefficient $k = 6.6\text{e-}5$ m/s,
- relative density $D_r = 40\%$ (loose sand according to ASTM)

The artificial peak wave composed of vertical accelerations (see Fig. 9.3b) was used as dynamic loading. Numerical analysis of dynamic consolidation problem modeled in a manner presented above, can be running longer than the excitation lasts. Thus assuming the period of analysis 10 time longer than shock duration (of about 2 s) and the time increment equal to 0.002 s we obtained 1,024 time steps. The choice of impulse wave loading was dictated of the need for complete frequency spectrum generation.

The shear stress calculated at each Gauss point for each time step, gave almost 100,000 values. The SWANDYNE postprocessor is very simple and does not allow for advanced visualizations. In order to generate the 3D surfaces and contour maps for every 10th step the Surfer program was used. Figure 9.4 shows some specific distributions of tangential stresses. Shadings depict stress accumulation thus slip plane and liquefaction location. Comparing stress distribution for this 6 time steps we can detect how the diagonal shear bands appear and how they propagate during excitation. Observing the peaks location (at 3D visualizations), we can notice that these bands cross each over which proves that the phenomena well known under static and quasi-static tests takes place during dynamic excitation either. In the post impulse time shear stresses travel only in the plane perpendicular to the load direction (see 411th and 421st time step at Fig. 9.4), indicating the shear wave propagation.

Finally in order to assess the results by means of Wavelet Transform the shear stresses in the reference point P (see Fig. 9.3a) were plotted and treated as an input signal (see Fig. 9.5).

9.5 Wavelet Analysis

During geodynamic, earthquake driven processes we are dealing with a non-stationary problem, which means that the frequencies vary with time. Then classical signal processing tools such as Fourier transform or spectrograms are insufficient.

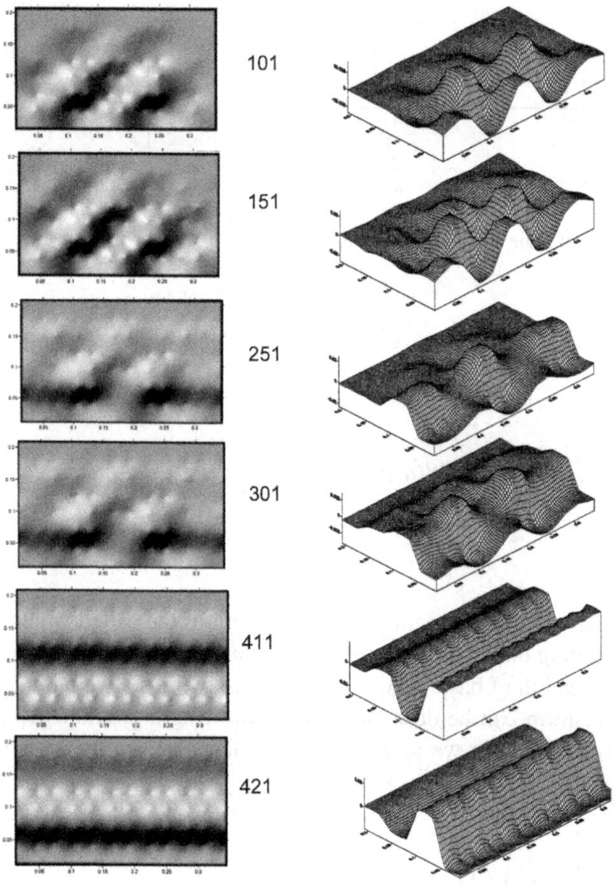

Fig. 9.4 Shear stress changes in selected time steps: (**a**) maps, (**b**) 3D visualization

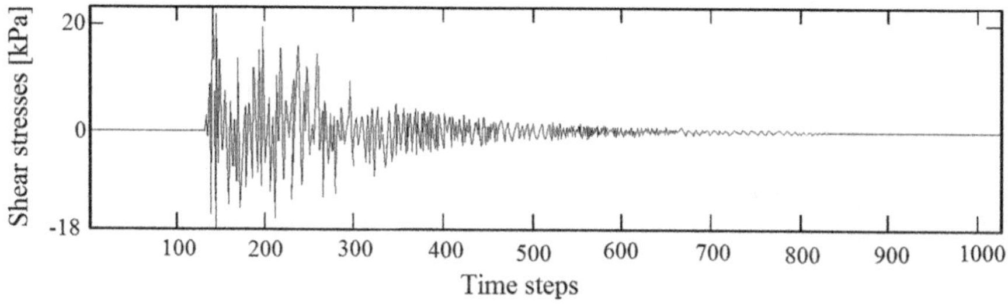

Fig. 9.5 Tangential stress in the reference point P

The wavelet analysis is being used mainly in processing of signals (continuous transform) and pictures (discrete transform). In general the idea is similar to Fourier analysis only that we are using whole family of wave function instead of sine. Wavelet transform is define as [6]:

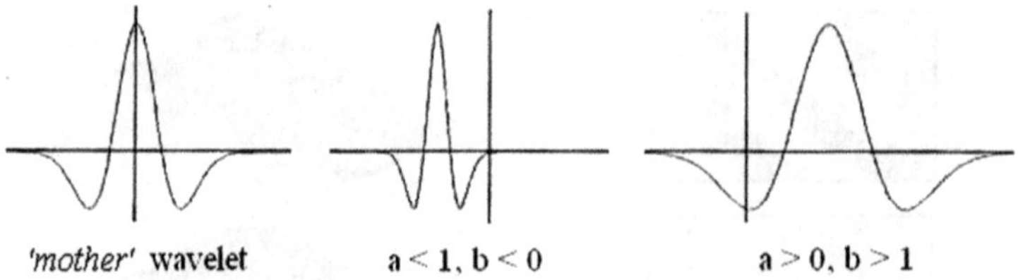

Fig. 9.6 Influence of scale and phase parameters

$$W[x(t)] = W(a,b) = \frac{1}{\sqrt{a}} \int_{-\infty}^{+\infty} x(t) \Psi \left[\frac{t-b}{a} \right] dt \quad (9.13)$$

where Ψ is the domain of orthogonal wavelet functions (so called "mother" or basic wavelets) depending on two parameters: scale a and phase b. Figure 9.6 shows how the shape and location of basic wavelet is influence by those two parameters.

Wavelet transform can be describe by the following algorithm. After choosing the basic wavelet function we scan the signal, starting from the signal beginning ($t = 0$).

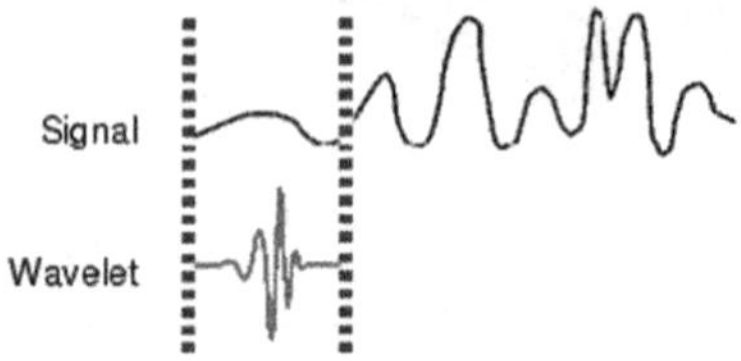

Then we continue to change the phase parameter b while the scale parameter a remains constant.

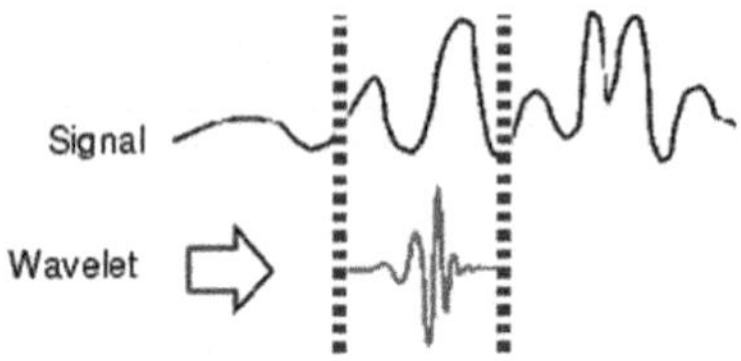

After scanning the whole signal we change the scale parameter a and repeat the whole procedure for as many scales as wide is the frequency range we want to check – compare with (9.13).

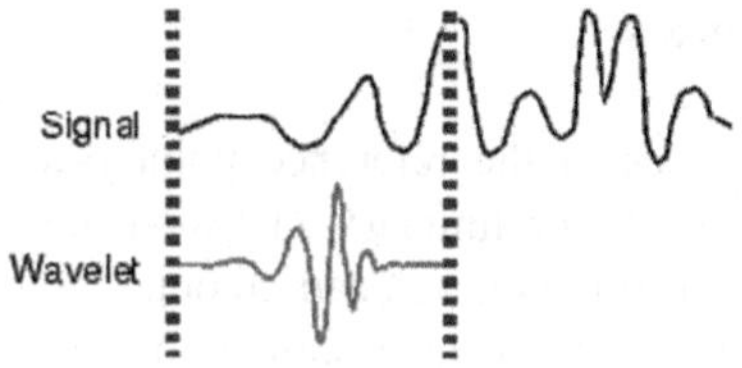

The main idea is to fit tightly the wave within the original signal. The measure of this adjustment is so called fitting function W(a, $b(t)$), the higher values of W the more similar is the wavelet to the signal.

According to the wavelets theory there are some restrictions impose on basic wavelet. The function $\varphi(t) \in L^2(\Re)$ can not be use as the wavelet unless it fulfills the following condition:

$$C_\varphi = \int_0^\infty \frac{|\Psi(\omega)|^2}{\omega} d\omega < \infty \tag{9.14}$$

where $\Psi(\omega)$ is Fourier transform of function $\varphi(t)$. From the practical point of view, the relation between frequency and scale parameter a is of the great importance, it allows us to follow the frequency changes in time:

$$f = \frac{f_c}{a\Delta} \tag{9.15}$$

where f_c is so called central frequency (unique for every basic wavelet) and Δ is a sampling period. For purposes presented in the paper the complex Morlet basic wavelet is the most suitable. It can be presented in the following form (Fig. 9.7):

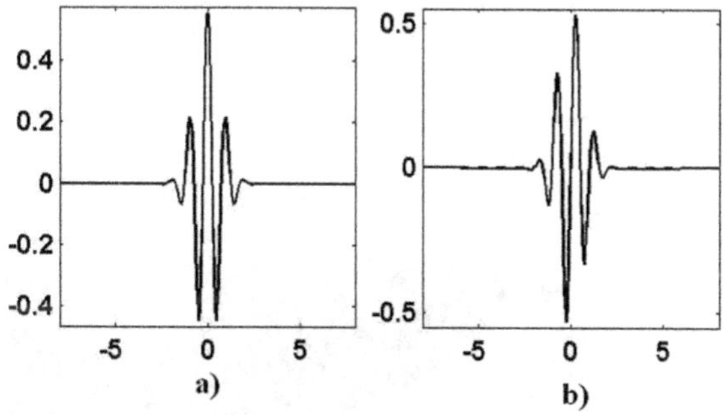

Fig. 9.7 Complex Morlet wavelet: (**a**) real part, (**b**) imaginary part

Central frequency for Morlet wavelet is equal to one ($f_c = 1$) but could be also presented in form of graphs [7].

9.6 Wavelet Analysis of the Shear Stresses

Results from SWANDYNE, here the shear stresses in the reference point (see Fig. 9.5), were processed with help of MathCAD. The preliminary analysis using Fourier Transform first and detailed Wavelet Transform after, was carried out.

Those two analysis are presented at Figs. 9.8 and 9.9, in a similar manner. The frequency axis is a horizontal one. We can easily conclude that they give quantitatively comparable frequency spectrum.

Introducing the wavelet transform we gain the information about the frequency location in time. Predominant frequency of 45 Hz appears three times in the response history in about 0.7 0.9 and 1.1 s after the start of the impulse. The second most influential frequency of 80 Hz appears only once during the response that is at 0.9 s.

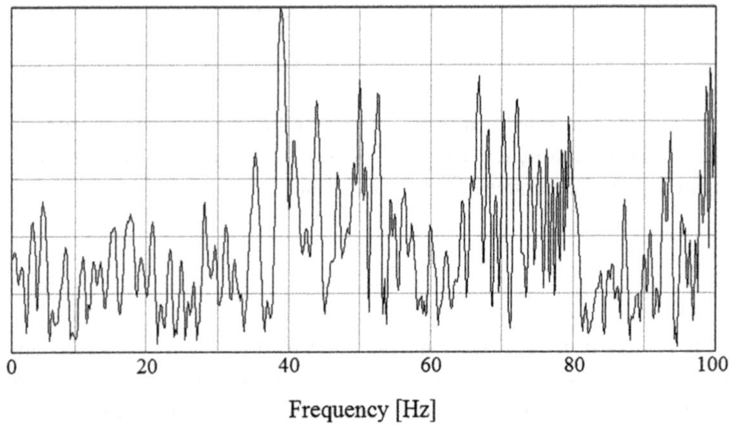

Fig. 9.8 Fourier transform of the shear stresses in the reference point P

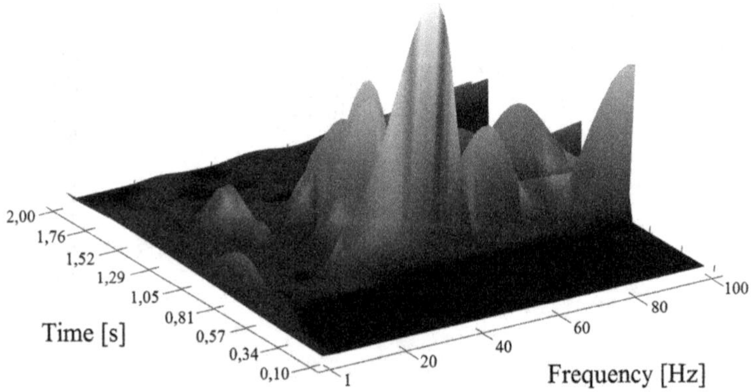

Fig. 9.9 Wavelet transform of the shear stresses in reference point P – 3D surface

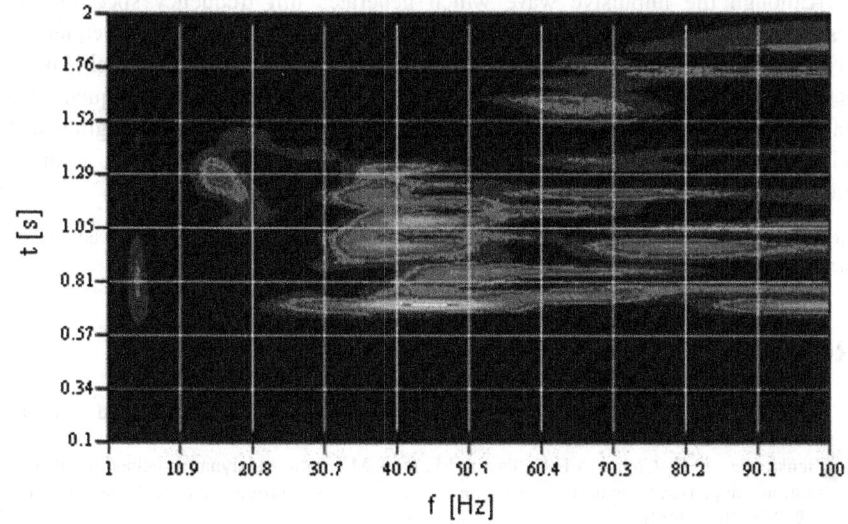

Fig. 9.10 Wavelet transform of the shear stresses – contour map

Figure 9.10 illustrates peaks location in time more precisely. Not only the 45 Hz frequency but also this of 65 Hz repeats in time. In the other hand low frequencies (below 20 Hz) appear very late or at all. We can also observe that the shear wave reaches the middle of the layer (reference point P) suddenly at about 0,7 s which is 0,5 s past the end of impulse loading.

9.7 Conclusions

Examining the behavior of tangential stresses (see Fig. 9.4) we can conclude that:

- diagonal slip bands relocate with time – proving that the phenomena known in static takes place in soil-dynamics too,
- after fade of input we still have the response – what indicates to dynamic consolidation of the sand.

Spectrum calculated using Morlet wavelet contains similar frequencies as the one obtained from Fourier. However thanks to wavelet transform (Figs. 9.9 and 9.10) we can retrieve following information:

- the time when the predominant frequency (highest peak) is reached,
- whether predominant frequency repeats or fades,
- which frequencies appears and/or disappears in time.

Although the impulsive wave which generates full frequency spectrum the response spectrum in the middle of the layer is quit thin. This filtering of frequencies is connected with dumping – well know phenomena in real soils. Commonly known is that the use of elastic models in geodynamics is problematic but frequent. The another purpose of the paper is to use less popular, non-commercial program with advanced constitutive relation like Pastor – Zienkiewicz mark III which, as shows the wavelet analysis, models the dynamic behavior of soil properly.

Acknowledgements To professor A.H.C. Chan for sharing the SWANDYNE source code under the academic license.

References

1. Biot, M.A.: Theory of propagation of elastic waves in a fluid – saturated porous solid. J. Acoust. Soc. Am. **28**(2), 168–178 (1956)
2. Zienkiewicz, O.C., Chana, A.H.C., Pastor, M., Paul, M.: Static and dynamic behaviour of soils: a rational approach to quantitative solutions. I Fully saturated problems. Proc. R. Soc. Lond. A **429**, 285–309 (1990)
3. Biot, M.A.: Theory of elasticity and consolidation for a porous anisotropic soil. J. Appl. Phys. **26**, 182–185 (1955)
4. Zienkiewicz, O.C., Chana, A.H.C., Pastor, M., Schrefler, B.A., Shiomi, T.: Computational Geomechanics with Special Reference to Earthquake Engineering. Wiley, New York, NY (2000)
5. Arulmoli, K., et al.: VELACS Laboratory Testing Program Soil Data Report. National Science Foundation, Washington, DC (1993)
6. Daubechies, I.: Ten lectures on wavelets. CBMS – NSF Conference in Applied Mathematics, SIAM Ed, Philadelphia, PA (1992)
7. Białasiewicz, T.: Falki i aproksymacje. Wydawnictwo Naukowo. Techniczne, Warszawa (2000)

Chapter 10
Strength of Composites with Fibres

E. Kormaníková, D. Riecky, and M. Žmindák

Abstract The laminate analogy is very useful for the calculation of the strength of composite materials with short fibres. In connection with the maximum strain criterion, the lamination theory is suitable for predicting the strength of composites with short fibres. The prediction of the laminate strength is carried out by evaluating the stress state within each layer of the laminate based on the classical lamination theory. Finite Element Method (FEM) is used as a tool to predict the laminate strength. Numerical simulation has been prepared by using a commercially available ANSYS code.

Keywords Composite · Short fiber · Lamination theory · Failure criteria

10.1 Introduction

Composite materials have become common engineering materials due to their good mechanical and electro-chemical properties. Specifically fibre-reinforced composites are one of the most widely used man-made composite materials; they are constituted by reinforcing fibres embedded in a matrix material.

Modelling can play an important role in the analysis and design of fibre-reinforced composite materials. Their mechanical properties and possible failure modes can be predicted early during the design stage using effective modelling techniques [1]. Recent developments in commercial FEA packages along with their user-friendliness and pre- and post-processing, as well as their powerful user-programmable features, have made the detailed analysis of composites quite accessible to the designer.

In a laminate, stresses in the individual layers with different orientations are generally different. Therefore, some of the layers probably reach their limiting stresses before the other remaining layers and they fail first. This is generally

E. Kormaníková (✉)
Department of Structural Mechanics, Faculty of Civil Engineering, Technical University of Košice, Košice, Slovakia
e-mail: eva.kormanikova@tuke.sk

J. Murín et al. (eds.), *Computational Modelling and Advanced Simulations*, Computational Methods in Applied Sciences 24, DOI 10.1007/978-94-007-0317-9_10,

referred to as first-ply failure [2, 3]. A fibre-reinforced laminate may or may not be able to carry loads except during the failure initiation, depending on the nature of the first failure. There are two factors contributing to this behaviour. First, the constituent materials are brittle in nature and do not tolerate local failures. The second factor is a large difference in stiffness and strength between the two principal material directions in a layer.

This paper pertains to the micromechanical analysis of composite materials with short fibres. The micromechanical analysis takes into account the nature of the constituents and their distribution. It can be used to evaluate the overall properties of composites. Failure criteria are used to calculate a failure index (FI) from the computed stresses and user-supplied material strengths. The micromechanical analysis has been carried out using the authors' own programme in MATLAB. Finite Element Method (FEM) is used as a tool to predict the laminate strength based on the classical lamination theory. Numerical simulation has been prepared by using a commercially available ANSYS code.

10.2 Micromechanics of Composite Materials with Short Fibres

The classical laminate theory is the most commonly used theory for analysing composites with randomly-oriented short fibres [4–6]. The laminates with the orientation of angles [0/±45/90] and [0/±60] are particularly very suitable for practical applications. In order to predict the strength of this type of composite, it is best to use the maximum strain criterion and then the strength of a composite with randomly-oriented short fibres can be determined by using the properties of unidirectionally reinforced composites with short fibres.

The longitudinal and transverse moduli of these composites can be expressed with the help of so-called Halphin-Tsai equations [7]

$$E_1 = E^{(m)} \frac{1 + \frac{l}{d}\zeta_E \eta_L \xi}{1 - \eta_L \xi} \qquad E_2 = E^{(m)} \frac{1 + \zeta_E \eta_T \xi}{1 - \eta_T \xi}$$

$$G_{12} = G^{(m)} \frac{1 + \zeta_E \eta_G \xi}{1 - \eta_G \xi} \qquad \nu_{12} = \nu^{(m)} \frac{1 + \zeta_E \eta_\nu \xi}{1 - \eta_\nu \xi} \tag{10.1}$$

where
$$\eta_L = \frac{\dfrac{E^{(f)}}{E^{(m)}} - 1}{\dfrac{E^{(f)}}{E^{(m)}} + \zeta_E \dfrac{l}{d}} \qquad \eta_T = \frac{\dfrac{E^{(f)}}{E^{(m)}} - 1}{\dfrac{E^{(f)}}{E^{(m)}} + \zeta_E}$$

$$\eta_G = \frac{\dfrac{G^{(f)}}{G^{(m)}} - 1}{\dfrac{G^{(f)}}{G^{(m)}} + \zeta_E} \qquad \eta_\nu = \frac{\dfrac{\nu^{(f)}}{\nu^{(m)}} - 1}{\dfrac{\nu^{(f)}}{\nu^{(m)}} + \zeta_E} \tag{10.2}$$

while

the superscripts $^{(m)}$ and $^{(f)}$ refer to matrix and fibre, respectively, ζ_E is a reinforcing factor.

It depends on the geometry of the fibres in a composite, the packing arrangement of the fibres and its loading conditions. It ranges in value between 1 and 2. However, only when a reliable experimental value of the E_2 is available for a composite, the factor ζ_E can be derived and then applied to predict the E_2 for a range of fibre-volume ratios of the same composite.

A random microstructure results in transversely isotropic properties on a mezzo-scale. A simple alternative is to assume that the random microstructure is well-approximated by a periodic microstructure model (Fig. 10.1). Periodic microstructure mechanics exploits the geometric periodicity of the system in order to simplify mechanical field variables, such as stress, strain, and stiffness. In general, there is a correlation between all of these terms and the position inside the representative volume element (RVE) [8].

A simpler alternative is to assume that the random microstructure is well approximated by the hexagonal microstructure displayed in Fig. 10.2.

The elastic properties of a homogenized material can be computed by [9], i.e. the longitudinal and transversal Young's moduli E_1 and E_2, the longitudinal and transversal Poisson's ratios ν_{12} and ν_{21} and the longitudinal shear modulus G_{12}, as follows

$$\begin{aligned} E_1 &= C_{11} - 2C_{12}^2/\left(C_{22} + C_{23}\right) \\ \nu_{12} &= C_{12}/\left(C_{22} + C_{23}\right) \\ E_2 &= \left(C_{11}\left(C_{22} + C_{23}\right) - 2C_{12}^2\right)\left(C_{22} - C_{23}\right)/\left(C_{11}C_{22} - C_{12}^2\right) \\ G_{12} &= C_{66} \\ \nu_{21} &= \nu_{12}\frac{E_2}{E_1} \end{aligned} \tag{10.3}$$

In order to evaluate the elastic matrix $\boldsymbol{C}$ of a composite, the RVE (Fig. 10.3) is subjected to an average strain.

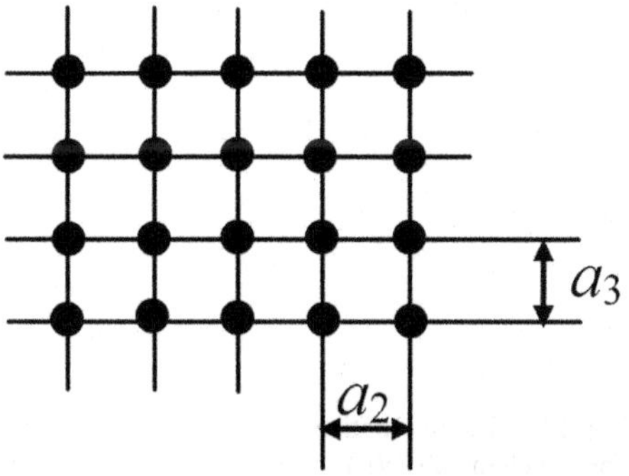

Fig. 10.1 A periodic microstructure model

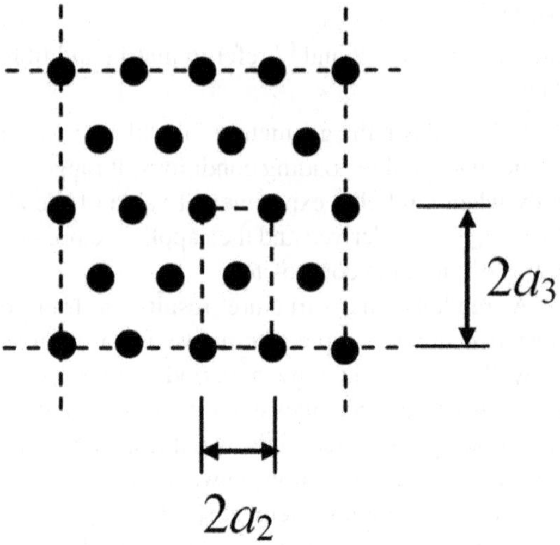

Fig. 10.2 A hexagonal microstructure model

Then the volume average of the strain in the RVE equals to the applied strain

$$\bar{\varepsilon}_{ij} = \frac{1}{V} \int_V \varepsilon_{ij} dV \tag{10.4}$$

The components of the tensor $\boldsymbol{C}$ are determined by solving three elastic models of the RVE with its parameters (a_1, a_2, a_3) subjected to the boundary conditions (BC). The unit strain applied to the boundary results in a complex state of stress in the RVE. Subsequently, the volume average of stress in the RVE equals the required components of the elastic matrix as follows

$$C_{ij} = \bar{\sigma}_i = \frac{1}{V} \int_V \sigma_i dV \tag{10.5}$$

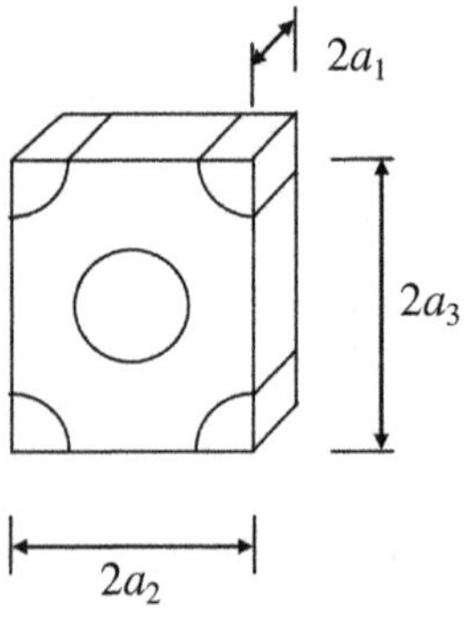

Fig. 10.3 Representative volume element (RVE)

The coefficients in $\boldsymbol{C}$ are specified by setting a different problem for each column of $\boldsymbol{C}$. Then the components C_{ij} are determined in three steps:

1. For the components C_{i1} $(i = 1,\ 2,\ 3)$, the strain is applied to stretch the RVE in the fibre direction x_1

$$\varepsilon_1^0 = 1$$
$$\varepsilon_2^0 = \varepsilon_3^0 = \gamma_4^0 = \gamma_5^0 = \gamma_6^0 = 0 \tag{10.6}$$

   and the applied BC are

$$\begin{array}{l} u_1\,(a_1, x_2, x_3) = a_1,\ u_2\,(x_1, a_2, x_3) = 0,\ u_3\,(x_1, x_2, a_3) = 0 \\ u_1\,(0, x_2, x_3) = 0, \quad u_2\,(x_1, 0, x_3) = 0, \quad u_3\,(x_1, x_2, 0) = 0 \end{array} \tag{10.7}$$

   The coefficients C_{i1} are specified by using the expression

$$C_{i1} = \bar{\sigma}_i \tag{10.8}$$

2. For the components C_{i2} $(i = 1,\ 2,\ 3)$, the strain is applied to stretch the RVE in the direction x_2.

$$\varepsilon_2^0 = 1$$
$$\varepsilon_1^0 = \varepsilon_3^0 = \gamma_4^0 = \gamma_5^0 = \gamma_6^0 = 0 \tag{10.9}$$

   and the applied BC are

$$\begin{array}{l} u_1\,(a_1, x_2, x_3) = 0,\ u_2\,(x_1, a_2, x_3) = a_2,\ u_3\,(x_1, x_2, a_3) = 0 \\ u_1\,(0, x_2, x_3) = 0, \quad u_2\,(x_1, 0, x_3) = 0, \quad u_3\,(x_1, x_2, 0) = 0 \end{array} \tag{10.10}$$

   Again, the coefficients C_{i2} are specified by using

$$C_{i2} = \bar{\sigma}_i \tag{10.11}$$

3. For the components C_{i3} $(i = 1,\ 2,\ 3)$, the following strain is applied to stretch the RVE in the direction x_3.

$$\varepsilon_3^0 = 1$$
$$\varepsilon_1^0 = \varepsilon_2^0 = \gamma_4^0 = \gamma_5^0 = \gamma_6^0 = 0 \tag{10.12}$$

   and the applied BC are

$$\begin{array}{l} u_1\,(a_1, x_2, x_3) = 0,\ u_2\,(x_1, a_2, x_3) = 0,\ u_3\,(x_1, x_2, a_3) = a_3 \\ u_1\,(0, x_2, x_3) = 0, \quad u_2\,(x_1, 0, x_3) = 0, \quad u_3\,(x_1, x_2, 0) = 0 \end{array} \tag{10.13}$$

Equally, the coefficients C_{i3} are specified by using

$$C_{i3} = \bar{\sigma}_i \tag{10.14}$$

The coefficient C_{44} can be determined as follows

$$C_{44} = \frac{1}{2}\left(C_{22} - C_{23}\right) \tag{10.15}$$

The fibre-volume fraction is expressed as

$$v_f = \frac{\pi}{\sqrt{3}}\left(\frac{d^2}{2a^2}\right) \tag{10.16}$$

10.3 Classical Lamination Theory

Similarly to the Euler-Bernoulli beam theory and the plate theory, the classical lamination theory (CLT) can be applied to thin laminates only (the span a and $b > 10 \times$ thickness t) with small displacements w in the transverse direction ($w << t$). In CLT displacements are considered to be continuous throughout the total thickness of the laminate and the constitutive equations are linear.

The strains relations in the condensed form can be noted as [6]

$$\boldsymbol{\varepsilon}(\mathbf{x}) = \bar{\boldsymbol{\varepsilon}}(x, y) + z\boldsymbol{\kappa}(x, y) \tag{10.17}$$

where $\bar{\boldsymbol{\varepsilon}}$ is the vector of mid-plane strains and $\boldsymbol{\kappa}$ is the vector of curvatures that are constant throughout the thickness.

In modelling a laminate it is assumed that each individual layer of the laminate behaves as a linear elastic material. All layers are bonded together with a perfect bond and each lamina of a composite material behaves macroscopically as a homogeneous orthotropic material (Fig. 10.4).

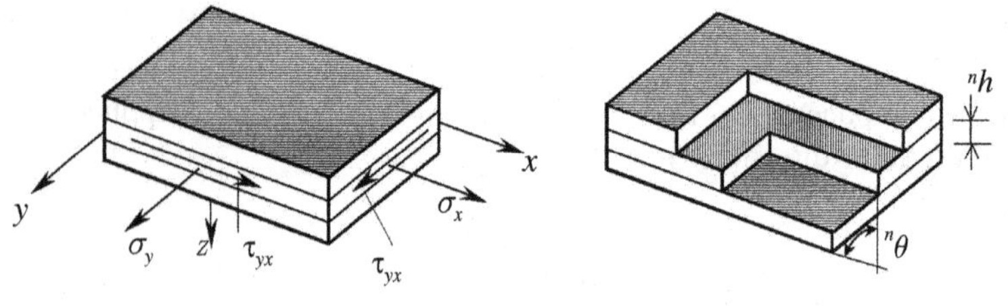

Fig. 10.4 Global and local coordinate system of a laminate

The constitutive equations can be written in the condensed hyper-matrix form

$$\begin{pmatrix} \boldsymbol{N} \\ \boldsymbol{M} \end{pmatrix} = \begin{pmatrix} \boldsymbol{A} & \boldsymbol{B} \\ \boldsymbol{B} & \boldsymbol{D} \end{pmatrix} \begin{pmatrix} \bar{\boldsymbol{\varepsilon}} \\ \boldsymbol{\kappa} \end{pmatrix} \tag{10.18}$$

with

$$\{A,B,D\} = \int_{-h/2}^{+h/2} E(z)\left\{1, z, z^2\right\} dz \tag{10.19}$$

The elasticity matrix can be expressed as

$$E = \sum_{n=1}^{N} \frac{{}^{n}h}{h} {}^{n}E$$

$${}^{n}E = \bar{T}^{T}\left({}^{n}\theta\right)^{(n)} E_L \bar{T}\left({}^{n}\theta\right), \quad \bar{T}(\theta) = \left(T^{T}(\theta)\right)^{-1} \tag{10.20}$$

where T is the transformation matrix [3].

10.4 Failure Criteria for Fibre-Reinforced Orthotropic Layers

The strength of unidirectional fibres composite materials depends on the direction of fibres on a macroscopic scale. Composite layers are much stronger in the fibres direction than in the direction perpendicular to their fibres. For loads that are primarily parallel to the fibres, either in tension or compression, the material strength is generally determined by the failure of the fibres. For loads transverse to the fibres, failure is controlled by the failure of the much weaker matrix material.

The strength of a composite layer in any other direction is based on various failure criteria [3, 7].

10.4.1 Maximum Stress and Maximum Strain Criteria

The basic assumption in predicting the failure of fibre-reinforced layers using the maximum stress and maximum strain criteria is the same as for any other isotropic material. Failure is assumed when the maximum stress along the fibre or transverse to the fibre directions exceeds the strengths in tension or compression.

The failure surface is defined as

$$\sigma_1 < X_t, \quad \sigma_2 < Y_t \quad \text{for} \quad \sigma_1, \ \sigma_2 > 0$$

$$\sigma_1 > -X_c, \quad \sigma_2 > -Y_c \quad \text{for} \quad \sigma_1, \ \sigma_2 < 0 \tag{10.21}$$
$$|\tau_{12}| < S$$

where

X and Y represent the ultimate strengths along and transverse to the fibre directions respectively,

indexes $_t$ and $_c$ refer to tension and compression, respectively,
S is the ultimate in-plane shear strength of a specimen under pure shear loading.

Similarly, the maximum strain criterion states that failure occurs when one of the following inequalities is violated

$$\begin{gathered}\varepsilon_1 < \frac{X_t}{E_1}, \quad \varepsilon_2 < \frac{Y_t}{E_2}, \quad \varepsilon_1, \varepsilon_2 > 0 \\ \varepsilon_1 > -\frac{X_c}{E_1}, \varepsilon_2 > -\frac{Y_c}{E_2}, \quad \varepsilon_1, \varepsilon_2 < 0 \\ |\gamma_{12}| < \frac{S}{G_{12}}\end{gathered} \tag{10.22}$$

10.4.2 Tsai-Wu Criterion

A more general form of the failure criterion for orthotropic materials under plane stress is expressed as [7]

$$F_{01}\sigma_1 + F_{11}\sigma_1^2 + 2F_{12}\sigma_1\sigma_2 + F_{02}\sigma_2 + F_{22}\sigma_2^2 + F_{44}\tau_{12}^2 < 1 \tag{10.23}$$

where

$$F_{01} = \frac{1}{X_t} - \frac{1}{X_c}, \quad F_{11} = \frac{1}{X_t X_c}, \quad F_{02} = \frac{1}{Y_t} - \frac{1}{Y_c}, \quad F_{22} = \frac{1}{Y_t Y_c}$$
$$F_{12} = -\frac{1}{2}\frac{1}{\sqrt{X_t X_c Y_t Y_c}}, F_{44} = \frac{1}{S^2}$$

The failure criterion for an orthotropic material under a strain is expressed as

$$G_{01}\varepsilon_1 + G_{11}\varepsilon_1^2 + G_{12}\varepsilon_1\varepsilon_2 + G_{02}\varepsilon_2 + G_{22}\varepsilon_2^2 + G_{44}\gamma_{12}^2 < 1 \tag{10.24}$$

where

$$\begin{gathered}G_{01} = F_{01}E_{11} + F_{02}E_{12} \quad G_{02} = F_{02}E_{22} + F_{01}E_{12} \\ G_{11} = F_{11}E_{11}^2 + F_{22}E_{12}^2 + F_{12}E_{11}E_{12} \\ G_{22} = F_{22}E_{22}^2 + F_{11}E_{12}^2 + F_{12}E_{22}E_{12} \\ G_{12} = 2E_{12}(F_{11}E_{11} + F_{22}E_{22}) + 2F_1(E_{12}^2 + E_{11}E_{22}) \\ G_{44} = F_{44}E_{44}^2\end{gathered}$$

When $F_{12} = \frac{-1}{2X_t^2}$, the Tsai-Wu criterion is reduced to the Tsai-Hill criterion, and when $F_{12} = \frac{-1}{2X_t X_c}$ the Tsai-Wu criterion is reduced to the Hoffman criterion [3, 7].

These failure criteria are used to calculate a failure index (FI) from the computed stresses and user-supplied material strengths. The failure index as a response of quantity is used for several FEA packages and it is defined as

$$I_F = \frac{stress}{strength} \tag{10.25}$$

Failure criteria predict the first occurrence of failure in one of the laminate layers. A value less than 1 denotes no failure and failure is predicted when $I_F \geq 1$. The strength ratio is the inverse of the failure index.

It is important to distinguish between the fibre failure (FF) and the inter-fibre failure (IFF). In the case of shear plane stress, the IFF criteria discriminates three different modes [3]. The IFF Mode *A* is when perpendicular transversal cracks appear in the lamina under transverse tensile stress with or without in-plane shear stress. The IFF Mode *B* denotes perpendicular transversal cracks, but in this case they appear under in-plane shear stress with small transverse compression stress. The IFF Mode *C* indicates the start of oblique cracks when the material is under significant transversal compression.

The FF and the three IFF modes yield separate failure indices. The failure index for FF is defined as

$$I_{FF} = \begin{cases} \sigma_1/X_t & if \quad \sigma_1 > 0 \\ -\sigma_1/X_c & if \quad \sigma_1 < 0 \end{cases} \tag{10.26}$$

For IFF with positive transverse stress, Mode A is active. The failure index in this case is defined as

$$I_{IFF,A} = \sqrt{\left(\frac{\tau_{12}}{S}\right)^2 + \left(1 - p_{6t}\frac{Y_t}{S}\right)^2 \left(\frac{\sigma_2}{Y_t}\right)^2} + p_{6t}\frac{\sigma_2}{S} \quad if \quad \sigma_2 \geq 0 \tag{10.27}$$

where $p_{6t} = 0.3$.

Under negative transverse stress, either Mode *B* or Mode *C* is active, depending on the relationship between in-plane shear stress and transversal shear stress. The failure indices are defined as

$$I_{IFF,B} = \frac{1}{S}\left[\sqrt{\tau_{12}^2 + (p_{6c}\sigma_2)^2} + p_{6c}\sigma_2\right] \quad if \quad \begin{cases} \sigma_2 < 0 \\ \left|\frac{\sigma_2}{\tau_{12}}\right| \leq \frac{F_{2A}}{F_{6A}} \end{cases} \tag{10.28}$$

$$I_{IFF,C} = -\frac{Y_c}{\sigma_2}\left[\sqrt{\left(\frac{\tau_{12}}{2\,(1+p_{2c})\,S}\right)^2 + \left(\frac{\sigma_2}{Y_c}\right)^2}\right] \quad if \quad \begin{cases} \sigma_2 < 0 \\ \left|\frac{\sigma_2}{\tau_{12}}\right| \geq \frac{F_{2A}}{F_{6A}} \end{cases} \tag{10.29}$$

where $p_{6c} = 0.2$.

The limit between Mode *B* and Mode *C* is defined by the relation F_{2A}/F_{6A}, where

$$F_{2A} = \frac{S}{2p_{6c}}\left[\sqrt{1 + 2p_{6c}\frac{Y_c}{S}} - 1\right] \tag{10.30}$$

$$F_{6A} = S\sqrt{1 + 2p_{2c}} \quad p_{2c} = p_{6c}\frac{F_{2A}}{S} \tag{10.31}$$

10.5 Finite Element Formulation

Among the computational techniques implemented for layered-plate and shell analyses, a predominant role has been played by Finite Element Method (FEM). The extensively used FEM formulation is based on the variation of the total potential energy. It leads to

$$\boldsymbol{K}\,\boldsymbol{v} = \boldsymbol{f}_p + \boldsymbol{f}_q \tag{10.32}$$

where, $\boldsymbol{K}$ is a stiffness matrix, $\boldsymbol{f}_p$ and $\boldsymbol{f}_q$ are the vectors of volume and surface forces.

The through-thickness numerical integration is carried out by modifying the variable ζ to ${}^n\zeta$ in the nth layer so that ${}^n\zeta$ varies from –1 to 1 in that layer

$$d\zeta = \frac{{}^n h}{h} d^n\zeta \tag{10.33}$$

Then the resultant relation for the element stiffness matrix is

$$\boldsymbol{K}_E = \sum_{n=1}^{N} \int_{-1}^{1}\int_{-1}^{1}\int_{-1}^{1} \boldsymbol{B}^T \boldsymbol{E} \boldsymbol{B}\, |\boldsymbol{J}| \frac{{}^n h}{h} d\xi\, d\eta\, d^n\zeta \tag{10.34}$$

It should be noted that the elasticity matrix $\boldsymbol{E}$ is obtained through the suitable transformations in two stages, firstly from the principal material directions to the element local directions and secondly to the global directions.

10.6 Numerical Examples

10.6.1 Example 1

In this example, a composite with randomly oriented fibres is assumed with its material characteristics for the fibres: $E_f = 210$ GPa, $\nu_f = 0.3$ and for the matrix $E_m = 31$ GPa, $\nu_m = 0.15$. The geometrical characteristics of the fibres are $L = 6$ cm, $d = 0.75$ mm.

The material characteristics of the composite material given for a variable amount of fibres are shown in Tables 10.1, 10.2 and 10.3, including their Young's Moduli E and Poisson's Ratios. For the given amount of fibres of 30 and 60 kg/m^3, there

Table 10.1 Material characteristics calculated according to Halphin Tsai

	30 kg/m^3	60 kg/m^3
E Young's Moduli [GPa]	31.644	31.795
ν Poisson's Ratio	0.151	0.151

Table 10.2 Material characteristics calculated using the periodic microstructure model

	30 kg/m^3	60 kg/m^3
E Young's Moduli [GPa]	31.239	31.426
ν Poisson's Ratio	0.150	0.151

Table 10.3 Material characteristics calculated using the classical laminate theory

	30 kg/m^3	60 kg/m^3
E Young's Moduli [GPa]	31.470	31.960
ν Poisson's Ratio	0.1470	0.1480

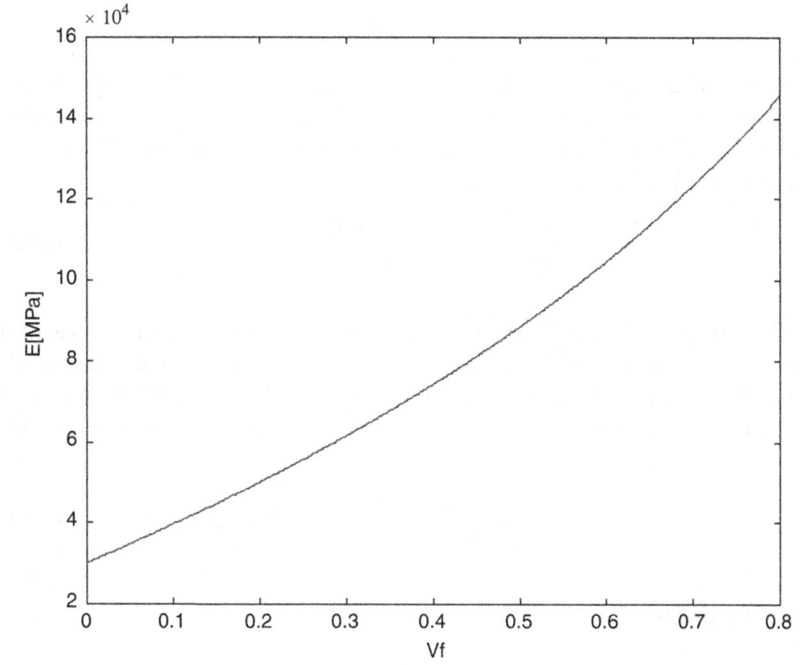

Fig. 10.5 Correlation between the modulus of elasticity and the fibre volume fraction

are only minor differences in the material characteristics regardless of the method used, whether the Halphin Tsai method, the periodic microstructure model or the classical laminate theory. The correlation between the modulus of elasticity and the fibre volume fraction is given in Fig. 10.5.

The moduli of elasticity – the first-ply E_{1FAIL}, the second-ply E_{2FAIL} and the ultimate failure E_{3FAIL} in a fictitious laminate $[0/45/{-45}/90]_S$ – are used instead of these in a composite with short fibres. Their values calculated using the classical laminate theory are given in Table 10.4.

Table 10.4 Moduli of elasticity of a composite calculated using the classical laminate theory

	30 kg/m^3	60 kg/m^3
$E_{1\mathrm{FAIL}}$ [GPa]	31.47	31.96
$E_{2\mathrm{FAIL}}$ [GPa]	23.66	24.10
$E_{3\mathrm{FAIL}}$ [GPa]	15.75	16.02

Table 10.5 Flexural tensile strength and characteristic flexural tensile strength for the fibres

	30 kg/m^3	30 kg/m^{3}RC80/60-BN	60 kg/m^3	60 kg/m^{3}ZC30/0.50
$f_{\mathrm{fctm},eq}$ [MPa]	2.448	2.26	3.519	3.40
$f_{\mathrm{fctk},eq}$ [MPa]	1.714	1.83	2.463	2.40

The main characteristic essentially changed is the strength of a composite with short fibres. In the absence of more accurate information, the average and characteristic value of an equivalent flexural tensile strength for steel wire fibres can be calculated as follows [10]

$$R_{e_{m,150}} = \frac{180 W_f \lambda_f d_f^{\,1/3}}{180C + W_f \lambda_f d_f^{\,1/3}} \tag{10.35}$$

with $C = 20$ for hooked-end steel fibres under the trade name of Dramix, where W_f is the fibre content (in kg/m^3), d_f is the diameter of steel fibres, and λ_f is the ratio between the length and the diameter of steel fibres. The flexural tensile strength $f_{\mathrm{fctm},eq}$ and characteristic flexural tensile strength $f_{\mathrm{fctk},eq}$ are given as (Table 10.5)

$$f_{\mathrm{fctm},eq} = \frac{R_{e_{m,150}} f_{\mathrm{fctm},fl}}{100}, \quad f_{\mathrm{fctk},eq} = 0.7 f_{\mathrm{fctm},eq} \tag{10.36}$$

10.6.2 Example 2

In this example, a simply supported square plate was analysed with a side length of 3 m, a thickness of 0.2 m, and the flexural uniform loading $q = 50$ kPa. The material is a composite with randomly oriented fibres and the material characteristics of the fibres are $E_f = 210$ GPa, $\nu_f = 0.3$ and of the matrix $E_m = 31$ GPa, $\nu_m = 0.15$. The amount of Dramix fibres RC-80/60-BN is 30 kg/m^3.

A simply supported square steel-fibre-reinforced concrete plate with a fictitious laminate $[0/45/\text{-}45/90]_S$ instead of a composite with short fibres is used as the tested example. The maximum stress criterion is used for the calculation of the failure index using the FEM ANSYS code. There is used the quadrilateral finite element SHELL91. In Figs. 10.6 and 10.7, the maximum failure indices for a reinforced plate and an unreinforced plate are shown. The maximum failure index appears in the middle of the plate. The equivalent flexural tensile strength for RC-80/60-BN fibres is $f_{\mathrm{fctk},eq} = 1.83$ MPa and the characteristic flexural tensile strength is $f_{\mathrm{fctk},fl} = 2.26$ MPa.

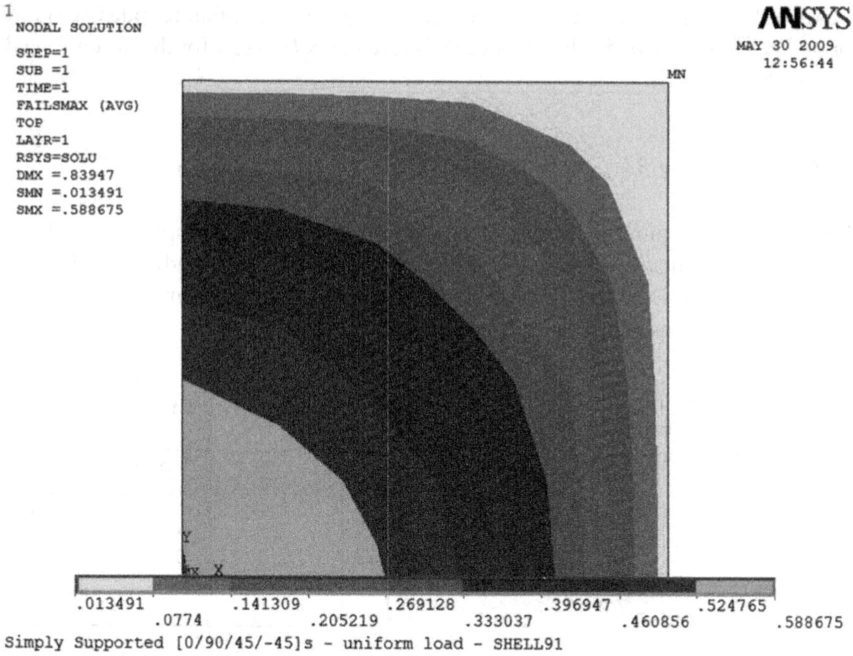

Fig. 10.6 The maximum failure index in a composite plate

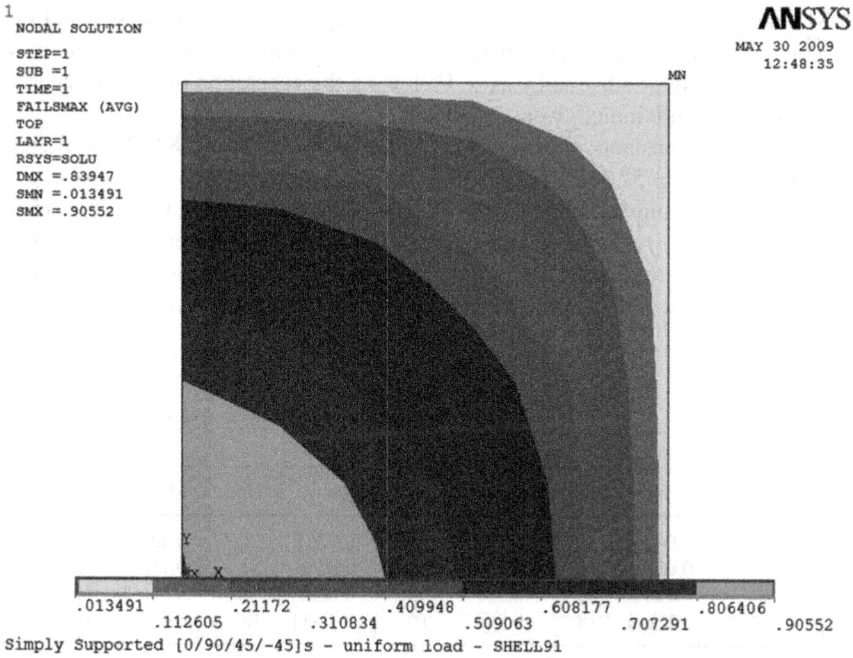

Fig. 10.7 The failure index in an unreinforced plate

According to the maximum stress criterion, the maximum failure index $I_{F,\mathrm{MAX}\,S}$ for 30 kg/m^3 is 0.588675. The maximum failure index $I_{F,\mathrm{MAX}\,S}$ for the unreinforced plate is 0.90552.

10.6.3 Example 3

Similarly to Example 2, failure indices were determined in Example 3 as well. In this case the laminate consists of four equally-oriented layers made from the uniform material AS4D/9110 [11] with the uniform thickness. The dimensions and the boundary conditions are described in Fig 10.8.

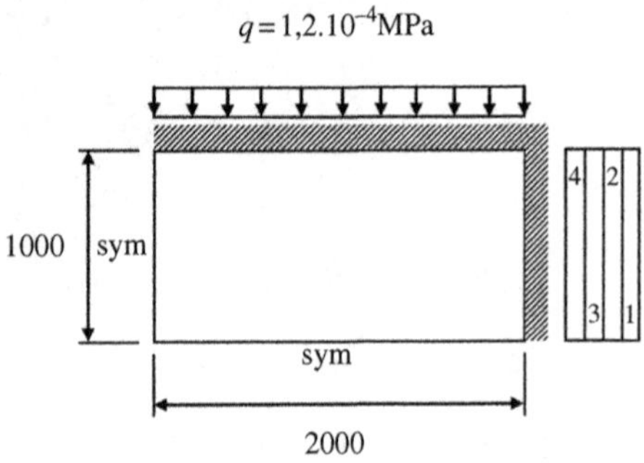

Fig. 10.8 Loads and the geometry of the laminate and the numbering of its layers

The model is obtained by using the ANSYS input command sequence and the element SHELL99. These elements possess a local coordinate system that is axially rotated by 30° from both x and y axes. Following the generation of model material strengths, the failure indices were calculated.

The strength constants of AS4D/9110 material are: $F_{1t} = 1,830$ MPa, $F_{1c} = -1,096$ MPa, $F_{2t} = 57$ MPa, $F_{2c} = -228$ MPa, $c_6 = -1$, $c_5 = -1$, $c_4 = -1$.

The results are summarized in Table 10.6, where $I_{F,\mathrm{MAX}\,S}$ is the maximum stress failure index and $I_{F,TW}$ is the Tsai-Wu failure index. As can be seen from the given results, the maximum FIs are in Layer 1. The distribution of failure indices is given in Fig. 10.9. In the following two layers the maximum FIs occurred in the middle

Table 10.6 Failure indices $I_{F,\mathrm{MAX}S}$, $I_{F,TW}$

$I_{F,\mathrm{MAX}\,S}$					$I_{F,TW}$			
No. lamina	Max.	Node	Min.	Node	Max	Node	Min	Node
1.	0.0424	1	0.22e-3	38	0.0251	1	0.23e-2	40
2.	0.0283	1	0.15e-3	38	0.0166	1	0.15e-2	40
3.	0.0141	1	0.75e-4	38	0.0083	1	0.76e-3	40
4.	0.11e-5	18	0.88e-8	16	0.1e-11	18	0.4e-16	16

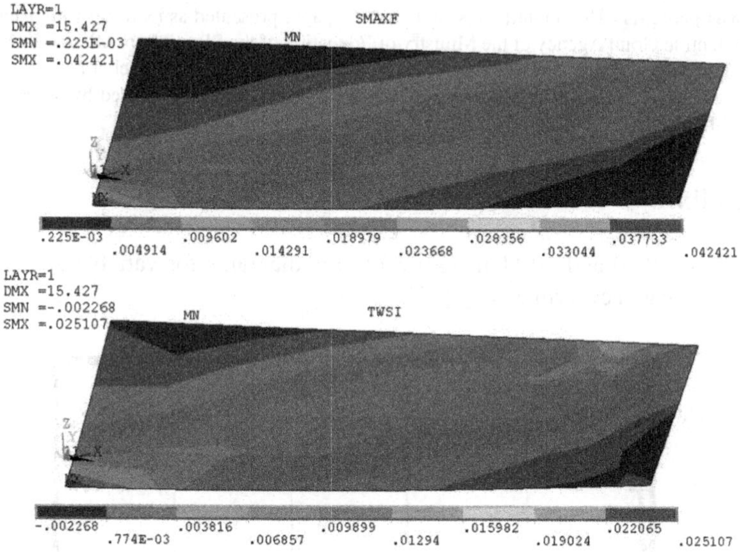

Fig. 10.9 The distribution of failure indices $I_{F,\,\mathrm{MAX\,S}}$, $I_{F,TW}$ in the first lamina

of the laminate. Moreover, the distribution of the FIs in Layer 4 is different, the FI values are the smallest and their locations changed. The minimum values occurred in the corner of the laminate where it is fixed.

10.7 Conclusion

Some calculation methods for the calculation of the modulus of elasticity and the strength of randomly reinforced composite materials are derived in the first example. There are no significant differences in the modulus of elasticity when using the Halphin Tsai method, the periodic microstructure model or the classical laminate theory. The main characteristic that changed essentially is the strength of a composite with short fibres. This characteristic is determined theoretically or by using experimental results. The deformation energy that can be absorbed is a measure of the toughness of the material.

Safety is one of the most important aspects in designing structures. The deformation behaviour under loading is of primary importance too. In the paper presented the maximum failure index was calculated using the maximum stress criterion. A simply supported square steel-fibre-reinforced concrete plate with a fictitious laminate $[0/45/\text{-}45/90]_S$ instead of a composite with short fibres was used as the second tested example. A laminate plate made from AS4D/9110 material was used as the third tested example and the maximum stress failure index $I_{F,\,\mathrm{MAX}\,S}$ and the Tsai-Wu failure index $I_{F,TW}$ were computed.

Acknowledgements The scientific research and the paper presented as its result were supported by the Scientific Grant Agency of the Ministry of Education of the Slovak Republic and the Slovak Academy of Science under Project 1/4202/07, by the Slovak Science and Technology Assistance Agency registered under No. APVV-0169-07 and NFP 26220220051 supported by the European structural funds.

Appendix

In the Figs. 10.10 and 10.11 there are $F(\sigma)$-w diagrams for variable amounts of fibres investigated experimentally [12].

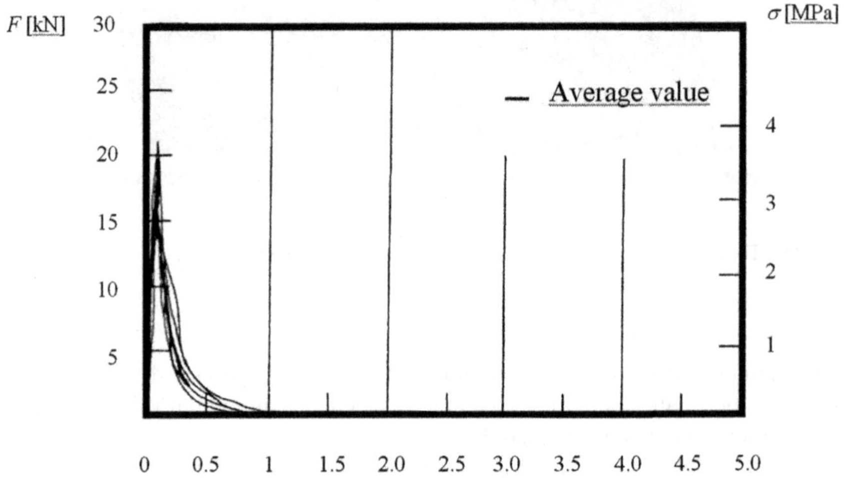

Fig. 10.10 Diagram $F(\sigma)$ - w for an unreinforced composite

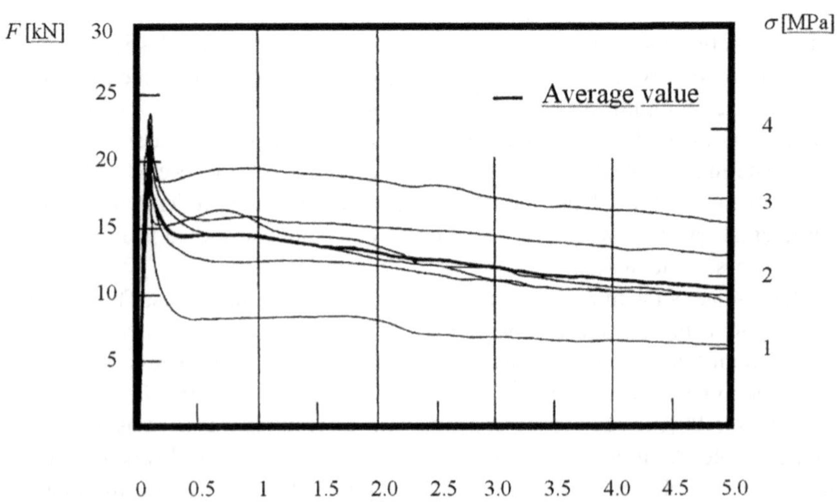

Fig. 10.11 Diagram $F(\sigma)$ - w for a composite with the fibre content of 30 kg/m^3

References

1. Kompiš, V., Štiavnický, M., Žmindák, M., Murčinková, Z.: Trefftz radial basis functions. Proceedings of Leuven Symposium on Applied Mechanics in Engineering (LSAME.08), Part I, Proceeding of Trefftz.08, 5th International Workshop on Trefftz Methods, Leuven (2008)
2. Mallick, P.K.: Fiber-Reinforced Composites: Materials, Manufacturing and Design. CRC Press, Taylor & Francis, Boca Raton, FL (2007)
3. Altenbach, H., Altenbach, J., Kissing. W.: Structural Analysis of Laminate and Sandwich Beams and Plates. Lubel- skie Towarzystwo Naukowe, Lublin (2001)
4. Agarwal, B.D., Broutman, L.J.: Fibre Composites (in Czech). SNTL, DT, Prague (1987)
5. Carrera, E.: Theories and finite elements for multilayered, anisotropic, composite plates and shells. Arch. Comput. Methods Eng. **9**(2) 87–140 (2002)
6. Carrera, E.: Theories and finite elements for multi-layered plates and shells. Arch. Comput. Methods. Eng. **10**(3), 215–296 (2003)
7. Laš, V.: Mechanics of Composite Materials (in Czech). West Bohemia University, Pilsen (2008)
8. Qu, J., Cherekaoiu M.: Fundamentals of Micromechanics of Solids. Wiley, Hoboken (2006)
9. Luciano, R., Barbero, E.J.: Formulas for the stiffness of composites with periodic microstructure. Int. J. Solids Struct. **31**(21), 2933–2944 (1995)
10. Vandewalle, L., Var Nieuwerburg, D., Var Gysel, A., Vyncke, J., Deforche, E.: DRAMIX Guideline: Design of Concrete Structures, No. 4 (1995)
11. Barbero, E.J.: Finite Element Analysis of Composite Materials. CRC Press, Taylor & Francis Group, Boca Raton, FL (2008)
12. Falkner, H., Teutsch, M., Klinkert, H.: Power Class of Iron Fibre-Concrete. TU Braunschweig, Booklet 143 (1999)

Chapter 11
A Direct Boundary Element Formulation for the First Plane Problem in the Dual System of Micropolar Elasticity

Gy. Szeidl and J. Dudra

Abstract This paper studies the plane strain problem of micropolar elastostatics assuming that the governing equations are given in terms of stress functions of order one. We clarify the conditions of single valuedness and construct the fundamental solution for the dual basic equations. We then establish the integral equations of the direct method. Numerical examples illustrate the applicability of these integral equations.

Keywords Micropolar elasticity · The first plane problem · Dual formulation · Boundary element method

11.1 Introduction

In their book [1], the Cosserat brothers assume that the motion of a material particle is described by a displacement field and an independent rotation field. According to this assumption a material particle of the body behaves as if it were a very small rigid body. Under this assumption a correct description of the interaction on the inner surfaces of a solid body requires the existence not only of force stresses but of couple stresses as well. The corresponding stress and couple stress tensors are, however, not symmetric. The Cosserat brothers did not deal with the issue of the constitutive equations.

Eringen [2] who supplemented the theory with the constitutive equations. Books The theory was reinvented among others by Eringen and Suhibi [3, 4] and [5] and [6] by Nowaczky provide an excellent overview on the linear theory of micropolar elasticity.

According to the famous Tonti scheme [7, 8] the variables in the equations of mathematical physics are categorized as fundamental variables, intermediate variables of the first and second kind and source variables. Problems of mathematical

Gy. Szeidl (✉)
Department of Mechanics, University of Miskolc, 3515 Miskolc, Hungary
e-mail: gyorgy.szeidl@uni-miskolc.hu

J. Murín et al. (eds.), *Computational Modelling and Advanced Simulations*,
Computational Methods in Applied Sciences 24, DOI 10.1007/978-94-007-0317-9_11,

physics can be set up in a primal formulation and in a dual one. In [a primal]{a dual} formulation the set of field equations involve the defining equations, which relate the fundamental variables to the intermediate variables of the first kind, the constitutive equations, which connect the intermediate variables of the second kind to those of the first kind, and the balance equations, which relate the intermediate variables of the second kind to the source variables. The intermediate variables of the second kind in the [primal]{dual} formulation (system) coincide with the intermediate variables of the first kind in the [dual]{primal} formulation (system). In addition the [primal]{dual} defining equations identically satisfy the [dual]{primal} balance equations.

As regards the primal formulation in micropolar elasticity the displacement and rotation vectors (together are referred to as displacements) are the fundamental variables, the asymmetric strain tensor and the curvature twist tensor (together strains) are the intermediate variables of the first kind while the asymmetric force stresses and couple stresses (together stresses) are the intermediate variables of the second kind. The body forces and couples constitute the source variables [9, 10]. Problems in the primal formulation are governed by the kinematic equations (defining equations), which give the asymmetric strain tensor and curvature twist tensor (together the strains) in terms of the displacement and rotation vectors (together the displacements), Hooke's law which connects the force and couple stress tensors to the two strain tensors, and the equilibrium equations (the balance equations) which relate the force and couple stresses to the body forces and couples (the source variables).

General solutions to the primal equilibrium equations in terms of stress functions of order one have been established independently of each other by Schaefer [11, 12] and Carlson [13]. It is worth mentioning that their solutions are equivalent to each other.

For the dual formulation of micropolar elasticity stress functions of order one are the fundamental variables, the force and couple stresses are the intermediate variables of the first kind, the asymmetric strain tensor and the curvature twist tensor are the intermediate variables of the second kind and the tensors of incompatibility (they are equal to zero in real problems) constitute the source variables. Problems in the dual formulation are governed by the representation of stresses in terms of stress functions of order one (dual kinematic or defining equations), the inverse form of Hooke's law (dual constitutive equations), which relates the dual intermediate variables of the second kind (the strains) to that of the first kind (to the stresses), and the compatibility equations (dual balance equations).

The first plane problem of micropolar elasticity was formulated in the primal system by [5, 14]. Assuming isotropy, Iesan derived the associated integral equations [15]. Schiavone made these results more accurate and precise for outer regions, and also investigated mixed boundary value problems in [16]. Fuang and Liang [17] developed a boundary element analysis of stress concentration in a micropolar plate to determine the stress concentration factors for some cases.

Sladek and Sladek [18–20] study the boundary element formulations in microploar elasticity provided that thermal effects are included. Numerical solutions are also presented.

A dual formulation for the first plane problem was published in paper [21]. To the best of our knowledge, the direct boundary element formulation for the first plane problem in the dual system has not yet been developed. In unpublished work, Szeidl and Ivan [22] dealt with the formulation of indirect boundary integral equations and presented some existence theorems.

In this paper, we first investigate the conditions of single valuedness for the first plane problem of micropolar elasticity. Using these conditions we develop the fundamental solutions of order one and two. Then we derive the Somigliana relations both for inner and outer regions. Finally we set up a direct boundary element formulation and present an algorithm for numerical solutions.

The paper is organized into 10 sections. Section 11.2 presents some preliminaries, and Section 11.3 clarifies the supplementary conditions of single valuedness. The basic equations and the fundamental solutions of order one are established in Sections 11.4 and 11.5 deals with the determination of the fundamental solutions of order two. The dual Somigliana relations for inner and outer regions are derived in Sections 11.6 and 11.7. Section 11.8 clarifies how to determine the stresses on the boundary. The last two sections present some numerical examples and a conclusion.

11.2 Preliminaries

Throughout this paper $x_1 = x$, $x_2 = y$ and $x_3 = z$ are rectangular Cartesian coordinates with origin O. The coordinate plane (x_1, x_2) coincides with the plane of strains. The ordered pair (x_1, x_2) is denoted by x. {Greek}[Latin subscripts] are assumed to have the range {(1,2)}[(1,2,3)], and summation over repeated subscripts is implied unless explicitly suspended. The triple connected plane region under consideration (Fig. 11.1) is denoted by A^+ –inner region– and is bounded by the outer contour.

$$\mathcal{L}_0 = \mathcal{L}_{t1} \cup \mathcal{L}_{u2} \cup \mathcal{L}_{t3} \cup \mathcal{L}_{u2}$$

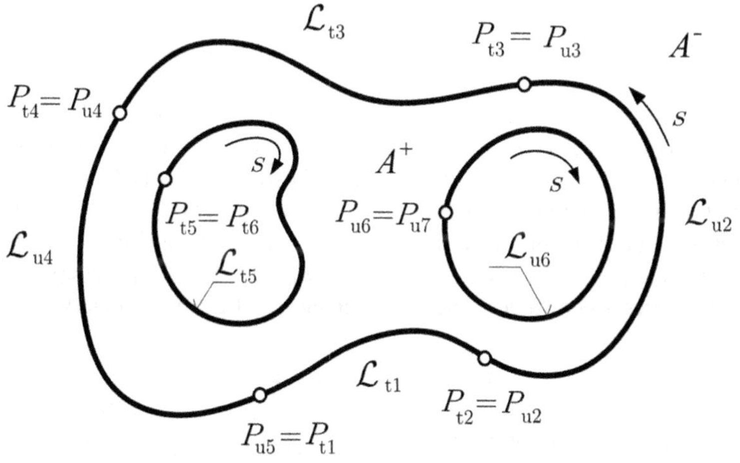

Fig. 11.1 The region under consideration

and the inner contours

$$\mathcal{L}_0 = \mathcal{L}_{t5} \text{ and } \mathcal{L}_2 = \mathcal{L}_{u6}$$

which – partly or wholly – consists of the arcs $\mathcal{L}_{t1}$, $\mathcal{L}_{t3}$, $\mathcal{L}_{t5}$ and $\mathcal{L}_{u1}$, $\mathcal{L}_{u3}$, $\mathcal{L}_{u5}$ Here the first {t}[u] in the subscripts expresses that {displacements}[tractions] are prescribed on the arc considered. The second subscript counts the arcs which constitute the contours. Note that the inner contours $\mathcal{L}_1$ and $\mathcal{L}_2$ lie wholly in the interior of the outer contour $\mathcal{L}_0$ and they have no points in common. We assume that each contour has a continuously turning unit tangent τ_κ and admits a nonsingular parametrization in terms of its arc length s. The outer normal is denoted by n_π. Let $\delta_{\kappa\lambda}$ denote the Kronecker symbol, ∂_α the derivatives with respect to x_α and ε_{pkl} the permutation symbol.

Let u_κ and φ_3 be the displacement and microrotation vectors (together the displacements); $\gamma_{\pi\rho}$ and $\kappa_{\rho 3}$ the asymmetric strain tensor and curvature twist tensor (together the strains); and $t_{\pi\rho}$ and $\mu_{\nu 3}$, $\mu_{\nu 3}$ the asymmetric stress tensor and couple-stress tensor (together the stresses). Body forces and couples are denoted by b_ρ and c_3. Displacements and strains are assumed to be small. {The symmetric} [The skew] part of a tensor, say the tensor $t_{\kappa\lambda}$, is denoted by $\{t_{(\kappa\lambda)}\}[t_{<\kappa\lambda>}]$.

In the primal system the first plane problem is governed by

- the kinematic equations

$$\kappa_{\rho 3} - \varphi_3 \partial_\rho = 0, \quad \gamma_{\pi\rho} - u_\rho \partial_\pi - \varepsilon_{\rho\pi 3}\varphi_3 = 0 \tag{11.1}$$

- Hooke's law

$$t_{\kappa\lambda} = 2\mu\left(\gamma_{(\kappa\lambda)} + \gamma_{\phi\phi}\delta_{\kappa\lambda}\nu/(1-2\nu)\right) + 2\alpha\gamma_{\langle\kappa\lambda\rangle}, \qquad \mu_{\nu 3} = (\gamma + \varepsilon)\,\kappa_{\nu 3}, \tag{11.2}$$

where μ, α, γ, ν and ε are the material constants of an isotropic body, and

- the primal balance equations

$$\partial_\nu t_{\nu\rho} + b_\rho = 0, \qquad \partial_\nu \mu_{\nu 3} + \varepsilon_{3\nu\rho} t_{\nu\rho} + c_3 = 0. \tag{11.3}$$

Field equations (11.1), (11.2) and (11.3) should be associated with appropriate boundary conditions. If a contour is not divided into parts then either tractions or displacements can be imposed on it. If a contour is divided into parts then it is assumed to consist of an even number of arcs, with alternating prescriptions of displacements and tractions. For the inner region shown in Fig. 11.1 {tractions}[displacements] are given on the arc $\{\mathcal{L}_t = \mathcal{L}_{t1} \cup \mathcal{L}_{t3} \cup \mathcal{L}_{t5}\}[\mathcal{L}_u = \mathcal{L}_{u1} \cup \mathcal{L}_{u3} \cup \mathcal{L}_{u5}]$. Letters with hats stand for the prescribed values. The displacement and traction boundary conditions are given by

$$u_\kappa = \hat{u}_\kappa, \qquad \varphi_3 = \hat{\varphi}_3 \tag{11.4}$$

$$n_\nu t_{\nu\rho} = \hat{t}_\rho, \qquad n_\nu \mu_{\nu_3} = \hat{\mu}. \tag{11.5}$$

In the dual system the first plane problem of micropolar elastostatics [23] is governed by

- the dual kinematic equations

$$t_{\pi\rho} = \varepsilon_{\pi\mu 3}\partial_\mu \mathcal{F}_\rho + \overset{o}{t}_{\pi\rho}, \qquad \mu_{\nu 3} = \varepsilon_{\nu\pi 3}\left(\partial_\pi \mathcal{H} + \varepsilon_{3\pi\rho}\mathcal{F}_\rho\right) + \overset{o}{\mu}_{\nu 3} \tag{11.6}$$

which give the stresses $t_{\pi\rho}$ and $\mu_{\nu 3}$ in terms of stress functions of order one $\mathcal{F}_\rho$ and $\mathcal{H}$ while $\overset{o}{t}_{\pi\rho}$ and $\overset{o}{\mu}_{\nu 3}$ are particular solutions of the primal equilibrium equations,

- the inverse form of Hooke's law (dual constitutive equations)

$$\gamma_{\kappa\lambda} = \frac{1}{2\mu}t_{(\kappa\lambda)} + \frac{1}{2\alpha}t_{\langle\kappa\lambda\rangle} - \frac{\nu}{2\mu}t_{\phi\phi}\delta_{\kappa\lambda}, \qquad \kappa_{\nu 3} = \frac{1}{\gamma+\varepsilon}\mu_{\nu 3}; \tag{11.7}$$

- and the compatibility differential equations (dual balance equations)

$$E = \varepsilon_{3\pi\rho}\kappa_{\rho 3}\partial_\pi = 0, \qquad D_\rho = \varepsilon_{3\nu\pi}\left(\gamma_{\pi\rho}\partial_\nu + \varepsilon_{\pi\rho 3}\kappa_{\nu 3}\right) = 0. \tag{11.8}$$

Remark 1. If $\mu > 0$, $\gamma + \varepsilon > 0$, $\alpha >$, $\nu\,\varepsilon$ (0,0.5) then the strain energy density

$$e\left(t_{11},\ldots,\mu_{23}\right) = \frac{1}{2}\left\{\frac{1}{2\mu}\left(t_{(\kappa\lambda)}t_{(\kappa\lambda)} - \nu t_{\phi\phi}t_{\rho\rho}\right) + \frac{1}{4\alpha}t_{\langle\kappa\lambda\rangle}t_{\langle\kappa\lambda\rangle} + \frac{1}{\gamma+\varepsilon}\mu_{\rho 3}\mu_{\rho 3}\right\} \tag{11.9}$$

is strictly positive provided that at least one stress component is different from zero.

Remark 2. For plane problems there are four independent material constants: μ, ν, α and $\gamma + \varepsilon$. It is customary to introduce the characteristic lengths l, l_l and the coupling number N which are defined by

$$1^2 = \frac{\gamma+\varepsilon}{4\mu}, \qquad l_1^2 = \frac{(\gamma+\varepsilon)(\mu+\alpha)}{4\mu\alpha}, \qquad N^2 = \frac{1^2}{l_1^2} = \frac{\alpha}{\mu+\alpha}. \tag{11.10}$$

If l and l_l tend to zero, the equations of micropolar elasticity simplify to those of classical elasticity. If N tends to 1 we obtain the equations of the couple stress theory.

Remark 3. The stress representations (11.6) (the dual kinematic equations) are solutions to the primal equilibrium equations. The particular solutions $\overset{o}{t}_{\pi\rho}$ and $\overset{o}{\mu}_{\nu 3}$ have the form

$$\overset{o}{t}_{\pi\rho} = \partial_\pi p_\rho, \qquad \overset{o}{\mu}_{\nu 3} = \varepsilon_{\nu 3\eta}p_\eta + \partial_\nu q_3,$$

in which

$$\Delta p_\rho = -b_\rho, \qquad \Delta q_3 = c_3.$$

These representations were found independently of each other by Schaefer [12] and Carlson [13].

Remark 4. There belong no stresses to the stress functions

$$\mathcal{F}_\rho = \overset{o}{\mathcal{F}}_\rho \text{ and } \mathcal{H} = \overset{o}{\mathcal{H}} - \varepsilon_{3\pi\rho} x_\pi \overset{o}{\mathcal{F}}_\rho \tag{11.11}$$

where $\overset{o}{\mathcal{F}}_\rho$ and $\overset{o}{\mathcal{H}}$ are arbitrary constants. These stress functions correspond to the rigid body motion which causes no strains in the region under consideration and are referred to as dual rigid body motion.

Remark 5. There are no strains due to the rigid body motion

$$\mu_\rho = \overset{o}{\mu}_\rho + \varepsilon_{3\pi\rho} x_\pi \overset{o}{\varphi}_3 \text{ and } \varphi_3 = \overset{o}{\varphi}_3 \tag{11.12}$$

where $\overset{o}{\mu}_\rho$ and $\overset{o}{\varphi}_3$ are arbitrary constants. These displacements correspond to those stress functions causing no stresses in the region under consideration.

Remark 6. The primal kinematic equations (11.1) are solutions to the dual balance equations (11.8).

Remark 7. If the compatibility differential equations or, equivalently, the dual balance equations are satisfied, then the strains $\gamma_{\pi\rho}$ and $\kappa_{\nu 3}$ are compatible on a single connected domain, i.e. the primal kinematic equations have one solution for the displacements u_ρ and φ_3 provided that we disregard rigid body motions.

Field equations (11.6), (11.7) and (11.8) should be associated with appropriate boundary conditions. Kozák and Szeidl [24] have shown that the strain boundary conditions

$$\tau_\pi \gamma_{\pi\rho} = d\hat{u}_\rho/ds + \tau_\pi \varepsilon_{\rho\pi 3} \hat{\varphi}_3, \quad s \in \mathcal{L}_u \tag{11.13}$$

$$\tau_\pi \kappa_{\pi 3} = \frac{d\hat{\varphi}_3}{ds}, \quad s \in \mathcal{L}_u \tag{11.14}$$

correspond to the displacement boundary conditions of the primal system.

Consider now the boundary conditions on arc $\mathcal{L}_t$. Using (11.5) and (11.6) we have

$$\hat{t}_\rho - \overset{o}{t}_\rho = n_\pi \varepsilon_{\pi\nu 3} \left(\mathcal{F}_\rho \partial_\nu\right) = \frac{d\mathcal{F}_\rho}{ds}, \quad s \in \mathcal{L}_t \tag{11.15}$$

$$\hat{\mu} - \overset{o}{\mu} = n_\pi \varepsilon_{\pi\nu 3} \left(\mathcal{H}\partial_\nu + \varepsilon_{3\nu\rho} \mathcal{F}_\rho\right) = \frac{d\mathcal{H}}{ds} - n_\rho \mathcal{F}_\rho, \quad s \in \mathcal{L}_t \tag{11.16}$$

where

$$\overset{o}{t}_\rho = n_\sigma \left(p_\rho \partial_\sigma\right), \qquad \overset{o}{\mu} = n_\rho \left(\varepsilon_{\rho 3\eta} p_\eta + q\partial_\rho\right). \tag{11.17}$$

To solve the differential equation (11.15), let

$$\hat{\mathcal{F}}_\rho(s) = \int_{P_{ti}}^{s} \left[\hat{t}_\rho(\sigma) - \overset{o}{t}_\rho(\sigma) \right] d\sigma, \quad s \in \mathcal{L}_{ti}, \quad i = 1,3,5, \tag{11.18}$$

and let $\underset{(ti)}{C_\rho}(i = 1,3,5)$ be constants of integration. Then any solution of (11.15) can be written as

$$\mathcal{F}_\rho(s) = \hat{\mathcal{F}}_\rho(s) + \underset{(ti)}{C_\rho}, \quad s \in L_{ti}, \quad i = 1,3,5, \tag{11.19}$$

This equation is one of the boundary conditions for the arc $\mathcal{L}_t$.

Let $\hat{\mathcal{H}}(s)$ be a solution to the differential equation

$$\hat{\mu} - \overset{o}{\mu} = d\hat{\mathcal{H}}/ds - n_\rho \hat{\mathcal{F}}_\rho \tag{11.20}$$

Integration yields

$$\hat{\mathcal{H}}(s) = \int_{P_{ti}}^{s} \left[\hat{\mu}(\sigma) - \overset{o}{\mu}(\sigma) + n_\rho \hat{\mathcal{F}}_\rho \right] d\sigma. \tag{11.21}$$

Subtracting (11.20) from (11.16) and then using (11.19) we obtain the differential equation

$$d(\mathcal{H} - \hat{\mathcal{H}})/ds - n_\rho \underset{(ti)}{C_\rho} = 0 \tag{11.22}$$

which implies the boundary condition

$$\mathcal{H}(s) = \hat{\mathcal{H}}(s) + \underset{(ti)}{C} - \varepsilon_{3\kappa\rho} \left[x_\kappa(s) - x_\kappa(P_{ti}) \right] \underset{(ti)}{C_\rho}, \quad s \in \mathcal{L}_{ti} \;\; i = 1,\,3,\,5 \tag{11.23}$$

where $\underset{(ti)}{C}$ is a further constant of integration.

11.3 Supplementary Compatibility Conditions

The compatibility field equations (11.8) do not guarantee single valuedness of the displacements if the region is multiply connected or if tractions are prescribed on more than one boundary arc. In this section we derive compatibility conditions for these cases as well from the principle of minimum complementary energy. As is well known, the total complementary energy functional is

$$K = -\frac{1}{2} \int_{A^+} \left(t_{\pi\rho} \gamma_{\pi\rho} + \mu_{\nu 3} \kappa_{\nu 3} \right) dA + \int_{\mathcal{L}_u} \left(n_\pi t_{\pi\rho} \hat{u}_\rho + n_\nu \mu_{\nu 3} \hat{\varphi}_3 \right) ds. \tag{11.24}$$

The stationarity condition

$$\delta K = -\int_{A^+} (\gamma_{\pi\rho}\delta t_{\pi\rho} + \kappa_{\nu 3}\delta\mu_{\nu 3})dA + \int_{\mathcal{L}_u} (n_\pi \delta t_{\pi\rho}\hat{u}_\rho + n_\nu \delta\mu_{\nu 3}\hat{\varphi}_3)ds = 0. \quad (11.25)$$

ensures that the strains $\gamma_{\pi\rho}$ and $\kappa_{\nu 3}$ satisfy the conditions to be kinematically admissible. In functional (11.24) $\gamma_{\pi\rho}$ and $\kappa_{\nu 3}$ are given in terms of the stresses $t_{\kappa\lambda}$ and $\mu_{\nu 3}$ via Hooke's law while the stresses $t_{\kappa\lambda}$ and $\mu_{\nu 3}$ should satisfy the equilibrium equations and the traction boundary conditions. Consequently, the variations of stresses can not be arbitrary but should meet the side conditions

$$\partial_\pi \delta t_{\pi\rho} = 0, \quad \partial_\nu \delta\mu_{\nu 3} + \varepsilon_{3\nu\rho}\delta t_{\nu\rho} = 0, \quad x \in A^+ \quad (11.26)$$

$$n_\pi \delta t_{\pi\rho} = 0, \quad n_\nu \delta\mu_{\nu 3} = 0, \quad s \in \mathcal{L}_t. \quad (11.27)$$

Both (11.26) and (11.27) are satisfied if $\delta t_{\pi\rho}$ and $\delta\mu_{\nu 3}$ are expressed with the stress functions and the integration constants introduced in (11.19) and (11.23) as follows

$$\begin{gathered} \delta t_{\pi\rho} = \varepsilon_{\pi\mu 3}\partial_\mu \delta\mathcal{F}_\rho, \quad \delta\mu_{\nu 3} = \varepsilon_{\nu\pi 3}\left(\partial_\pi \delta\mathcal{H} + \varepsilon_{3\pi\rho}\delta\mathcal{F}_\rho\right)\mathcal{F}_\rho, \quad x \in A^+ \\ \delta\mathcal{F}_\rho(s) = \delta \underset{(ti)}{C_\rho}, \quad i = 1, 3, 5 \quad s \in \mathcal{L}_{ti} \end{gathered} \quad (11.28)$$

$$\delta\mathcal{H}(s) = \delta \underset{(ti)}{C} - \varepsilon_{3\kappa\rho}\left[x_\kappa(s) - x_\kappa(P_{ti})\right]\delta \underset{(ti)}{C_\rho}, \quad i = 1, 3, 5 \quad s \in \mathcal{L}_{ti}. \quad (11.29)$$

Substituting (11.28) into the extremum condition and applying the Green theorem we have

$$\delta K = \delta K_A + \delta K_{\mathcal{L}} + \delta K_u = 0, \quad (11.30)$$

where

$$\delta K_A = -\int_{A^+} \left[D_\rho \delta\mathcal{F}_\rho + E\delta\mathcal{H}\right] dA = 0 \quad (11.31)$$

and

$$\delta K_{\mathcal{L}} = \int_{\mathcal{L}_u \cup \mathcal{L}_t} \left(n_\mu \varepsilon_{\mu\pi 3}\gamma_{\pi\rho}\,\delta\mathcal{F}_\rho + n_\pi \varepsilon_{\pi\nu 3}\kappa_{\nu 3}\,\delta\mathcal{H}\right) ds, \quad (11.32)$$

$$\delta K_u = \int_{\mathcal{L}_u} \left[n_\pi \varepsilon_{\pi\mu 3}\delta\mathcal{F}_\rho \partial_\mu \hat{u}_\rho + n_\nu \varepsilon_{\nu\psi 3}\left(\delta\mathcal{H}\partial_\psi + \varepsilon_{3\psi\rho}\delta\mathcal{F}_\rho\right)\hat{\varphi}_3\right] ds. \quad (11.33)$$

To derive the final expression for $\delta K_{\mathcal{L}} + \delta K_u$, we combine the following observations:

- $n_\pi \varepsilon_{\pi\mu 3} = \tau_\mu$ and $\mathcal{L} = \mathcal{L}_\mathrm{u} \cup \mathcal{L}_\mathrm{t}$;
- the variations of stress functions on $\mathcal{L}_t$ are given by (11.28) and (11.29);

- the variations of stress functions are continuous on $\mathcal{L}$ including the points P_{t1}, P_{t2}, P_{t3} and P_{t4} as well;
- the desired result can be achieved if we perform partial integrations when transforming δK_u and take the continuity of the stress functions into account.

Building on these steps, we can integrate by parts the expression for δK_u and use the continuity of the stress functions to obtain from (11.32) and (11.33) the following result:

$$\begin{aligned}
&\delta K_{\mathcal{L}} + \delta K_{\mathcal{L}} = \\
&= \int_{\mathcal{L}_u} \left[\tau_\pi \gamma_{\pi\rho} - \frac{d\hat{u}_\rho}{ds} - \tau_\pi \varepsilon_{\rho\pi 3} \hat{\varphi}_3 \right] \delta \mathcal{F}_\rho ds + \int_{\mathcal{L}_u} \left[\tau_\pi \kappa_{\pi 3} - \frac{d\hat{\varphi}_3}{ds} \right] \delta \mathcal{H} ds \\
&\quad + \oint_{\mathcal{L}_{t5}} \tau_\pi \kappa_{\pi 3} ds\, \delta \underset{(t5)}{C} + \oint_{\mathcal{L}_{t5}} \tau_\pi \left[\gamma_{\pi\rho} - \kappa_{\pi 3} \varepsilon_{\rho 3\sigma} \left(x_\sigma - x_\sigma(P_{t5}) \right) \right] ds\, \delta \underset{(t5)}{C_\rho} \\
&\quad + \sum_{i=1,3} \left\{ \int_{L_{ti}} \tau_\pi \kappa_{\pi 3} ds - \hat{\varphi}_3 \Big|_{P_{ti}}^{P_{t,i+1}} \right\} \delta \underset{(ti)}{C} \\
&\quad + \sum_{i=1,3} \left\{ \int_{L_{ti}} \tau_\pi \left[\gamma_{\pi\rho} - \kappa_{\pi 3} \varepsilon_{\rho 3\sigma} \left(x_\sigma - x_\sigma(P_{t5}) \right) \right] ds \right. \\
&\quad \left. - \hat{u}_\rho \Big|_{P_{ti}}^{P_{t,i+1}} + \hat{\varphi}_3 \varepsilon_{\rho 3\sigma} \left(x_\sigma(P_{t,i+1}) - x_\sigma(P_{ti}) \right) \right\} \delta \underset{(ti)}{C_\rho} .
\end{aligned} \tag{11.34}$$

Since the variations are arbitrary in (11.31) and (11.34), we obtain four groups of equations:

- the compatibility conditions on A^+:

$$E(x) = 0, \qquad D_\rho(x) = 0; \tag{11.35}$$

- the strain boundary conditions on $\mathcal{L}_u$:

$$\tau_\pi \gamma_{\pi\rho} - \frac{d\hat{u}_\rho}{ds} - \tau_\pi \varepsilon_{\rho\pi 3} \hat{\varphi}_3 = 0, \qquad \tau_\pi \kappa_{\pi 3} - \frac{d\hat{\varphi}_3}{ds} = 0; \tag{11.36}$$

- the compatibility conditions at large on $\mathcal{L}_{t5}$:

$$\oint_{\mathcal{L}_1} \tau_\pi \left[\gamma_{\pi\rho} - \kappa_{\pi 3} \varepsilon_{\rho 3\sigma} \left(x_\sigma - x_\sigma(P_{t5}) \right) \right] ds = 0, \qquad \oint_{\mathcal{L}_1} \tau_\pi \kappa_{\pi 3} ds = 0; \tag{11.37}$$

and

- the supplementary boundary conditions on $\mathcal{L}_0$:

$$\oint_{\mathcal{L}_{ti}} \tau_\pi \left[\gamma_{\pi\rho} - \kappa_{\pi 3}\varepsilon_{\rho 3\sigma}\left(x_\sigma - x_\sigma(P_{t5})\right)\right] ds - \hat{u}_\rho\Big|_{P_{ti}}^{P_{t,i+1}} + \\ +\hat{\varphi}_3\varepsilon_{\rho 3\sigma}\left(x_\sigma\left(P_{t,i+1}\right) - x_\sigma\left(P_{ti}\right)\right) = 0, \quad i = 1,3; \tag{11.38}$$

$$\int_{\mathcal{L}_{ti}} \tau_\pi \kappa_{\pi 3} ds - \hat{\varphi}_3 \Big|_{P_{ti}}^{P_{t,i+1}} = 0. \tag{11.39}$$

Remark 8. It can be shown that only two of the three macro conditions of compatibility – the compatibility condition at the large (11.37) and the supplementary conditions of single valuedness (11.38) and (11.39) are independent of each other. Since each of these conditions involves three equations, we can set three of the nine undetermined constants – for example, $\underset{(t1)}{C}$ and $\underset{(t1)}{C_\rho}$– to zero. This follows because no stresses belong to the corresponding stress functions

$$\mathcal{F}_\rho(s) = \underset{(ti)}{C_\rho} \text{ and } \mathcal{H}(s) = \underset{(ti)}{C} - \varepsilon_{3\kappa\rho}\left[x_\kappa(s) - x_\kappa(P_{ti})\right] \underset{(ti)}{C_\rho}.$$

As a result, the number of independent macro conditions for single valuedness equals the number of undetermined constants of integration.

11.4 Basic Equations and Fundamental Solutions of Order One

Here and in the sequel we shall assume that there are no body forces. Substituting the dual kinematic equation (11.6) into Hooke's law (11.7) and the result into the compatibility equations (11.8) we have

$$\left[-\left(\frac{1}{4\mu} + \frac{1}{4\alpha}\right)\partial_1\partial_1 - \frac{1-\nu}{2\mu}\partial_2\partial_2 + \frac{1}{\gamma+\varepsilon}\right] \\ \mathcal{F}_1 + \left(\frac{1}{4\mu} - \frac{1}{4\alpha} - \frac{\nu}{2\mu}\right)\partial_1\partial_2\mathcal{F}_2 - \frac{1}{\gamma+\varepsilon}\partial_2\mathcal{H} = 0, \tag{11.40}$$

$$\left(\frac{1}{4\mu} - \frac{1}{4\alpha} - \frac{\nu}{2\mu}\right)\partial_2\partial_1\mathcal{F}_1 + \left[-\left(\frac{1}{4\mu} + \frac{1}{4\alpha}\right)\partial_2\partial_2 - \frac{1-\nu}{2\mu}\partial_1\partial_1 + \frac{1}{\gamma+\varepsilon}\right] \\ \mathcal{F}_2 - \frac{1}{\gamma+\varepsilon}\partial_1\mathcal{H} = 0, \tag{11.41}$$

$$\frac{1}{\gamma+\varepsilon}\partial_2\mathcal{F}_1 - \frac{1}{\gamma+\varepsilon}\partial_1\mathcal{F}_2 - \frac{1}{\gamma+\varepsilon}\Delta\mathcal{H} = 0. \tag{11.42}$$

The quantities that can be prescribed on the arcs constituting $\mathcal{L}_u$ are obtained from (11.13) and (11.14):

$$\tau_1\gamma_{11} + \tau_2\gamma_{21} = \left\{\frac{1-\nu}{2\mu}\tau_1\partial_2 - \left(\frac{1}{4\mu} + \frac{1}{4\alpha}\right)\tau_1\partial_1\right\} \mathcal{F}_1 + \left\{\left(\frac{1}{4\mu} - \frac{1}{4\alpha}\right)\tau_2\partial_2 + \frac{\nu}{2\mu}\tau_1\partial_1\right\}\mathcal{F}_2 \tag{11.43}$$

$$\tau_1\gamma_{12} + \tau_2\gamma_{22} = \left\{\left(\frac{1}{4\mu} - \frac{1}{4\alpha}\right)\tau_1\partial_1 - \frac{\nu}{2\mu}\tau_2\partial_2\right\} \mathcal{F}_1 - \left\{\frac{1+\nu}{2\mu}\tau_2\partial_1 - \left(\frac{1}{4\mu} + \frac{1}{4\alpha}\right)\tau_1\partial_2\right\}\mathcal{F}_2 \tag{11.44}$$

$$\tau_1\kappa_{13} + \tau_2\kappa_{23} = \frac{1}{\gamma+\varepsilon}\{-\tau_1\mathcal{F}_1 - \tau_2\mathcal{F}_2 + \tau_1\partial_2\mathcal{H} - \tau_2\partial_1\mathcal{H}\}. \tag{11.45}$$

Let

$$\mathfrak{a} = (\gamma+\varepsilon)(1-\nu)/2\mu \tag{11.46}$$

be a new constant. It is obvious – see Remark 1. for details – that $\mathfrak{a} > 0$. After multiplying (11.40), ..., (11.42) with $\gamma + \varepsilon$ and straightforward manipulations we arrive at the basic equations in the dual system of the first plane problem:

$$(1 - \mathfrak{l}^2\Delta)\mathcal{F}_1 + (\mathfrak{l}^2 - \mathfrak{a})(\mathcal{F}_1\partial_2 - \mathcal{F}_2\partial_1)\partial_2 - \partial_2\mathcal{H} = 0, \tag{11.47}$$

$$(\mathfrak{l}^2 - \mathfrak{a})(\mathcal{F}_1\partial_2 - \mathcal{F}_2\partial_1)\partial_1 + (1 - \mathfrak{l}^2\Delta)F_2 + \partial_1\mathcal{H} = 0, \tag{11.48}$$

$$\partial_2\mathcal{F}_1 - \partial_1\mathcal{F}_2 - \Delta\mathcal{H} = 0, \tag{11.49}$$

or in matrix form

$$\mathfrak{D}_{lk}\mathfrak{u}_k = 0, \tag{11.50}$$

where

$$\mathfrak{D}_{lk} = \begin{bmatrix} 1 - \mathfrak{l}^2\Delta + (\mathfrak{l}^2 - \mathfrak{a})\partial_2\partial_2 & -(\mathfrak{l}^2 - \mathfrak{a})\partial_1\partial_2 & -\partial_2 \\ -(\mathfrak{l}^2 - \mathfrak{a})\partial_1\partial_2 & 1 - \mathfrak{l}^2\Delta + (\mathfrak{l}^2 - \mathfrak{a})\partial_1\partial_1 & \partial_1 \\ \partial_2 & -\partial_1 & -\Delta \end{bmatrix} \tag{11.51}$$

is the corresponding differential operator and

$$\mathfrak{u}_k = (\mathcal{F}_1|\mathcal{F}_2|\mathcal{H}) \tag{11.52}$$

is the vector of unknowns, i.e., the vector of stress functions or the dual displacement field.

Remark 9. The system of differential equations (11.50) is elliptic if the material constants meet our preconditions – see Remark 1. for details.

Let D_{kj} be the cofactor of $\mathcal{D}_{jk}$:

$$[D_{kl}]=\begin{bmatrix} -\left[1-\mathfrak{l}^2\Delta+(\mathfrak{l}^2-\mathfrak{a})\partial_1^2\right]\Delta+\partial_1^2 & -(\mathfrak{l}^2-\mathfrak{a})\Delta\partial_1\partial_2+\partial_1\partial_2 & (1-\mathfrak{l}^2\Delta)\partial_2 \\ -(\mathfrak{l}^2-\mathfrak{a})\Delta\partial_2\partial_1+\partial_2\partial_1 & -\left[1-l^2\Delta+(\mathfrak{l}^2-\mathfrak{a})\partial_2^2\right]\Delta+\partial_2^2 & -(1-\mathfrak{l}^2\Delta)\partial_1 \\ -(1-\mathfrak{l}^2\Delta)\partial_2 & (1-\mathfrak{l}^2\Delta)\partial_1 & (1-\mathfrak{l}^2\Delta)(1-\mathfrak{a}\Delta) \end{bmatrix}. \tag{11.53}$$

It is obvious that

$$D_{ik}\mathfrak{D}_{kl}=\mathfrak{D}_{ik}D_{kl}=\det(\mathfrak{D}_{jl})\,\delta_{il}, \tag{11.54}$$

where

$$\det(\mathfrak{D}_{jl})=\mathfrak{a}(1-\mathfrak{l}^2\Delta)\Delta\Delta\,. \tag{11.55}$$

If we introduce a new unknown χ_l [25, 26] defined by the equation

$$\mathfrak{u}_k=D_{kl}\chi_l \tag{11.56}$$

and substitute it back into (11.26), we obtain an uncoupled system of differential equations

$$\mathfrak{D}_{ik}\mathfrak{u}_k=\mathfrak{D}_{ik}D_{kl}\chi_l=\det(\mathfrak{D}_{jl})\chi_i=0. \tag{11.57}$$

Let $Q(\xi_1,\xi_2)$ and $M(x_1,x_2)$ be two points in the plane of strain (the source point and the field point). Further let $\mathbf{e}$ with components e_i be a unit vector at Q. We shall assume temporarily that the point Q is fixed. The distance between Q and M is R, the position vector of M relative to Q is r_κ. We call the solution of the differential equation

$$\mathfrak{D}_{lk}\mathfrak{u}_k+(\gamma+\varepsilon)\delta(M-Q)e_l=0 \tag{11.58}$$

in which $\delta(M-Q)$ is the Dirac function the fundamental solution. It follows from our earlier arguments – see in particular (11.55), (11.56) and (11.58) – that the fundamental solution can be obtained from the fundamental solution for the Galjorkin functions χ_l, i.e., from the solution of the differential equation

$$\mathfrak{D}_{ik}\mathfrak{u}_k+(\gamma+\varepsilon)\delta(M-Q)e_i=\mathfrak{a}(1-\mathfrak{l}^2\Delta)\Delta\Delta\chi_i+(\gamma+\varepsilon)\delta(M-Q)e_i=0. \tag{11.59}$$

Let

$$k^2=1/\mathfrak{l}^2 \quad \text{and} \quad a=(\gamma+\varepsilon)/\mathfrak{a}\mathfrak{l}^2=2\mu k^2/(1-\nu) \tag{11.60}$$

be further constants. With this notation, (11.59) implies that

$$\left(\Delta - k^2\right) \Delta\Delta\chi_i = a\delta(M - Q)e_i) \tag{11.61}$$

Consequently

$$\chi_i(M, Q) = \chi(R)e_i, \ \ \chi(R) = -\frac{a}{8\pi k^4}\left[k^2R^2\ln R + 4\ln R + 4K_o(kR)\right] \tag{11.62}$$

are the solutions for the Galjorkin functions – see [15, 27] – where K_o is the modified Bessel function of order zero. As is well known [28]

$$K_o(z) = -\ln z - \frac{z^2}{4}\ln z - \frac{z^4}{64}\ln z - ..., \ \ K_1(z) = \frac{1}{z} + \frac{z}{2}\ln z + \frac{z^2}{16}\ln z + ... \tag{11.63}$$

and

$$K_o(z) = e^{-z}\sqrt{\pi/2z} + ..., \ \ K_1(z) = e^{-z}\sqrt{\pi/2z} + ... \tag{11.64}$$

are the expansions in series if z → 0 and the asymptotic expansions if z → ∞. In addition

$$\begin{gathered}\overset{M}{\Delta} = \overset{Q}{\Delta} = \frac{d^2}{dR^2} + \frac{1}{R}\frac{d}{dR}, \quad \frac{dK_o(kR)}{dR} = -kK_1(kR),\\ \frac{dK_1(kR)}{dR} = -k\left[K_o(kR) + \frac{1}{kR}K_1(kR)\right].\end{gathered} \tag{11.65}$$

Here the Q or M superscript means that the differentiation applies with respect to the coordinates of the point Q or M. We will continue to use this notation below. Let

$$\varepsilon(z) = (K_1(z) - 1/z)/z \text{ and } \mathcal{D}(z) = K_o(z) + 2\,(K_1(z) - 1/z)/z \tag{11.66}$$

be further functions that simplify notation. Substituting the expansions in series (11.63) we have

$$\lim_{z\to 0}\varepsilon(z) = \frac{1}{2}\ln z, \ \ \lim_{z\to 0} D(z) = -\frac{z^2}{8}\ln z. \tag{11.67}$$

Using equations and we obtain

$$\frac{d\varepsilon(kR)}{dR} = -\frac{\mathcal{D}(kR)}{R} \text{ and } \frac{d\mathcal{D}(kR)}{dR} = -kK_1(kR) - \frac{2}{R}\mathcal{D}(kR). \tag{11.68}$$

It is also not too difficult to check that the following equations hold:

$$\partial_\alpha = \frac{r_\alpha}{R}\frac{d}{dR}, \quad \partial_\alpha\partial_\beta = \frac{1}{R^2}\left[r_\alpha r_\beta\frac{d^2}{dR^2} + \frac{\delta_{\alpha\beta}R^2 - r_\alpha r_\beta}{R}\frac{d}{dR}\right],$$

$$\partial_\alpha\chi = -\frac{a}{8\pi k^4}\left\{k^2 r_\alpha\,(2\ln R + 1) - \frac{4kr_\alpha}{R}\left[K_1(kR) - \frac{1}{kR}\right]\right\} =$$

$$= -\frac{a}{8\pi k^4}\left\{k^2 r_\alpha\,(2\ln R + 1) - \frac{4kr_\alpha}{R}\mathcal{E}(kR)\right\},$$

$$\partial_\alpha\partial_\beta\chi = \frac{a}{8\pi k^4}\left\{k^2\left[2\delta_{\alpha\beta}\ln R + \frac{\delta_{\alpha\beta}R^2 + 2r_\alpha r_\beta}{R^2}\right] - \right.$$

$$\left. - \frac{4k}{R^3}\left[R^2\delta_{\alpha\beta} - 2r_\alpha r_\beta\right]\left[K_1(kR) - \frac{1}{kR}\right] + \frac{4k^2 r_\alpha r_\beta}{R^2}K_0(kR)\right\}, \tag{11.69}$$

$$\Delta\chi = -\frac{a}{2\pi k^2}\left[\ln R + K_0(kR) + 1\right], \ \partial_\alpha\Delta\chi = \frac{a}{2\pi}r_\alpha\mathcal{E}(kR),$$

$$\partial_\alpha\partial_\beta\Delta\chi = \frac{a}{2\pi}\left[\delta_{\alpha\beta}\mathcal{E}(kR) - \frac{r_\alpha r_\beta}{R^2}\mathcal{D}(kR)\right], \ \Delta\Delta\chi = -\frac{a}{2\pi}K_0(kR).$$

Combining relations (11.53), (11.56) and (11.62) we have

$$\mathfrak{u}_1 = e_1\underbrace{D_{11}\chi}_{\mathfrak{U}_{11}} + e_2\underbrace{D_{12}\chi}_{\mathfrak{U}_{12}} + e_3\underbrace{D_{13}\chi}_{\mathfrak{U}_{12}} = e_1\left\{\left[1 - \mathfrak{l}^2\Delta + (\mathfrak{l}^2 - \mathfrak{a})\partial_1^2\right]\Delta - \partial_1^2\right\}\chi$$

$$+e_2\left[(\mathfrak{l}^2 - \mathfrak{a})\Delta\partial_1\partial_2 - \partial_1\partial_2\right]\chi + e_3\left\{(1 - \mathfrak{l}^2\Delta)\partial_2\right\}\chi, \tag{11.70}$$

$$\mathfrak{u}_2 = e_1\underbrace{D_{21}\chi}_{\mathfrak{U}_{21}} + e_2\underbrace{D_{22}\chi}_{\mathfrak{U}_{22}} + e_3\underbrace{D_{23}\chi}_{\mathfrak{U}_{23}} = e_1\left[(\mathfrak{l}^2 - \mathfrak{a})\Delta\partial_2\partial_1 - \partial_2\partial_1\right]\chi \tag{11.71}$$

$$+e_2\left\{\left[1 - \mathfrak{l}^2\Delta + (\mathfrak{l}^2 - \mathfrak{a})\partial_1^2\right]\Delta - \partial_2^2\right\}\chi - e_3(1 - \mathfrak{l}^2\Delta)\partial_1\chi,$$

$$\mathfrak{u}_3 = e_1\underbrace{D_{31}\chi}_{\mathfrak{U}_{31}} + e_2\underbrace{D_{32}\chi}_{\mathfrak{U}_{32}} + e_3\underbrace{D_{33}\chi}_{\mathfrak{U}_{33}} =$$

$$= e_1\left\{-(1 - \mathfrak{l}^2\Delta)\partial_2\right\}\chi + e_2(1 - \mathfrak{l}^2\Delta)\partial_1\chi + e_3\left\{(1 - \mathfrak{l}^2\Delta)(1 - \mathfrak{a}\Delta)\right\}\chi. \tag{11.72}$$

We can check using the derivatives (11.65) and (11.69) and the asymptotic relations (11.63) that only $\Delta\Delta\chi$ and $\Delta\partial^2\chi/\partial x_\alpha\partial x_\beta$ are singular among the derivatives $\Delta\Delta\chi$, $\Delta\partial^2\chi/\partial x_\alpha\partial x_\beta$, $\Delta\partial\chi/\partial x_\alpha$, $\Delta\chi$, $\Delta\partial^2\chi/\partial x_\alpha\partial x_\beta$ in (11.70),…,(11.72) if $R \to 0$:

$$\Delta\Delta\chi = \frac{a}{2\pi}\ln R, \quad \Delta\partial^2\chi/\partial x_\alpha\partial x_\beta = \frac{a}{2\pi}\delta_{\alpha\beta}\ln R. \tag{11.73}$$

Omitting the formal transformations, we obtain from (11.56), (11.62), (11.65) and (11.66) that

$$\mathfrak{u}_l = e_k(Q)\mathfrak{U}_{kl}(M, Q), \tag{11.74}$$

in which

$$\mathfrak{U}_{\alpha\beta}(M,Q) = \frac{a}{2\pi k^2}\left\{\delta_{\alpha\beta}\left[\frac{1}{2}\ln R + \frac{3}{4} + \mathfrak{a}k^2\varepsilon(kR)\right] - \frac{r_\alpha r_\beta}{R^2}\left[\frac{1}{2} + \mathfrak{a}k^2\mathcal{D}(kr)\right]\right\}, \tag{11.75}$$

$$\mathfrak{U}_{3\alpha}(M, Q) = -\mathfrak{U}_{\alpha 3}(M, Q) = \frac{a}{2\pi k^2}(-1)^{(\alpha)} r_{3-\alpha}\left(\frac{1}{2}\ln R + \frac{1}{4}\right), \tag{11.76}$$

$$\mathfrak{U}_{33}(M, Q) = -\frac{a}{2\pi k^2}\left\{\frac{1}{4}R^2\ln R - \left(\frac{1}{k^2} + \mathfrak{a}\right) - \mathfrak{a}\ln R\right\} \tag{11.77}$$

are the elements of the matrix of fundamental solutions.

Remark 10. The fundamental solution $\mathfrak{U}_{kl}(M,Q)$ satisfies the symmetry and asymmetry conditions

$$\mathfrak{U}_{\alpha\beta}(M, Q) = \mathfrak{U}_{\beta\alpha}(M, Q), \quad \mathfrak{U}_{3\alpha}(M, Q) = -\mathfrak{U}_{\alpha 3}(M, Q). \tag{11.78}$$

Remark 11.: The [rows]{columns} of the fundamental solution $\mathfrak{U}_{kl}(M,Q)$ as three dimensional vectors satisfy the fundamental equation (11.50) if $[M]\{Q\}$ is the independent variable and $\{Q\}[M]$ is fixed.

Remark 12.: Using (11.67), which gives the singular part of $\varepsilon(z)$, and the formulas providing the material constants (11.46) and (11.60) we obtain the decomposition of $\mathfrak{U}_{kl}(M,Q)$ into singular and nonsingular parts:

$$\mathfrak{U}_{kl}(M, Q) = \overset{S}{\mathfrak{U}}_{kl}(M, Q) + \overset{N}{\mathfrak{U}}_{kl}(M, Q), \tag{11.79}$$

where

$$\overset{S}{\mathfrak{U}}_{kl}(M, Q) = \frac{1}{2\pi}\begin{bmatrix} b & 0 & 0 \\ 0 & b & 0 \\ 0 & 0 & d \end{bmatrix}\ln R \quad \text{in which} \quad b = \frac{\mu}{1-\nu} + \frac{2\alpha\mu}{\alpha+\mu}, \quad d = \gamma + \varepsilon. \tag{11.80}$$

11.5 Fundamental Solutions of Order Two

11.5.1 Calculating Dual Stresses from the Fundamental Solution of Order One

The dual stresses are defined by equations

$$\mathfrak{t}_\lambda = \tau_1\gamma_{1\lambda} + \tau_2\gamma_{2\lambda} \quad \text{and} \quad \mathfrak{t}_3 = \tau_1\kappa_{13} + \tau_2\kappa_{23}. \tag{11.81}$$

Note that these quantities constitute the left sides of the strain boundary conditions (11.13) and (11.14). We now introduce the following additional notations

$$\Phi_{\rho 1} = \frac{1}{2}\left[(1-2\nu) - (-1)^{(\rho)} 2\frac{r_\rho^2}{R^2}\right] + (-1)^{(\rho)} \mathfrak{a} k^2 \left[\left(1 - 4\frac{r_\rho^2}{R^2}\right)\mathcal{D}(kR) - \frac{r_\rho^2}{R} k K_1(kR)\right], \tag{11.82}$$

$$\Phi_{\rho 2} = \frac{1}{2}\left[(1-2\nu) + (-1)^{(\rho)} 2\frac{r_\rho^2}{R^2}\right] - (-1)^{(\rho)} \mathfrak{a} k^2 \left[\left(1 - 4\frac{r_\rho^2}{R^2}\right)\mathcal{D}(kR) - \frac{r_\rho^2}{R} k K_1(kR)\right], \tag{11.83}$$

$$\Phi_{3\kappa} = (1-2\nu)R^2\left(\frac{1}{2}\ln R + \frac{1}{4}\right) - \nu\frac{R^2}{2} + \frac{r_{3-\kappa}^2}{2}, \tag{11.84}$$

$$\Psi_{\rho\kappa} = \frac{1}{2}\left\{\frac{r_2^2 - r_1^2}{R^2} - (-1)^{(\rho)} 2\mathfrak{a} k^2 \left(1 - (-1)^{(\rho)} 2\frac{r_2^2 - r_1^2}{R^2}\right)\mathcal{D}(kR) + \mathfrak{a} k^2 \frac{r_2^2 - r_1^2}{R} k K_1(kR)\right\} + (-1)^{(\kappa)} \frac{\mu}{2\alpha} \mathfrak{a} k^2 R k K_1(kR), \tag{11.85}$$

$$\mathcal{K}_{\rho 3} = -\mathfrak{a} k^2 \left[R^2 \mathcal{E}(kR) - r_\rho^2 \mathcal{D}(kR)\right], \tag{11.86}$$

$$\Psi_{13} = \Psi_{23} = r_1 r_2 \mathfrak{a} k^2 \mathcal{D}(kR). \tag{11.87}$$

Omitting the long handmade calculations and making use of the notations (11.82),...,(11.87), we obtain

$$\mathfrak{t}_k = e_l(Q)\mathcal{T}_{lk}(M_\circ, Q), \tag{11.88}$$

where

$$\mathcal{T}_{lk}(M_\circ, Q) = \frac{1}{2\pi(1-\nu)R^2} \begin{bmatrix} -n_2 r_2 \Phi_{11} + n_1 r_1 \Psi_{11} & n_1 r_2 \Phi_{12} - n_2 r_1 \Psi_{12} & \dfrac{2\mu}{(\gamma+\varepsilon)}(-n_2 \mathcal{K}_{13} + n_1 \Psi_{13}) \\ n_2 r_1 \Phi_{21} + n_1 r_2 \Psi_{21} & -n_1 r_1 \Phi_{22} - n_2 r_2 \Psi_{22} & -\dfrac{2\mu}{(\gamma+\varepsilon)}(-n_1 \mathcal{K}_{23} + n_2 \Psi_{23}) \\ n_2 \Phi_{31} + \dfrac{1}{2} n_1 r_1 r_2 & -n_1 \Phi_{32} - \dfrac{1}{2} n_2 r_1 r_2 & -\dfrac{2\mu}{(\gamma+\varepsilon)}\mathfrak{a}\,(n_1 r_1 + n_2 r_2) \end{bmatrix}. \tag{11.89}$$

The last equation is the formula for calculating the dual stresses $\mathfrak{t}_\kappa$ (brought into existence by the incompatibility $e_l = e_l(Q)$) at the point $M_\circ$ on the boundary where the outward unit normal is $n_l = n_l(M_\circ)$.

Remark 13. Matrix $\mathcal{T}_{lk}(M_\circ, Q)$ is that of the fundamental solutions of order two.

Remark 14. Matrix $\mathcal{T}_{lk}(M_\circ, Q)$ can also be resolved into a strongly singular and a non-strongly singular part:

$$\mathcal{T}_{lk}(M_\circ, Q) = \overset{S}{\mathcal{T}}_{lk}(M_\circ, Q) + \overset{N}{\mathcal{T}}_{lk}(M_\circ, Q), \tag{11.90}$$

where

$$\overset{S}{\mathcal{T}}_{lk}(M_O, Q) = -\frac{1}{2\pi}\begin{bmatrix} \dfrac{\partial \ln R}{\partial n_{Mo}} & g\dfrac{\partial \ln R}{\partial s_{Mo}} & 0 \\ -g\dfrac{\partial \ln R}{\partial s_{Mo}} & \dfrac{\partial \ln R}{\partial n_{Mo}} & 0 \\ 0 & 0 & \dfrac{\partial \ln R}{\partial n_{Mo}} \end{bmatrix} \quad \text{and} \quad g = \frac{1}{2(1-\nu)} - \frac{2\mu}{\alpha+\mu}. \tag{11.91}$$

The singularity of matrix $\overset{N}{\mathcal{T}}_{kl}(M_\circ, Q)$ is weaker than that of matrix $\overset{S}{\mathcal{T}}_{kl}(M_\circ, Q)$.

11.6 Dual Somigliana Formulae for Inner Regions

11.6.1 The Dual Somigliana Identity

In this section we assume that the region under consideration is the simple connected inner region A^+ depicted in Fig. 11.2.

The functions $\mathcal{F}_\psi$, $t_{\kappa\lambda}$, $\mu_{\nu 3}$, $\gamma_{\kappa\lambda}$ and $\kappa_{\nu 3}$ are called an elastic state of the region A^+ if they satisfy the field equations (11.6),. . .,(11.8). Let

$$\mathcal{F}_\psi,\ \mathcal{H},\ t_{\kappa\lambda},\ \mu_{\nu 3},\ \gamma_{\kappa\lambda},\ \kappa_{\nu 3}$$

and

$$\mathcal{F}_\psi,^*\ \mathcal{H},^*\ t^*_{\kappa\lambda},\ \mu_{\nu 3},\ ^*\gamma_{\kappa\lambda},^*\ \kappa^*_{\nu 3}$$

be two elastic states of the region A^+. Integrating by parts the compatibility differential equations (the weights are $\mathcal{F}^*_\rho$ and $\mathcal{H}^*$), using (11.6) and noting that because there are no body forces the particular solutions for the stresses and couple stresses are equal to zero, we obtain

$$\begin{aligned} &\int_{A^+} \varepsilon_{3\mu\pi}\left(\gamma_{\pi\rho}\partial_\mu + \varepsilon_{\pi\rho 3}\kappa_{\mu 3}\right)\mathcal{F}^*_\rho\, dA + \int_{A^+} \varepsilon_{3\psi\nu}\,\kappa_{\nu 3}\partial_\psi \mathcal{H}^* dA \\ &\quad \oint_{\mathcal{L}_o} \tau_\pi \gamma_{\pi\rho}\mathcal{F}^*_\rho\, ds + \oint_{\mathcal{L}_o} \tau_\nu \kappa_{\nu 3}\mathcal{H} * ds \\ &\int_{A^+} \varepsilon_{\pi\mu 3}(\mathcal{F}^*_\rho \partial_\mu)\gamma_{\pi\rho}\, dA + \int_{A^+} \varepsilon_{\nu\psi 3}(\mathcal{H} * \partial_\psi + \varepsilon_{\psi\rho 3}\mathcal{F}^*_\rho)\kappa_{\nu 3}\, dA. \end{aligned} \tag{11.92}$$

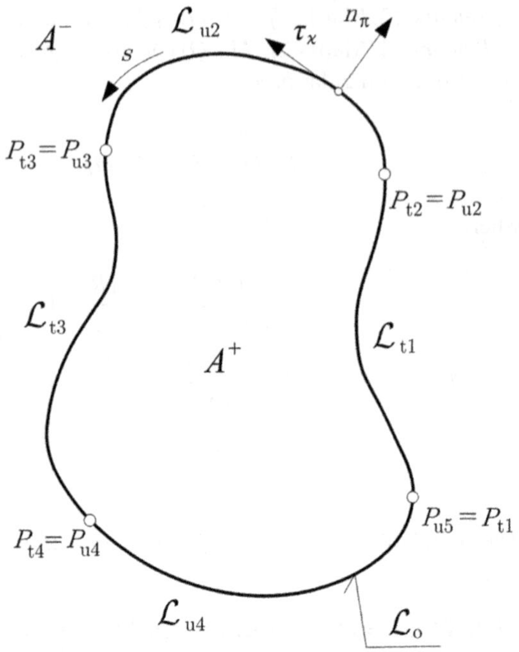

Fig. 11.2 The simple connected inner region

Remark 15. The value of the area integrals on the right side does not depend on whether the asterisk is over the first or the second variable in the products. This can be seen if we express the strains $\gamma_{\pi\rho}$ and $\kappa_{\nu 3}$ in terms of stress functions using (11.6) which provide the stresses in terms of stress functions and then the Hooke law.

The dual Somigliana identity is obtained if we move the asterisks from their original position to the variables which belong to the first elastic state, and then subtract the original (11.92):

$$\int_{A^+}\underbrace{\mathcal{F}_\rho\,\varepsilon_{3\mu\pi}\left(\gamma^*_{\pi\rho}\partial_\mu+\varepsilon_{\pi\rho 3}\kappa^*_{\mu 3}\right)}_{u_\rho\left({}'\mathfrak{D}_{\rho l}u^*_l\right)}dA-\int_{A^+}\underbrace{\mathcal{H}\,\varepsilon_{3\psi\nu}\left(\kappa^*_{\mu 3}\partial_\psi\right)}_{u_3\left({}'\mathfrak{D}_{3l}u^*_l\right)}dA$$

$$-\int_{A^+}\varepsilon_{3\mu\pi}\left(\gamma_{\pi\rho}\partial_\mu+\varepsilon_{\pi\rho 3}\,\kappa_{\mu 3}\right)\mathcal{F}^*_\rho dA+\int_{A^+}\varepsilon_{3\psi\nu}\left(\kappa_{\nu 3}\partial_\psi\right)\mathcal{H}^*\,dA$$

$$=\oint_{\mathcal{L}_o}\underbrace{\mathcal{F}_\rho\,\tau_\pi\gamma^*_{\pi\rho}}_{u_\rho t^*_\rho}\,ds-\oint_{\mathcal{L}_o}\underbrace{\mathcal{H}\,\tau_\nu\kappa^*_{\nu 3}}_{u_3 t^*_3}\,ds\;-\oint_{\mathcal{L}_o}\tau_\pi\gamma_{\pi\rho}\mathcal{F}^*_\rho ds+\oint_{\mathcal{L}_o}\tau_\nu\kappa_{\nu 3}\mathcal{H}^*\,ds. \tag{11.93}$$

On the right hand side we have the weighted integrals of the basic equations with the dual displacements as weights. The left hand side involves the products of the quantities on which boundary conditions can be imposed. Consequently the Somigliana identity can be cast into a form similar to the Green identity of the potential theory.

Introduce first the notation

$$'\mathfrak{D}_{kl} = \mathfrak{D}_{kl}/(\gamma + \varepsilon). \tag{11.94}$$

where $'\mathfrak{D}_{kl}$ is the operator of the basic equations (11.40),. . .,(11.42). Using (11.51) and (11.52) with (11.93), after some manipulations we obtain

$$\int_{A^+} \left[\mathfrak{u}_k \left('\mathfrak{D}_{kl}\mathfrak{u}_l^*\right) - \mathfrak{u}_k^* \left('\mathfrak{D}_{kl}\mathfrak{u}_l\right)\right] dA = \oint_{\mathcal{L}_o} \left[\mathfrak{u}_l t_l^* - t_l \mathfrak{u}_l^*\right] ds. \tag{11.95}$$

which has the same structure as the Green identity [29].

Remark 16. In the argument leading to (11.94) we did not use that the quantities $\mathfrak{u}_k^*$ and $\mathfrak{u}_k$ are elastic states of the region A^+. Consequently (11.94) always holds provided that $\mathfrak{u}_k^*$ and $\mathfrak{u}_k$ are arbitrary functions that can be differentiated as many times as required.

11.6.2 The Dual Somigliana Formulae for Inner Region

To derive the dual Somigliana formulae we shall assume that $\mathfrak{u}_l(M)$ is an elastic state of the region A^+. Suppose that the other elastic state, denoted by *, is the one which belongs to the fundamental solutions:

$$\mathfrak{u}_l^*(M) = e_k(Q)\mathfrak{U}_{kl}(M, Q)\,, \quad \mathfrak{t}_l^*(M) = e_k(Q)\mathfrak{T}_{kl}(M, Q). \tag{11.96}$$

The latter is singular at the point Q. Depending on the position of the point Q relative to the region A^+, we distinguish three cases – two of them are shown in Fig. 11.3.

1. If $Q \in A^+$, then we first remove the neighborhood of Q with radius R_ε, denoted A_ϵ and assumed to lie wholly in A^+, from A^+; and then we apply the dual Somigliana identity to the double connected domain $A' = A^+|A_\varepsilon$. Note that the contour $\mathcal{L}_\varepsilon$ of A_ε and the arc $\mathcal{L}'_\varepsilon$, which is assumed to be the part of the contour $\mathcal{L}_\varepsilon$ lying within A^+ coincide with each other.
2. If $Q = Q_o \in \partial A = \mathcal{L}_o$, then the part $A^+ \cap A_\varepsilon$ of the neighborhood A_ε of Q is removed from A^+ and we then apply the dual Somigliana identity to the simply connected region $A' = A^+ \backslash (A^+ \cap A_\varepsilon)$. In this case, the contour of the simply connected region just obtained consists of two arcs, the arc $\mathcal{L}'_\mathrm{o}$ left from $\mathcal{L}_\mathrm{o}$ after the removal of A_ε and the arc $\mathcal{L}'_\varepsilon$, i.e., the part of $\mathcal{L}_\varepsilon$ that lies within A_i.
3. If $Q \notin (A^+ \cup \mathcal{L}_o)$ we apply the dual Somigliana identity to the original region A^+.

   Since both $\mathfrak{u}_k^*$ and $\mathfrak{u}_k$ are elastic states the surface integrals in (11.95) are identically equal to zero.

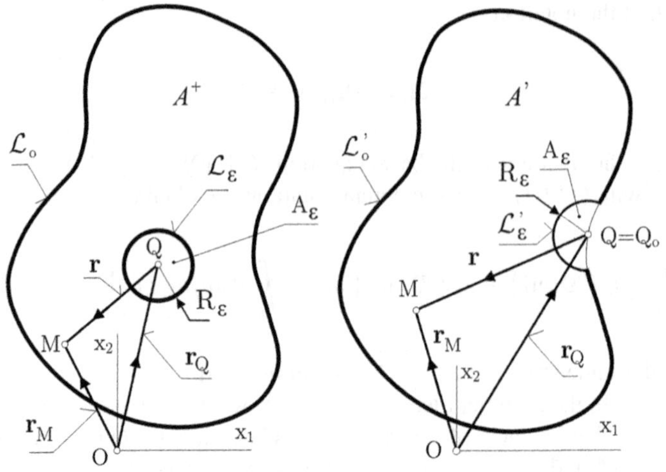

Fig. 11.3 The region under consideration depending on the position of the point Q

We now consider each of these three cases, focusing on the main steps of the argument.

1. If $Q \in A_i$ then substitute (11.96) into (11.95). This yields

$$\oint_{\mathcal{L}_o} [\mathfrak{T}_{kl}(M_\circ, Q)\mathrm{u}_l(M_\circ) - \mathfrak{U}_{kl}(M_\circ, Q)\mathrm{t}_l(M_\circ)]\, ds_{M_\circ} + \oint_{\mathcal{L}_\varepsilon} [\mathfrak{T}_{kl}(M_\circ, Q)\mathrm{u}_l(M_\circ) - \mathfrak{U}_{kl}(M_\circ Q)\mathrm{t}_l(M_\circ)]\, ds_{M_\circ} = 0. \tag{11.97}$$

Substituting the singular and non-singular parts of the fundamental solutions $\mathfrak{U}_{kl}(M_o, Q)$ and $\mathfrak{T}_{kl}(M_o, Q)$ into the left sides of (11.98) and (11.99), using (11.80), (11.90) and (11.91) and noting that the limit of integrals involving the non-singular parts is zero, we obtain

$$\lim_{R_\varepsilon \to 0} \oint_{\mathcal{L}_\varepsilon} \mathfrak{T}_{kl}(M_\circ, Q)\,[\mathrm{u}_l(M_\circ) - \mathrm{u}_l(Q)]\; ds_{M_\circ} = 0\,, \tag{11.98}$$

$$\lim_{R_\varepsilon \to 0} \oint_{\mathcal{L}_\varepsilon} \mathfrak{T}_{kl}(M_\circ, Q)\, ds_{M_\circ} = \delta_{kl} \quad \text{and} \quad \lim_{R_\varepsilon \to 0} \oint_{\mathcal{L}_\varepsilon} \mathfrak{U}_{kl}(M_\circ, Q)\mathrm{t}_l(M_\circ)\, ds_{M_\circ} = 0. \tag{11.99}$$

Substituting these limits into (11.97) yields the first dual Somigliana formula

$$\mathrm{u}_k(Q) = \oint_{\mathcal{L}_o} \mathfrak{U}_{kl}(M_\circ, Q)\mathrm{t}_l(M_\circ) ds_{M_o} - \oint_{\mathcal{L}_o} \mathfrak{T}_{kl}(M_\circ, Q)\mathrm{u}_l(M_\circ) ds_{M_o}. \tag{11.100}$$

Remark 17. Given the dual displacements (stress functions) $\mathfrak{u}_\lambda(M_o)$ and dual stresses (the corresponding displacement derivatives) $\mathfrak{t}_\lambda(M_o)$ on the contour $\mathcal{L}_o$, the first dual Somigliana formula provides the elastic state $\mathfrak{u}_k(Q)$ in terms of two determinate integrals.

2. If $Q = Q_o \in \partial A = \mathcal{L}_o$ then repeating the logic leading from (11.95), (11.96) and (11.97) we can write

$$\oint_{\mathcal{L}_o{}'}[\mathfrak{T}_{kl}(M_\circ, Q_\circ)\mathfrak{u}_l(M_\circ) - \mathfrak{U}_{kl}(M_\circ, Q_\circ)\mathfrak{t}_l(M_\circ)]\, ds_{M_\circ}$$
$$+\oint_{\mathcal{L}_\varepsilon{}'}[\mathfrak{T}_{kl}(M_\circ, Q_\circ)\mathfrak{u}_l(M_\circ) - \mathfrak{U}_{kl}(M_\circ, Q_\circ)\mathfrak{t}_l(M_\circ)]\, ds_{M_\circ} = 0, \tag{11.101}$$

where

$$\lim_{R_\varepsilon \to 0}\int_{\dot{\mathcal{L}}_\varepsilon}\mathfrak{T}_{kl}(M_\circ, Q_\circ)\, ds_{M_\circ} = c_{kl}(Q_\circ). \tag{11.102}$$

We remark that $c_{kl}(Q_o) = \delta_{kl}/2$ if the contour $\mathcal{L}_o$ is smooth at the point Q_o. If not then $c_{kl}(Q_o)$ depends on the angle formed by the tangents to the contour at Q_o. In addition

$$\lim_{R_\varepsilon \to 0}\int_{\dot{\mathcal{L}}_\varepsilon}\mathfrak{U}_{kl}(M_\circ, Q_\circ)\mathfrak{t}_l(M_\circ)\, ds_{M_\circ} = 0, \tag{11.103}$$

$$\lim_{R_\varepsilon \to 0}\int_{\dot{\mathcal{L}}_\varepsilon}\mathfrak{T}_{kl}(M_\circ, Q_\circ)\,[\mathfrak{u}_l(M_\circ) - \mathfrak{u}_l(Q_\circ)]\; ds_{M_\circ} = 0. \tag{11.104}$$

Combining (11.102),. . .,(11.104) for the limit of (11.101) when $R_\varepsilon \to 0$ we obtain

$$c_{kl}(Q_\circ)\mathfrak{u}_l(Q_\circ) = \oint_{\mathcal{L}_o}\mathfrak{U}_{kl}(M_\circ, Q_\circ)\mathfrak{t}_l(M_\circ)\, ds_{M_\circ} - \oint_{\mathcal{L}_o}\mathfrak{T}_{kl}(M_\circ, Q_\circ)\mathfrak{u}_l(M_\circ)\, ds_{M_\circ}. \tag{11.105}$$

This result is the second dual Somigliana formula for inner regions.

Remark 18. The integrals in (11.105) are to be taken in Cauchy principal value.

Remark 19. Since at a point of the contour $\mathcal{L}_o$ either the dual displacements (stress functions) $\mathfrak{u}_l(M_o)$ or the dual stresses (displacement derivatives) $\mathfrak{t}_l(M_o)$ can be prescribed, the above integral equation is suitable both for determining the missing $\mathfrak{u}_l(M_o)$ at a point Q where $\mathfrak{t}_l(M_o)$ is known, and for determining the missing $\mathfrak{t}_l(M_o)$ at a point Q where $\mathfrak{u}_l(M_o)$ is known. Given these quantities on the whole boundary $\mathcal{L}_o$, we can use the first dual Somigliana formula for computing the field variables (stresses and strains).

3. If $Q \in (A \cup L_o)$ then only the integral over $\mathcal{L}_o$ remains in (11.95), and following steps similar to above, we obtain the third dual Somogliana equation:

$$0 = \oint_{\mathcal{L}_o}\mathfrak{U}_{kl}(M_\circ, Q)\mathfrak{t}_l(M_\circ)\, ds_{M_\circ} - \oint_{\mathcal{L}_o}\mathfrak{T}_{kl}(M_\circ, Q)\mathfrak{t}_l(M_\circ)\, ds_{M_\circ}. \tag{11.106}$$

Remark 20. Assume that

$$\mathfrak{u}_l(M,Q) = o\left[\overset{o}{\mathcal{F}}_1 | \overset{o}{\mathcal{F}}_2 | \overset{o}{\mathcal{H}} + \overset{o}{\mathcal{F}}_1 r_2(M,Q) - \overset{o}{\mathcal{F}}_2 r_1(M,Q)\right], \tag{11.107}$$

where $\overset{o}{\mathcal{H}}$, $\overset{o}{\mathcal{F}}_1$ and $\overset{o}{\mathcal{F}}_2$ arbitrary constants while r_1 and r_2 are the coordinates of some pointwith respect to the fixed point Q, which may coincide with the origin. It can be checked neither stresses nor strains belong to this dual displacement vector. Hence

$$'\mathfrak{D}_{kl}\mathfrak{u}_l = \mathfrak{D}_{kl}\mathfrak{u}_l = 0, \quad \text{and} \quad \mathfrak{t}_k = 0. \tag{11.108}$$

Using these equations, identity (11.95) yields

$$\int_{A^+} \mathfrak{u}_k \left('\mathfrak{D}_{kl} \overset{*}{\underset{l}{\mathfrak{u}}}\right) \, dA = \oint_{\mathcal{L}_o} \mathfrak{u}_k \overset{*}{\underset{k}{\mathfrak{t}}} \, ds \,. \tag{11.109}$$

Assume that ${}^*\mathfrak{u}_l$ and ${}^*\mathfrak{t}_k$ belong to the fundamental solutions. Then a comparison of (11.58) and (11.94) gives

$$'\mathfrak{D}^*_{kl}\mathfrak{u}_l + \delta(M-Q)e_k(Q) = 0 \,. \tag{11.110}$$

Recalling that

$${}^*\mathfrak{u}_l(M,Q) = \mathfrak{U}_{ls}(M,Q)e_s(Q)\,, \quad {}^*\mathfrak{t}_k(Mo,Q) = e_s(Q)\mathfrak{T}_{sl}(Mo,Q)$$

and taking (11.110) into account we get from (11.109) that

$$\begin{aligned} -\int_{A^+} \mathfrak{u}_k \delta(M-Q)\delta_{kl}\, e_l(Q) dA &= -e_l(Q)\delta_{lk}\mathfrak{u}_k(Q,Q)\eta(Q) = \\ &\oint_{\mathcal{L}_o} e_l(Q)\mathfrak{T}_{lk}(Mo,Q)\mathfrak{u}_k(Mo)\, ds_{Mo}. \end{aligned} \tag{11.111}$$

Given the structure of the dual displacements (11.107), simple manipulations imply

$$\oint_{\mathcal{L}_o} [\mathfrak{T}_{l1} + r_2\mathfrak{T}_{l3} | \mathfrak{T}_{l2} - r_1\mathfrak{T}_{l3} | \mathfrak{T}_{l3}]\, ds_{\overset{o}{M}} = -\eta(Q)\,[\delta_{l1}|\delta_{l2}|\delta_{l3}]\,, \tag{11.112}$$

where

$$\eta(Q) = \begin{cases} 1 & \text{if } Q \in A^+ \\ 1/2 & \text{if } Q = Q_o \in \mathcal{L}_o \\ 1 & \text{if } Q \in A^- \end{cases}. \tag{11.113}$$

This result will allow us to compute the strongly singular integrals in a numerical implementation developed for solving boundary value problems in inner regions.

11.6.3 Formulae for Stresses

Substituting the first dual Somigliana formula (11.100) into (11.6) we obtain the stresses at the inner points Q of region A^+:

$$\mathfrak{s}_1 = t_{11} = \oint_{\mathcal{L}_o} \underbrace{\left[\mathfrak{U}_{1\,l}(M_o, Q)\,\overset{Q}{\partial}_2\right]}_{S_{1l}} \mathfrak{t}_l(M_o)\,ds_{M_\circ} - \oint_{\mathcal{L}_o} \underbrace{\left[\mathfrak{T}_{l1}(M_o, Q)\,\overset{Q}{\partial}_2\right]}_{-D_{1l}} \mathfrak{u}_l(M_o)\,ds_{M_\circ}, \tag{11.114}$$

$$\mathfrak{s}_2 = t_{12} = \oint_{\mathcal{L}_o} \underbrace{\left[\mathfrak{U}_{2\,l}(M_o, Q)\,\overset{Q}{\partial}_2\right]}_{S_{2l}} \mathfrak{t}_l(M_o)\,ds_{M_\circ} - \oint_{\mathcal{L}_o} \underbrace{\left[\mathfrak{T}_{2\,l}(M_o, Q)\,\overset{Q}{\partial}_2\right]}_{-D_{2l}} \mathfrak{u}_l(M_o)\,ds_{M_\circ}, \tag{11.115}$$

$$\mathfrak{s}_3 = t_{21} = -\oint_{\mathcal{L}_o} \underbrace{\left[\mathfrak{U}_{1\,l}(M_o, Q)\,\overset{Q}{\partial}_1\right]}_{-S_{3l}} \mathfrak{t}_l(M_o)\,ds_{M_\circ} + \oint_{\mathcal{L}_o} \underbrace{\left[\mathfrak{T}_{1\,l}(M_o, Q)\,\overset{Q}{\partial}_1\right]}_{D_{3l}} \mathfrak{u}_l(M_o)\,ds_{M_\circ}, \tag{11.116}$$

$$\mathfrak{s}_4 = t_{22} = -\oint_{\mathcal{L}_o} \underbrace{\left[\mathfrak{U}_{2\,l}(M_o, Q)\,\overset{Q}{\partial}_1\right]}_{-S_{4l}} \mathfrak{t}_l(M_o)\,ds_{M_\circ} + \oint_{\mathcal{L}_o} \underbrace{\left[\mathfrak{T}_{2\,l}(M_o, Q)\,\overset{Q}{\partial}_1\right]}_{D_{4l}} \mathfrak{u}_l(M_o)\,ds_{M_\circ}, \tag{11.117}$$

$$\mathfrak{m}_1 = \mu_{13} = \oint_{\mathcal{L}_o} \underbrace{\left[\mathfrak{U}_{3\,l}(M_o, Q)\,\overset{Q}{\partial}_2\right]}_{M_{1l}} \mathfrak{t}_l(M_\circ)\,ds_{M_\circ} - \oint_{\mathcal{L}_o} \underbrace{\left[\mathfrak{U}_{3\,l}(M_o, Q)\,\overset{Q}{\partial}_2\right]}_{N_{1l}}$$

$$\mathfrak{t}_l(M_o)\,ds_{M_\circ} - \oint_{\mathcal{L}_o} [\mathfrak{U}_{1\,l}(M_o, Q)]\;\mathfrak{t}_l(M_o)\,ds_{M_\circ} + \oint_{\mathcal{L}_o} [\mathfrak{U}_{1\,l}(M_o, Q)]\,(M_o)\mathfrak{t}_l\,ds_{M_\circ}, \tag{11.118}$$

$$\mathfrak{m}_2 = \mu_{23} = -\oint_{L_o} \underbrace{\left[\mathfrak{U}_{3\,l}(M_o, Q)\,\overset{Q}{\partial}_1\right]}_{-M_{2l}} \mathfrak{t}_l(M_o)\,ds_{M_\circ} + \oint_{K_o} \underbrace{\left[\mathfrak{T}_{3\,l}(M_o, Q)\,\overset{Q}{\partial}_1\right]}_{-N_{2l}}$$

$$\mathfrak{u}_l(M_o)\,ds_{M_\circ} - \oint_{L_o} [\mathfrak{U}_{2\,l}(M_o, Q)]\;\mathfrak{t}_l(M_o)\,ds_{M_\circ} + \oint_{L_o} [\mathfrak{T}_{2\,l}(M_o, Q)]\;\mathfrak{u}_l(M_o)\,ds_{M_\circ} \tag{11.119}$$

or in a more concise form

$$\mathfrak{s}_K(Q) = \oint_{\mathcal{L}_o} S_{Kl}(M_\circ, Q)\mathfrak{t}_l(M_\circ)\,ds_{M_\circ} + \oint_{\mathcal{L}_o} D_{Kl}(M_\circ, Q)\mathfrak{u}_l(M_\circ)\,ds_{M_\circ}, K = 1, \ldots, 4 \tag{11.120}$$

and

$$\mathfrak{m}_\kappa(Q) = \oint_{\mathcal{L}_o} M_{\kappa l}(M_\circ, Q)\mathfrak{t}_l(M_\circ)\, ds_{M_\circ} - \oint_{\mathcal{L}_o} N_{\kappa l}(M_\circ, Q)\mathfrak{u}_l(M_\circ)\, ds_{M_\circ}$$

$$- \oint_{\mathcal{L}_o} [\mathfrak{U}_{\kappa l}(M_\circ, Q)]\, \mathfrak{t}_l(M_\circ)\, ds_{M_\circ} + \oint_{\mathfrak{L}_o} [\mathfrak{T}_{\kappa l}(M_\circ, Q)]\, \mathfrak{u}_l(M_\circ)\, ds_{M_\circ}, \; \kappa = 1, 2. \tag{11.121}$$

Matrices $S_{Kl}(M_\circ, Q)$, $M_{\kappa l}(M_\circ, Q)$, $D_{\kappa l}(M_\circ, Q)$ and $N_{\kappa l}(M_\circ, Q)$ are all presented in thesis [30].

11.7 Dual Somiglian Formulae for Outer Regions

11.7.1 Stresses at Infinity

Let A^- denote the part of the plane that lies outside the contour $\mathcal{L}_o$ – see Fig. 11.4. Assume that the force stresses and couple stresses are constants at infinity. These are denoted by

$$t_{11}(\infty),\ t_{12}(\infty),\ t_{21}(\infty),\ t_{22}(\infty) \quad \text{and} \quad \mu_{13}(\infty),\ \mu_{23}(\infty).$$

By assumption the body forces and couples b_ρ and c_3 vanish at infinity.

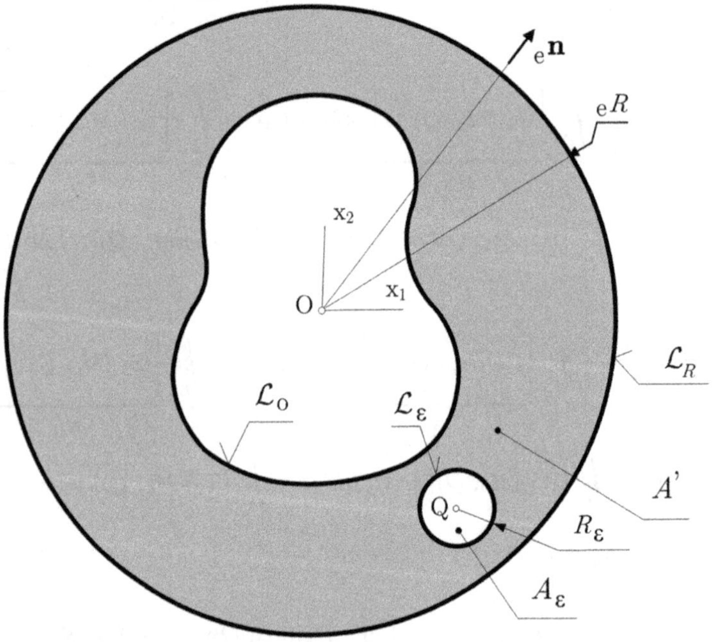

Fig. 11.4 The triple connected outer region

Since the couple stresses $\mu_{\nu 3}(\infty)$ are constants at infinity it follows from the equilibrium equation $(11.3)_2$ that the force stress tensor meets the symmetry condition

$$\varepsilon_{3\nu\rho} t_{\nu\rho}(\infty) = 0.$$

Consider the stress functions

$$\mathcal{F}_\rho = \varepsilon_{\pi\nu 3} t_{\rho\pi}(\infty)\xi_\nu + c_\rho(\infty). \tag{11.122}$$

and

$$\begin{aligned}\mathcal{H} = \mu_{13}(\infty)\xi_2 - \mu_{23}(\infty)\xi_1 + t_{11}(\infty)\frac{\xi_2^2}{2} - t_{12}(\infty)\xi_1\xi_2 \\ + t_{22}(\infty)\frac{\xi_1^2}{2} + c_1(\infty)\xi_2 - c_2(\infty)\xi_1\end{aligned} \tag{11.123}$$

where $c_\rho(\infty)$ is an arbitrary constant. If we substitute these stress functions into (11.6), in which we neglect the particular solutions since the body forces and couples are assumed to be zero, it follows that the stresses are constant on the entire plane. It is easy to see that the stress state

$$t_{11}(\infty),\ t_{12}(\infty),\ t_{21}(\infty),\ t_{22}(\infty) \quad \text{and} \quad \mu_{13}(\infty),\ \mu_{23}(\infty)$$

should be an elastic state of the whole plane (or of any subregion). As a result, the strains that belong to the above stress state should meet the dual balance equations (11.8) (the compatibility equations).

Since the couple stress tensor $\mu_{\nu 3}$ is constant it follows from the Hooke law (11.7) that the curvature twist tensor $\kappa_{\nu 3}$ is also constant, i.e., it satisfies the dual balance equation $(11.8)_1$.

If the force stress tensor $t_{\pi\rho}$ is constant then according to the Hooke law (11.7) so is the strain tensor $\gamma_{\pi\rho}$. The dual balance equation $(11.8)_2$ then implies that

$$D_\rho = \varepsilon_{3\nu\pi}\gamma_{\pi\rho}\partial_\nu - \kappa_{\rho 3} = -\kappa_{\rho 3} \neq 0 \tag{11.124}$$

which means that the second dual balance equation is satisfied if and only if $\kappa_{\rho 3} = 0$ or, equivalently given the Hooke law (11.2), if

$$\mu_{\nu 3}(\infty) = 0\,. \tag{11.125}$$

It then follows that the stress state at infinity is described by the stress functions

$$\begin{aligned}\mathcal{F}_1 = \tilde{u}_1(Q) = \xi_2 t_{11}(\infty) - \xi_1 t_{12}(\infty) + c_1(\infty), \\ \mathcal{F}_2 = \tilde{u}_2(Q) = \xi_2 t_{21}(\infty) - \xi_1 t_{22}(\infty) + c_2(\infty)\end{aligned} \tag{11.126}$$

and

$$\mathcal{H} = \tilde{\mathrm{u}}_3(Q) = t_{11}(\infty)\frac{\xi_2^2}{2} - t_{12}(\infty)\xi_1\xi_2 + t_{22}(\infty)\frac{\xi_1^2}{2} + c_1(\infty)\xi_2 - c_2(\infty)\xi_1. \quad (11.127)$$

which provide an elastic state on the whole plane (or on any part of it).

11.7.2 Derivation of the Dual Somigliana Formulae for Outer Regions

Figure 11.4 depicts a triple connected region A' bounded by the contours $\mathcal{L}_o$, $\mathcal{L}_\varepsilon$ and the circle $\mathcal{L}_R$ with radius ${}_eR$ and center O. Here $\mathcal{L}_\varepsilon$ is the contour of the neighborhood A_ε of Q with radius R_ε while ${}_e R$ is sufficiently large so that the region bounded by $\mathcal{L}_R$ covers both $\mathcal{L}_o$ and $\mathcal{L}_\varepsilon$. If ${}_eR \to \infty$ and $R_\varepsilon \to 0$ then clearly $A' \to A^-$.

Let $\mathrm{u}_k(M)$ and ${}^*\mathrm{u}_k(M)$ be sufficiently smooth elastic states (dual displacements) on A^-. The corresponding dual stresses on the contour are denoted by t_k and ${}^*\mathrm{t}_k$, respectively.

Equation

$$\begin{aligned}
&\int_{A'}\left[\mathrm{u}_k(M)\left(\overset{M}{\underset{kl}{\mathfrak{D}}}\, l\,\overset{*}{\mathrm{u}}_l(M)\right) - \underset{k}{\overset{*}{\mathrm{u}}}(M)\left(\overset{M}{\underset{kl}{\mathfrak{D}}}\,\mathrm{u}_l(M)\right)\right] dA_M \\
&\quad = \oint_{\mathcal{L}_o}\left[\mathrm{u}_l(M_o)\,\overset{*}{\mathrm{t}}_l(M_o) - \overset{*}{\mathrm{u}}_l(M_o)\mathrm{t}_l(M_o)\right] ds_{M_o} \\
&\quad + \oint_{\mathcal{L}_\varepsilon}\left[\mathrm{u}_l(M_o)\,\overset{*}{\mathrm{t}}_l(M_o) - \overset{*}{\mathrm{u}}_l(M_o)\mathrm{t}_l(M_o)\right] ds_{M_o} \qquad (11.128) \\
&\quad + \oint_{\mathcal{L}_R}\left[\mathrm{u}_l(M_o)\,\overset{*}{\mathrm{t}}_l(M_o) - \overset{*}{\mathrm{u}}_l(M_o)\mathrm{t}_l(M_o)\right] ds_{M_o}
\end{aligned}$$

is the dual Somigliana identity when it is applied to the triple connected region A'_e. Observe that M over a letter means that the corresponding derivatives are taken with respect to the coordinates of M. Let again

$$\overset{*}{\mathrm{u}}_l(M) = e_k(Q)\mathfrak{U}_{kl}(M, Q) \quad \text{and} \quad \overset{*}{\mathrm{t}}_l(M_o) = e_k(Q)\mathfrak{T}_{kl}(M_o, Q)$$

which provide a non singular elastic state of the plane in A'. We regard $\mathrm{u}_l(M)$ as a different elastic state in the region A^-. Further we assume that $\mathrm{u}_l(M)$ has the following far field pattern (asymptotic behavior)

$$\mathrm{u}_k = \tilde{\mathrm{u}}_k.$$

Without loss of generality, assume that the origin is an inner point of the region A^+. We need to consider three cases depending on the location of the point Q – Fig. 11.4 represents the first one.

1. If $Q \in A^-$ then region A' is the subject of our investigation. Substituting the above quantities into the Somigliana identity (11.128) and then noting that the surface integrals vanish and omitting $e_\kappa(Q)$ we have

$$\begin{aligned}
&\oint_{\mathcal{L}_o} [\mathfrak{T}_{kl}(M_o, Q)\mathfrak{u}_l(M_o) - \mathfrak{U}_{kl}(M_o, Q)\mathfrak{t}_l(M_o)]\, ds_{M_o} \\
&+ \oint_{\mathcal{L}_\varepsilon} [\mathfrak{T}_{kl}(M_o, Q)\mathfrak{u}_l(M_o) - \mathfrak{U}_{kl}(M_o, Q)\mathfrak{t}_l(M_o)]\, ds_{M_o} \\
&+ \oint_{\mathcal{L}_R} [\mathfrak{T}_{kl}(M_o, Q)\mathfrak{u}_l(M_o) - \mathfrak{U}_{kl}(M_o, Q)\mathfrak{t}_l(M_o)]\, ds_{M_o} = 0.
\end{aligned} \tag{11.129}$$

To compute the limit of the first two integrals as $R_\varepsilon \to 0$ and ${}_eR \to \infty$, note that they coincide with (11.100) if in the latter all terms are moved to the left side. Consequently

$$\oint_{\mathcal{L}_o} \cdots + \lim_{R_\varepsilon \to 0} \oint_{\mathcal{L}_\varepsilon} \cdots = \mathfrak{u}_k(Q) + \oint_{\mathcal{L}_o} [\mathfrak{T}_{kl}(M_\circ, Q)\mathfrak{u}_l(M_\circ) - \mathfrak{U}_{kl}(M_\circ, Q)\mathfrak{t}_l(M_\circ)]\, ds_{M_o}. \tag{11.130}$$

We now need to determine the limit of the third integral denoted by I_R:

$$\lim_{{}_eR \to \infty} \underbrace{\left\{ \oint_{\mathcal{L}_o} [\mathfrak{U}_{kl}(M_\circ, Q)\mathfrak{t}_l(M_\circ) - \mathfrak{T}_{kl}(M_\circ, Q)\mathfrak{u}_l(M_\circ)]\, ds_{M_o} \right\}}_{I_R}. \tag{11.131}$$

In this equation

- I_R coincides with the right side of the first dual Somigliana formula which is valid since Q is an inner point of the region;
- if ${}_eR \to \infty$ then $\mathfrak{u}_l$ and $\mathfrak{t}_l$ on the circle belong to the elastic state of the plane for which the force stresses are constant and the couple stresses vanish everywhere.

Therefore we get

$$\lim_{{}_eR \to \infty} I_R = \tilde{\mathfrak{u}}_k(Q).$$

Comparing (11.129), (11.130) and (11.132) yields the first dual Somigliana formula for exterior regions:

$$\mathfrak{u}_k(Q) = \tilde{\mathfrak{u}}_k(Q) + \oint_{\mathcal{L}_o} \mathfrak{U}_{kl}(M_\circ, Q)\mathfrak{t}_l(M_\circ)\, ds_{M_o} - \oint_{\mathcal{L}_o} \mathfrak{T}_{kl}(M_\circ, Q)\mathfrak{u}_l(M_\circ)\, ds_{M_o}. \tag{11.133}$$

2. If $Q = M_o$ is on $\mathcal{L}_o$ nothing changes concerning the limit of the integral taken on $\mathcal{L}_R$. Consequently

$$c_{kl}(Q_\circ)\mathfrak{u}_l(Q_\circ) = \tilde{\mathfrak{u}}_k(Q_\circ) + \oint_{\mathcal{L}_o} \mathfrak{U}_{kl}(M_\circ, Q_\circ)\mathfrak{t}_l(M_\circ)\, ds_{M_o} - \oint_{\mathcal{L}_o} \mathfrak{T}_{kl}(M_\circ, Q_\circ)\mathfrak{u}_l(M_\circ)\, ds_{M_o}. \tag{11.134}$$

Remark 21. Equation (11.134) is an integral equation for the unknowns $\mathfrak{u}_l(M_\circ)M_o \in \mathcal{L}_u$ and $\mathfrak{t}_l(M_\circ)M_o \in \mathcal{L}_t$. This is the integral equation of the direct method.
Remark 22.: We do not detail the derivations leading to (11.134). This is because the integrals taken on $\mathcal{R}_R$ are computed exactly the same way as before, and the other terms can be obtained analogously to the integral equation (11.105) we have set up for inner regions.

3. If Q is inside the contour $\mathcal{L}_o$, i.e., it lies in the region A^+, then it is easy to show that

$$0 = \tilde{\mathfrak{u}}_k(Q) + \oint_{\mathcal{L}_o} \mathfrak{U}_{kl}(M_\circ, Q)\mathfrak{t}_l(M_\circ)\, ds_{M_\circ} - \oint_{\mathcal{L}_o} \mathfrak{T}_{kl}(M_\circ, Q)\mathfrak{u}_l(M_\circ)\, ds_{M_\circ}. \tag{11.135}$$

which is the third Somigliana formula for exterior regions.

Remark 23.: Let us assume again recalling Remark 20 that $\mathfrak{u}_k$ is taken from (11.107) – see Remark 20 for details – and apply first relation (11.109) to region A_R in Fig. 11.4 ($\mathcal{L}_R$ and A_R should stand for $\mathcal{L}_o$ and A^+ in the relation we have referred to). Further let Q be an inner point of region A_R. Assume again that $\mathfrak{u}_{lj}$ and $\overset{*}{\mathfrak{t}}_k$ belong to the fundamental solutions. Then repeating the line of thought resulting in (11.111) we obtain

$$\int_{A_R} \mathfrak{u}_k({}'\mathfrak{D}_{kl}\overset{*}{\mathfrak{u}}_l)\, dA = \oint_{\mathcal{L}_R} \mathfrak{u}_k \overset{*}{\mathfrak{t}}_k\, ds = -\mathfrak{u}_l(Q)e_l(Q). \tag{11.136}$$

Next apply (11.109) to region A' in Fig. 11.4. We get

$$\oint_{\mathcal{L}_o} \mathfrak{u}_k \overset{*}{\mathfrak{t}}_k\, ds + \oint_{\mathcal{L}_o} \mathfrak{u}_k \overset{*}{\mathfrak{t}}_k\, ds = \int_{A'} \mathfrak{u}_k({}'\mathfrak{D}_{kl}\overset{*}{\mathfrak{u}}_l)\, dA, \tag{11.137}$$

where using (11.111) we have

$$\int_{A'} \mathfrak{u}_k\left({}'\mathfrak{D}^*_{kl}\mathfrak{u}_l\right)\, dA = -\eta(Q)\, e_l(Q) \left[\overset{o}{\mathcal{F}}{}_1\delta_{l1}|\overset{o}{\mathcal{F}}{}_2\delta_{l2}|\overset{o}{\mathcal{H}}{}_1\delta_{l3}\right]$$

while the integral on contour $\mathcal{L}_R$ can be taken from (11.136). Consequently the following equation

$$\oint_{L_o} \mathfrak{u}_k \overset{*}{\mathfrak{t}}_k \, ds - \mathfrak{u}_l(Q) e_l(Q) = -\eta(Q)\, e_l(Q) \,[\overset{o}{\mathcal{F}}_1 \,\delta_{l1} | \overset{o}{\mathcal{F}}_2 \,\delta_{l2} | \overset{o}{\mathcal{H}} \,\delta_{l3}]$$

holds independently of R, and hence for $R \to \infty$ as well. Recalling the definition of $\mathfrak{u}_k$ we arrive at the final form of the above equation

$$\oint_{\mathcal{L}_o} [\mathfrak{T}_{l1} + r_2 \mathfrak{T}_{l3} | \mathfrak{T}_{l2} - r_1 \mathfrak{T}_{l3} | \mathfrak{T}_{l3}] \, ds_{\overset{o}{M}} - [\delta_{l1} | \delta_{l2} | \delta_{l3}] = -\eta(Q) \, [\delta_{l1} | \delta_{l2} | \delta_{l3}], \tag{11.138}$$

where

$$\eta(Q) = \begin{cases} 1 & \text{if } Q \in A^- \\ 1/2 & \text{if } Q = Q_o \in \mathcal{L}_o \,. \\ 0 & \text{if } Q \in A^+ \end{cases} \tag{11.139}$$

This result will allow us to compute the strongly singular integrals in a numerical implementation set up for exterior regions.

11.8 Calculations of the Stresses on the Boundary

After solving the integral equations of the direct method ((11.105) for inner regions, (11.134) for exterior regions) we know the dual displacement vector $\mathfrak{u}_l$ as well as the dual stresses $\mathfrak{t}_1$, which can be given in terms of displacements derivatives with respect to the arc coordinate, on the contour.

The next question is how to determine the stresses on the contour in terms of these quantities.

Using (11.15), (11.16) and taking (11.5) into account we have

$$\frac{d\mathcal{F}_\rho}{ds} = n_1 t_{1\rho} + n_2 t_{2\rho}, \tag{11.140}$$

$$\frac{d\mathcal{H}}{ds} - n_\rho \mathcal{F}_\rho = n_1 \mu_{13} + n_2 \mu_{23}. \tag{11.141}$$

The above equations are to be supplemented by the definition of the dual stresses (11.81):

$$\mathfrak{t}_\rho = \tau_\pi \gamma_{\pi\rho} \quad \text{and} \quad \mathfrak{t}_3 = \tau_\pi \gamma_{\pi 3} \tag{11.142}$$

The left sides of (11.140),. . .,(11.142) are all known as soon as we have solved the integral equations of the direct method. Since the strains can always be given in terms of stresses via the Hooke law (11.7) the right hand sides contain the unknown force and couple stresses

$$t_{11}, \; t_{12}, \; t_{21} \; t_{22}, \; \mu_{12} \text{ and } \mu_{23}.$$

If we consider (11.140),. . .,(11.142) in the coordinate system (xyz) by substituting the relations $n_x = \tau_y$ and $n_y = -\tau_x$ into (11.142) we obtain the system of linear equations

$$\begin{bmatrix} n_x & 0 & n_y & 0 & 0 & 0 \\ 0 & n_x & 0 & n_y & 0 & 0 \\ 0 & 0 & 0 & 0 & n_x & n_y \\ -\frac{1-\nu}{2\mu}n_y & \dfrac{\alpha-\mu}{4\mu\alpha}n_x & \dfrac{\alpha+\mu}{4\mu\alpha}n_x & \dfrac{\nu}{2\mu}n_y & 0 & 0 \\ -\dfrac{\nu}{2\mu}n_x & -\dfrac{\alpha+\mu}{4\mu\alpha}n_y & -\dfrac{\alpha-\mu}{4\mu\alpha}n_y & \dfrac{1-\nu}{2\mu}n_x & 0 & 0 \\ 0 & 0 & 0 & 0 & -\dfrac{1}{\gamma+\varepsilon}n_y & \dfrac{1}{\gamma+\varepsilon}n_x \end{bmatrix} \begin{bmatrix} t_{xx} \\ t_{xy} \\ t_{yx} \\ t_{yy} \\ \mu_{xz} \\ \mu_{yz} \end{bmatrix} = \begin{bmatrix} \dfrac{d\mathfrak{u}_x}{ds} \\ \dfrac{d\mathfrak{u}_y}{ds} \\ \dfrac{d\mathfrak{u}_3}{ds} - n_x\mathcal{F}_x - n_y\mathcal{F}_y \\ \mathfrak{t}_x \\ \mathfrak{t}_y \\ \mathfrak{t}_z \end{bmatrix} \tag{11.143}$$

which can be solved for the stresses t_{xx}, t_{xy}, t_{yx} t_{yy}, μ_{xz} and μ_{yz}.

11.9 Solution Algorithm and Numerical Examples

11.9.1 The Algorithm

To solve the integral equation (11.134), we divide the contour $\mathcal{L}_o$ into n_{be} arcs, which are the boundary elements. The endpoints and midpoint of an element are, in many cases, the nodes on the element – their number is denoted by n_{en} (number of element nodes). The coordinates of an element as well as the functions $\mathfrak{u}_l(M_\circ)$ and $\mathfrak{t}_l(M_\circ)$ over an element are approximated by quadratic polynomials based on the nodal points. The order of the polynomial depends on the number of element nodes.

A quadratic approximation over an element might also be based on a different choice for the location of the nodal points. The approximation is partly discontinuous if the [first] {last} nodal point is inside the element, the [last] {first} nodal point is an endpoint of the element and there is a third nodal point between the two.

The nodal points are numbered locally increasing in the positive direction of the arc coordinate s and the node numbers take the values 1, 2 and 3.

The global node numbering differs from the local one. Let Q_i be the i-th node on the contour $\mathcal{L}_o : i = 1, 2, \ldots, n_{bn}$, where n_{bn} is the number of the boundary nodes on the contour. These are also numbered starting form an arbitrary nodal point of the contour and increasing in the positive direction.

In the computations we apply continuous (or partly discontinuous) quadratic shape functions. The arc coordinate on the element is mapped onto the interval $\eta \in [-1, 1]$ while the shape functions $N^i(\eta)$ are Lagrange polynomials given in terms of the nodal coordinates $\eta^1 < \eta^2 < \eta^3$; $\eta_i \in [-1, 1]\, i = 1, 2, 3$:

$$\begin{aligned} N^1(\eta) &= \frac{1}{(\eta^1-\eta^2)(\eta^1-\eta^3)}(\eta-\eta^2)(\eta-\eta^3), \\ N^2(\eta) &= \frac{1}{(\eta^2-\eta^3)(\eta^2-\eta^1)}(\eta-\eta^3)(\eta-\eta^1), \\ N^3(\eta) &= \frac{1}{(\eta^3-\eta^1)(\eta^3-\eta^2)}(\eta-\eta^1)(\eta-\eta^2) \end{aligned} \tag{11.144}$$

where $\eta^1 = -1$ and $-1 < \eta^2 < \eta^3 < 1$ if there is a discontinuity at the point $\eta = 1$ while $\eta^3 = 1$ and $-1 < \eta^1 < \eta^2 < 1$ if a discontinuity occurs at the point $\eta = -1$

Let $\overset{e}{x}{}^q_\lambda$, $\overset{e}{\mathrm{u}}{}^q_l$ and $\overset{e}{\mathrm{t}}{}^q_l$ be the nodal coordinates, the dual displacements and the dual stresses at the q-th nodal point of the element e. Then

$$\overset{e}{x}_\lambda = \sum_{q=1}^{n_{en}} N^q(\eta)\, \overset{e}{x}{}^q_\lambda, \quad \overset{e}{\mathrm{u}}_l = \sum_{q=1}^{n_{en}} N^q(\eta)\, \overset{e}{\mathrm{u}}{}^q_l \quad \text{and} \quad \overset{e}{\mathrm{t}}_l = \sum_{q=1}^{n_{en}} N^q(\eta)\, \overset{e}{\mathrm{t}}{}^q_l \tag{11.145}$$

are the approximations of the contour, the dual displacements and the dual stresses on the e-th element.

Using (11.145) and the definition of the shape functions (11.144) we can set up a closed form relation between ds and $d\eta$:

$$\mathrm{d}s = J(\eta)\mathrm{d}\eta. \tag{11.146}$$

The equation that provides $J(\eta)$ is not detailed here.

11.9.2 The Equation System to be Solved

Substituting approximations (11.145) into integral equation (11.105) and integrating element by element we obtain

$$\begin{aligned} c_{kl}(Q_\circ)\mathrm{u}_l(Q_\circ) &= \sum_{e=1}^{n_{be}} \int_{\mathcal{L}_o} \mathfrak{U}_{kl}(\eta, Q_\circ) \sum_{q=1}^{n_{en}} N^q(\eta)\, J(\eta)\mathrm{d}\eta\, \overset{e}{\mathrm{t}}{}^q_l \\ &- \sum_{e=1}^{n_{be}} \int_{\mathcal{L}_o} \mathfrak{T}_{kl}(\eta, Q_\circ) \sum_{q=1}^{n_{en}} N^q(\eta)\, J(\eta)\mathrm{d}\eta\, \overset{e}{\mathrm{u}}{}^q_l, \quad Q = Q_\circ \in \mathcal{L}_o. \end{aligned} \tag{11.147}$$

Let

$$\mathfrak{u}_j = \begin{bmatrix} \mathfrak{u}_1^j \\ \mathfrak{u}_2^j \\ \mathfrak{u}_3^j \end{bmatrix} \quad \text{and} \quad \mathfrak{t}_j = \begin{bmatrix} \mathfrak{t}_1^j \\ \mathfrak{t}_2^j \\ \mathfrak{t}_3^j \end{bmatrix}, \quad j = 1, ..., n_{bn} \tag{11.148}$$

be the matrices of the dual displacements and stresses at the nodal point j. For the whole contour equations

$$\mathfrak{u}^T = [\underbrace{\mathfrak{u}_1^1\,\mathfrak{u}_2^1\,\mathfrak{u}_3^1}_{\mathfrak{u}_1^T} \mid \underbrace{\mathfrak{u}_1^2\,\mathfrak{u}_2^2\,\mathfrak{u}_3^2}_{\mathfrak{u}_2^T} \mid ... \mid \underbrace{\mathfrak{u}_1^{n_{bn}}\,\mathfrak{u}_2^{n_{bn}}\,\mathfrak{u}_3^{n_{bn}}}_{\mathfrak{u}_{n_{bn}}^T}] \tag{11.149}$$

and

$$\mathfrak{t}^T = [\underbrace{\mathfrak{t}_1^1\,\mathfrak{t}_2^1\,\mathfrak{t}_3^1}_{\mathfrak{t}_1^T} \mid \underbrace{\mathfrak{t}_1^2\,\mathfrak{t}_2^2\,\mathfrak{t}_3^2}_{\mathfrak{t}_2^T} \mid ... \mid \underbrace{\mathfrak{t}_1^{n_{bn}}\,\mathfrak{t}_2^{n_{bn}}\,\mathfrak{t}_3^{n_{bn}}}_{\mathfrak{t}_{n_{bn}}^T}] \tag{11.150}$$

define the matrices of the dual displacements $\mathfrak{u}$ and the dual stresses $\mathfrak{t}$ where ^{T} denotes the transpose of a matrix. The value of the function *a(j,e)* is the local node number that belongs to the global node number j on element e. To simplify later derivations, we introduce the 3 × 3 submatrices $\mathbf{h}_{ij}$ and $\mathbf{b}_{ij}$ defined as

$$\hat{\mathbf{h}}_{ij} = \left[\sum_{e \in j} \int_{\mathcal{L}_e} \mathfrak{T}_{kl}(Q_i, \eta) N^{a(j,e)}(\eta) J(\eta)\, d\eta \right] \tag{11.151}$$

and

$$\mathbf{b}_{ij} = \left[\sum_{e \in j} \int_{\mathcal{L}_e} \mathfrak{U}_{kl}(Q_i, \eta) N^{a(j,e)}(\eta) J(\eta)\, d\eta \right] \tag{11.152}$$

in which (a) Q_i is the i-t nodal point referred to as collocation point; (b) the summation is over those boundary elements having the nodal point j as their common nodal point; and (c) $N^{a(j,e)}(\eta)$ is the a-th shape function. Introducing the additional notation

$$\mathbf{c}_{ii} = [c_{\kappa\lambda}(Q_i)], \tag{11.153}$$

and

$$\mathbf{h}_{ij} = \begin{cases} \hat{\mathbf{h}}_{ii} + \mathbf{c}_{ii}, & \text{if } i = j \\ \hat{\mathbf{h}}_{ij}, & \text{if } i \neq j \end{cases} \tag{11.154}$$

and then assuming that $Q_o = Q_i$, we can rewrite (11.147) as

$$\left[\mathbf{h}_{i1}\ \mathbf{h}_{i2} \cdots \mathbf{h}_{in_{bn}}\right]\begin{bmatrix}\mathbf{u}_1\\ \mathbf{u}_2\\ \cdots\\ \mathbf{u}_{n_{bn}}\end{bmatrix} = \left[\mathbf{b}_{i1}\ \mathbf{b}_{i2} \cdots \mathbf{b}_{in_{bn}}\right]\begin{bmatrix}\mathbf{t}_1\\ \mathbf{t}_2\\ \cdots\\ \mathbf{t}_{n_{bn}}\end{bmatrix}, \quad i = 1, ..., n_{bn} \tag{11.155}$$

Unification of the above equations yields

$$\begin{bmatrix}\mathbf{h}_{11} & \mathbf{h}_{12} & \cdots & \mathbf{h}_{1n_{bn}}\\ \mathbf{h}_{21} & \mathbf{h}_{22} & \cdots & \mathbf{h}_{2n_{bn}}\\ \cdots\cdots & \cdots\cdots & \cdots & \cdots\cdots\\ \mathbf{h}_{n_{bn}1} & \mathbf{h}_{n_{bn}2} & \cdots & \mathbf{h}_{n_{bn}n_{bn}}\end{bmatrix}\begin{bmatrix}\mathbf{u}_1\\ \mathbf{u}_2\\ \cdots\\ \mathbf{u}_{n_{bn}}\end{bmatrix} = \begin{bmatrix}\mathbf{b}_{11} & \mathbf{b}_{12} & \cdots & \mathbf{b}_{1n_{bn}}\\ \mathbf{b}_{21} & \mathbf{b}_{22} & \cdots & \mathbf{b}_{2n_{bn}}\\ \cdots\cdots & \cdots\cdots & \cdots & \cdots\cdots\\ \mathbf{b}_{n_{bn}1} & \mathbf{b}_{n_{bn}2} & \cdots & \mathbf{b}_{n_{bn}n_{bn}}\end{bmatrix}\begin{bmatrix}\mathbf{t}_1\\ \mathbf{t}_2\\ \cdots\\ \mathbf{t}_{n_{bn}}\end{bmatrix}, \tag{11.156}$$

which can be cast into a more compact form as

$$\mathbf{H}\mathbf{u} = \mathbf{B}\mathbf{t}. \tag{11.157}$$

In the above equation either $\mathbf{u}_{k}$ or $\mathbf{t}_{k}$ is known from the boundary conditions. Consequently we have as many equations as there are unknowns. After determining the unknown nodal values both $\mathfrak{u}_l(M_o)$ and $\mathfrak{t}_l(M_o)$ can be regarded as known functions on the contour $\mathcal{L}_o$.

Field variables at inner points are then obtained from (11.100) and (11.114),...,(11.119) and (11.120), (11.121).

Determination of the stresses on the boundary part $\mathcal{L}_u$ requires the solution of equations.

Determination of the diagonal submatrices $\mathbf{h}_{ij}$, $i = 1, \ldots, n_{bn}$ requires the computation of strongly singular integrals. However we can avoid computing strongly singular integrals if we follow the following procedure.

Let Q_i be the collocation point and typeset [the elements of the matrix $\mathbf{h}_{ij}$]{the matrix $\mathbf{h}_{ij}$ in full} in the following way:

$$h^{ij}_{kl} \quad \text{and} \quad \begin{bmatrix}h_{11}\ h_{12}\ h_{13}\\ h_{21}\ h_{22}\ h_{23}\\ h_{31}\ h_{33}\ h_{33}\end{bmatrix}^{ij}, \quad i, j = 1, \ldots, n_{bn}, \quad k, l = 1, 2, 3. \tag{11.158}$$

First assume that $\overset{o}{\mathcal{F}}_1 = \overset{o}{\mathcal{F}}_2 = 0$ and $\overset{o}{\mathcal{H}} = 1$. It follows from Remark 20 that no stresses and strains belong to the dual displacements

$$\mathfrak{u}_l(M, Q_i) = [0|0|1]$$

obtained from (11.107). Substituting the above dual displacements into (11.155) we have

$$\sum_{j=1}^{j=n_{bn}}\begin{bmatrix}h_{11}\ h_{12}\ h_{13}\\ h_{21}\ h_{22}\ h_{23}\\ h_{31}\ h_{33}\ h_{33}\end{bmatrix}^{ij}\begin{bmatrix}0\\0\\1\end{bmatrix} = \begin{bmatrix}0\\0\\0\end{bmatrix}$$

and hence

$$h^{ii}_{k3} = -\sum_{\substack{j=1\\ j\neq i}}^{j=n_{bn}} h^{ij}_{k3}\,. \tag{11.159}$$

Second assume that $\overset{o}{\mathcal{F}}_1 = 1$ and $\overset{o}{\mathcal{F}}_2 = \mathcal{H} = 0$. Again, the logic of Remark 20 implies that no stresses and strains belong to the dual displacements

$$\mathrm{u}_l(M, Q_i) = [1\,|\,0\,|\,r_2]$$

obtained from (11.107). Substitution of the above dual displacements into (11.155) yields

$$\sum_{j=1}^{j=n_{bn}} \begin{bmatrix} h_{11} & h_{12} & h_{13} \\ h_{21} & h_{22} & h_{23} \\ h_{31} & h_{33} & h_{33} \end{bmatrix}^{ij} \begin{bmatrix} 1 \\ 0 \\ r_2 \end{bmatrix}^{j} = \begin{bmatrix} 0 \\ 0 \\ 0 \end{bmatrix}$$

from where

$$h^{ii}_{k1} = -\sum_{\substack{j=1\\ j\neq i}}^{j=n_{bn}} h^{ij}_{k1} - h^{ii}_{k3} r^i_2 - \sum_{\substack{j=1\\ j\neq i}}^{j=n_{bn}} h^{ij}_{k3} r^j_2 = -\sum_{\substack{j=1\\ j\neq i}}^{j=n_{bn}} h^{ij}_{k1} - \sum_{\substack{j=1\\ j\neq i}}^{j=n_{bn}} h^{ij}_{k3}(r^i_2 - r^j_2). \tag{11.160}$$

Third we assume that $\overset{o}{F}_1 = H = 0$ and $\overset{o}{F}_2 = 1$. Then repeating the logic leading to (11.160) we get

$$h^{ii}_{k2} = -\sum_{\substack{j=1\\ j\neq i}}^{j=n_{bn}} h^{ij}_{k2} - \sum_{\substack{j=1\\ j\neq i}}^{j=n_{bn}} h^{ij}_{k3}(r^i_1 - r^j_1). \tag{11.161}$$

Using (11.159),. . .,(11.161) we can avoid computing strongly singular integrals.

For an outer region, we have to add a further term to the equation system to be solved. Define the matrix $\tilde{\mathbf{u}}$ as

$$\tilde{\mathbf{u}}^T = [\,\underbrace{\tilde{\mathrm{u}}^1_1\,\tilde{\mathrm{u}}^1_2\,\tilde{\mathrm{u}}^1_3}_{\tilde{\mathbf{u}}^T_1}\,|\,\underbrace{\tilde{\mathrm{u}}^2_1\,\tilde{\mathrm{u}}^2_2\,\tilde{\mathrm{u}}^2_3}_{\tilde{\mathbf{u}}^T_2}\,|\ldots|\,\underbrace{\tilde{\mathrm{u}}^{n_{bn}}_1\,\tilde{\mathrm{u}}^{n_{bn}}_2\,\tilde{\mathrm{u}}^{n_{bn}}_3}_{\tilde{\mathbf{u}}^T_{n_{bn}}}\,] \tag{11.162}$$

where the matrix $\tilde{\mathbf{u}}_j$ has the dual displacements $\tilde{\mathrm{u}}_k$ taken at the points $Q_j(j = 1, \ldots, n_{bn})$ as its elements. With this notation the equation system to be solved for the unknown nodal values assumes the form

$$\mathbf{H}\mathbf{u} = \tilde{\mathbf{u}} + \mathbf{B}\mathbf{t}\,. \tag{11.163}$$

The next question is how to compute the strongly singular integrals for exterior regions. Our procedure is as follows: (a) determine h_{kl}^{ii} under the assumption that the region under consideration is the interior region A^+; (b) then apply the formula

$$\underbrace{h_{kl}^{ii}}_{for\ A^-} = \underbrace{h_{kl}^{ii}}_{for\ A^+} + \delta_{kl} \tag{11.164}$$

The proof is based on (11.138). The details are, however, omitted here.

11.9.3 Examples

We have solved two external boundary value problems. First we consider the coordinate plane with a circular hole (Fig. 11.5b); second the coordinate plane with a rigid inclusion (Fig. 11.5c). Figure 11.5a shows the region to use if we solve the integral equation of the direct method in its traditional form, i.e., if the exterior region is replaced by a bounded one.

We remark that there exists an analytical solution for the stress concentration problem around the hole in an infinite plate of isotropic material if the stress state at infinity is $t_{xx} = p = \text{constant} = 100\,\text{n/mm}^2$, $t_{xy} = t_{yx} = t_{yy} = \mu_{xz} = \mu_{yz} = 0$, i.e., the plate is in tension [31–33]. The stress concentration factor K for the above problem can be calculated from the following equations:

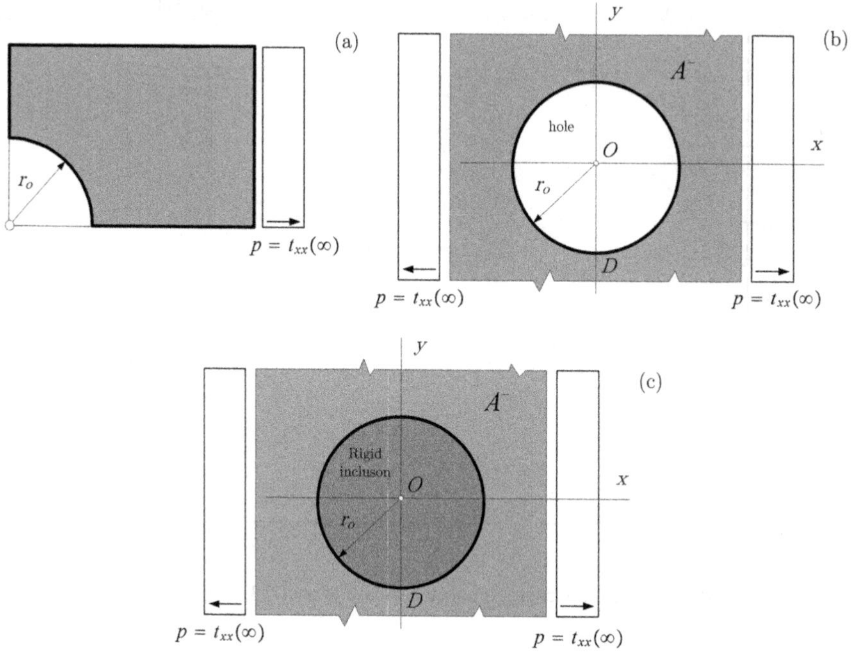

Fig. 11.5 Coordinate plane with circular hole and a rigid inclusion

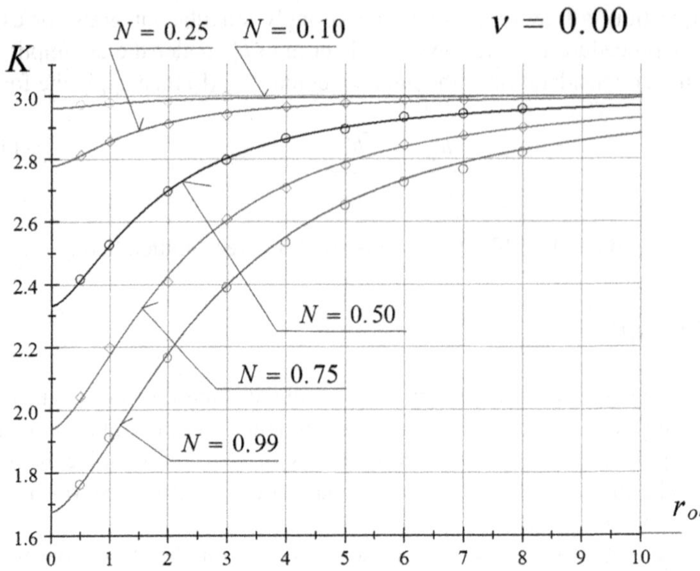

Fig. 11.6 The values of stress concentration factor in case of $\nu = 0.0$

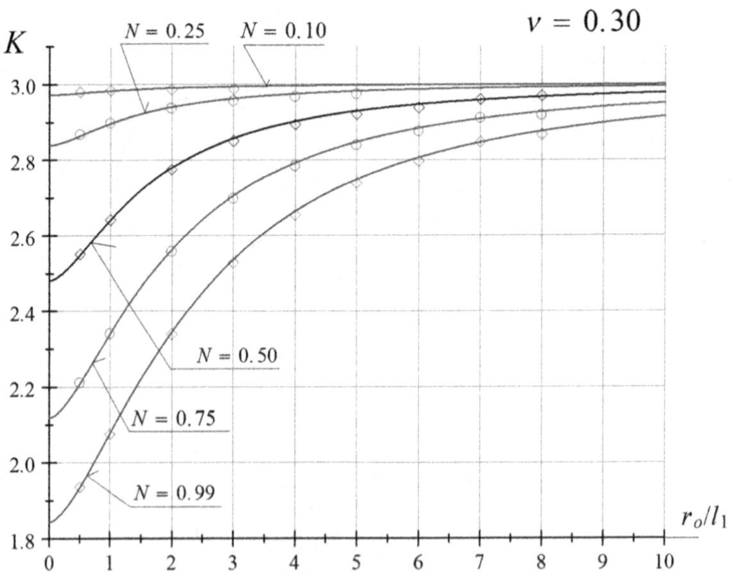

Fig. 11.7 The values of stress concentration factor in case of $\nu = 0.3$

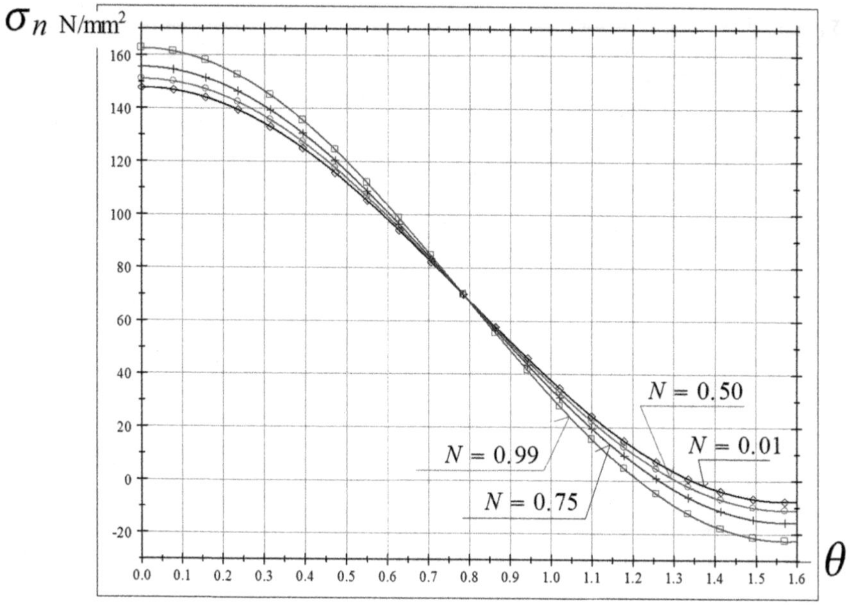

Fig. 11.8 The normal stress $\sigma_n = \sigma_r$ against the polar angle

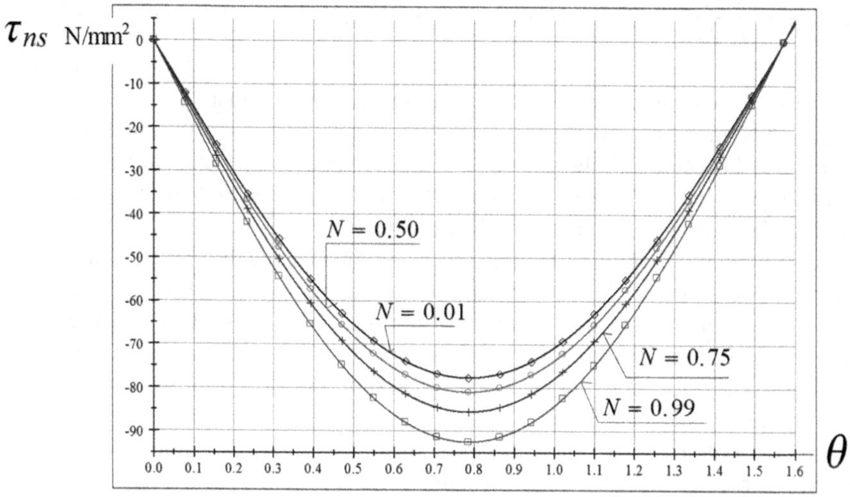

Fig. 11.9 The shear stress $\tau_{ns} = \tau_{r\theta}$ against the polar angle

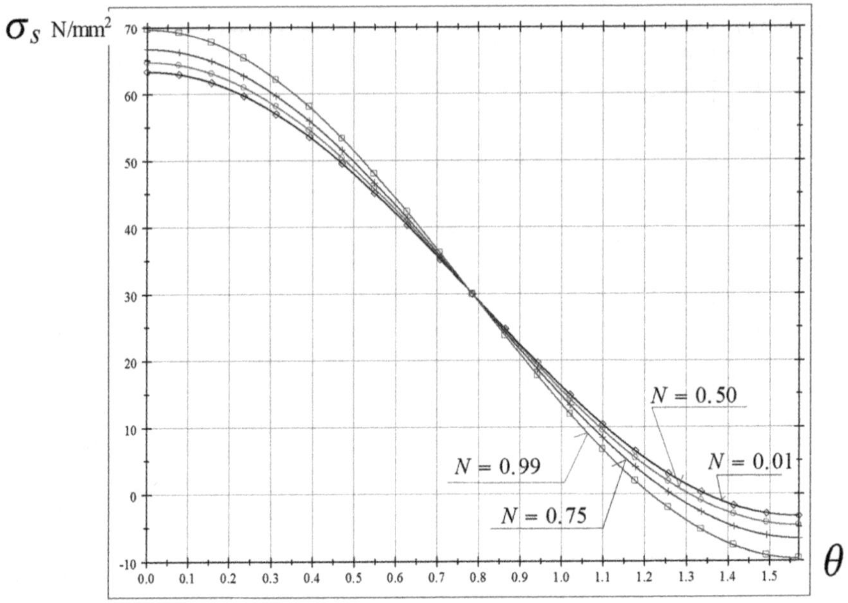

Fig. 11.10 The hoop stress $\sigma_s = \sigma_\theta$ .against the polar angle

$$K = \frac{t_{xx}(x = 0, y = r_o)}{t_{xx}(\infty)} = \frac{3 + F}{1 + F}, \quad F = 8(1 - \nu)N^2 \left[4 + \frac{r_o^2}{l_1^2} + \frac{2r_o}{l_1}\frac{K_0(r_o/l_1)}{K_1(r_o/l_1)}\right]^{-1} \tag{11.165}$$

We have carried out the computation under the assumption that
$\mu = 5\,\text{N/mm}^2$, $r_O = 0.36\,\text{mm}$, $\nu = 0.0$ (see Fig. 11.6) or $\nu = 0.0$ (see Fig. 11.7).

Figures 11.6 and 11.7 show the exact values for various couple numbers N (solid lines) as well as the values computed. The latter are denoted by diamonds and circles. These fit well the graphs of the exact values.

Results for the rigid inclusion have been computed using the same data as for the circular hole, however there are no computational results for $\nu = 0.0$. Figure 11.8 shows the normal stress $\sigma_n = \sigma_r$ against the polar angel $\theta \in [0, \pi/2]$. The black line represents the classical solution. This curve fits very well the results computed for $N = 0.01$. The graphs for $N = 0.50$, $N = 0.75$ and for $N = 0.99$ (couple stress theory) are also depicted in Fig. 11.8.

Figures 11.9 and 11.10 show the shear stress $\tau_{ns} = \tau_{r\theta}$ and the hoop stress $\sigma_s = \sigma_\theta$. The agreement with the classical solution for $N = 0.01$ is again very good.

11.10 Concluding Remarks

The present paper has dealt with the following issues:

1. Assuming isotropic and homogenous material we have presented the field equations and boundary conditions for the first plane problem of micropolar elasticity in a dual formulation. We have also clarified the supplementary conditions of

single valuedness that strains should meet for a class of mixed boundary value problems both for simply and multiply connected regions.
2. By applying Galjorkin functions and following the procedure presented in Kupradze's book [34] and elsewhere, we have derived the dual fundamental solutions of order one and two for the first plane problem of micropolar elasticity.[1]
3. We have set up the dual Somigliana relations both for inner and for exterior regions. A constant stress state at infinity is a part of our formulation for exterior regions. We also developed an integral representation for the stresses.
4. We developed a solution algorithm, coded it in Fortran 90, and used it to numerically solve simple boundary value problems in order to demonstrate the applicability of the solution algorithm.

We remark that the supplementary conditions of single valuedness should be incorporated into the algorithm if (a) the number of arcs on which tractions are prescribed is more than one or (b) if in addition to this the region under consideration is multiply connected. Work on these issues is in progress.

Acknowledgements The support provided by the Hungarian National Research Foundation within the framework of the project OTKA T 046834 is gratefully acknowledged.

References

1. Cosserat, E., Cosserat, F.: Théorie des corps déformables. Herman, Paris (1979)
2. Eringen, A.C.: Linear theory of micropolar elasticity. J. Math. Mech. **15**(6), 909–923 (1967)
3. Eringen, A.C., Suhubi, E.S.: Nonlinear theory of simple microelastic solids i. Int. J. Eng. Sci. **2**(2), 189–203 (1964)
4. Eringen, A.C., Suhubi, E.S.: Nonlinear theory of simple microelastic solids ii. Int. J. Eng. Sci. **2**(4), 389–404 (1964)
5. Nowaczky, W.: Theory of Micropolar Elasticity. Springer, Wien, NewYork, Udine (1970)
6. Nowaczky, W.: Theory of Elasticity. Mir, Moscow, in Russian (1970)
7. Tonti, E.: Variational principles in elastostatics. Meccanica **14**, 201–208 (1967)
8. Tonti, E.: A mathematical model for physical theories I., II. Rendiconti Accademia Nazionale dei Lincei. **LII** (2–3), 175–181, 351–356 (1972)
9. Szeidl, Gy.: Variational principles and solutions to some boundary value problems in the asymmetric elasticity [A nemszimmetrikus rugalmasságtan duál variációs elvei és egyes peremértékfeladatainak megoldása], PhD Thesis, Hungarian Academy of Sciences, Budapset (in Hungarian) (1985)
10. Kozák, I., Szeidl, Gy.: The field equations and the boundary conditions with force–stresses and couple–stresses in the linearized theory of micropolar elastostatics. Acta Techn. Hung. **90**(3/4), 57–80 (1980)
11. Schaefer, H.: Die Spannungsfunktionen des dreidimensionalen Kontinuums und des elastischen Körpers. Z. Angew. Math. Mech. **33**, 356–362 (1953)
12. Schaefer, H.: Die Spannungsfunktionen eines Kontinuums mit momentenspannungen i.–ii., Bulletin de l'Academie Polonaise des Sciences. Serie des sinces techniques. **15**(1), 63–67, 69–73, (1967)

[1] This procedure is attributed to Hörmander [26] by the boundary element community. To our knowledge however it is A.I. Lurie who applied this technique first to determine the fundamental solutions for the 3D problems of classical elasticity [25]

13. Carlson, D.E.: On Günthers stress functions for couple stresses. Q. Appl. Math. **25**, 139–146 (1967)
14. Nowaczky, W.: The plane problem of micropolar thermoelasticity. Arch. Mech. **22**, 3–26 (1970)
15. Iesan, D.: Existence theorems in the theory of micropolar elasticity. Int. J. Eng. Sci. **8**, 777–791 (1970)
16. Schiavone, P.: Integral equation methods in plane asymmetric elasticity. J. Elas. **43**, 31–43 (1996)
17. Huang, F.Y., Liang, K.Z.: Boundary element analysis of stress concentration in micropolar elastic plate. Int. J. Numer. Methods Eng. **40**, 1611–1622 (1997)
18. Sladek, V., Sladek, J.: Boundary element method in micropolar thermoelasticity. Part I. Boundary integral equations. Eng. Anal. Bound. Elem. **2**, 40–50 (1985)
19. Sladek, V., Sladek, J.: Boundary element method in micropolar thermoelasticity, Part II., Boundary integrodifferental equations. Eng. Anal. Bound. Elem. **2**, 81–91 (1985)
20. Sladek, V., Sladek, J.: Boundary element method in micropolar thermoelasticity, Part III. Numerical solution. Eng. Anal. Bound. Elem. **2**, 155–162 (1985)
21. Szeidl, Gy.: Dual variational principles for the first plane problem of micropolar elastostatics, Publ TUHI., Series D. Nat. Sci. **35**(3), 3–20 (1982)
22. Szeidl, Gy., Iván, I.: Fundamental solutions and boundary integral equations for the first plane problem of micropolar elastostatics in dual system. In Conference on Numerical Methods and Computational Mechanics, Abstracts,. University of Miskolc, Miskolc, p. 74, July 15–19 (1996)
23. Szeidl, Gy., Iván, I.: Macro conditions of compatibility and strain boundary condititons for somemixed plane boundary value problems of micropolar elastostatics, Publications of the University of Miskolc, Series D. Nat. Sci. Math. **36**(2), 35–45 (1996)
24. Kozák, I., Szeidl, Gy.: Contribution to the field equations and boundary conditions in terms of stresses of the first plane problem of micropolar elasticity. Publ TUHI., Series D. Nat. Sci. **34**(2), 135–146 (1981)
25. Lurie, A.I.: On theory of systems of linear differential equations with constant ceofficients, Transactions of the Leningrad Industrial Institute, 6, Section Phys. Math. **6**(3), 31–36 (1937)
26. Hörmander, L.: Liner Partial Differential Operators. Springer, Berlin (1964)
27. Erdélyi, A.: Higher Transcendental Functions, vol. 2. McGraw-Hill, NewYork, Toronto, London (1953)
28. Janke, E., Emde, F., Lösch, F.: Taflen Höheren Funktionen. B. G. Teubner Verlagsgesellschaft, Stuttgart (1960)
29. Jaswon, M.A., Symm, G.T.: Integral Equation Methods in Potential Theory and Elastostatics. Academic, London, NewYork, San Francisco (1977)
30. Dudra, J.: Boundary element method for plane problems of orthotropic and micropolar bodies in the primal and dual system of elasticity. PhD thesis, University of Miskolc, Miskolc (in Hungarian) (2009)
31. Sadd, M.H.: Elasticity, Theory, Applications and Numerics, pp. 395–397. Elsevier, New York, NY (2005)
32. Kaloni, P.N., Ariman, T.: Stress concentration effects in micropolar elasticity. J. Appl. Math. Phys. **18**, 136–141 (1967)
33. Kaloni, P.N., Ariman, T.: Stress concentration in micropolar elasticity. J. Appl. Math. Phys. **18**, 289–296 (1973)
34. Kupradze, V.D.: Potential Methods in Elasticity. Series Physico-Mathematical Library. State Publishing House, Moscow (in Russian) (1963)

Chapter 12
Implementation of Meshless Method for a Problem of a Plate Large Deflection

A. Uscilowska

Abstract Numerical methods for solving nonlinear problem of plate large deflection play an important role across many disciplines. In this article two alternative methods are proposed to solve problem of equations nonlinearity. Both methods require to solve linear biharmonic boundary value problem. For this purpose the Method of Fundamental solutions is implemented. Some numerical examples are presented to confirm that the proposed methods are good tools to solve nonlinear problems.

Keywords Large deflection of a plate · Nonlinear boundary value problem · Biharmonic equation · Method of fundamental solutions · Picard iterations · Homotopy analysis method

12.1 Introduction

This paper presents numerical solution to a problem of a large deflection of thin elastic plates. Some kinds of plate edges support are investigated. The phenomena is a boundary value problem [1]. It is described by von Karman equations, which are two highly nonlinear-coupled partial differential equations. These equations are in implicit form. It is not easy to solve these highly nonlinear-coupled equations analytically except for some special cases. Therefore, numerical methods are usually employed to deal with the large deflection problems. In the recent years, the boundary element methods are successfully applied to solve the problems of large deflection of thin elastic plate. The difficult point of using boundary element methods to solve such kind of problems is to find a fundamental solution of von Karman equations. Because of implicit form the system of equations is solved in an iterative fashion [2]. Therefore the technique for solving boundary value problem imposed

A. Uscilowska (✉)
Institute of Applied Mechanics, Poznan University of Technology, ul. Piotrowo 3, 60-965 Poznan, Poland
e-mail: anita.uscilowska@put.poznan.pl

J. Murín et al. (eds.), *Computational Modelling and Advanced Simulations*, Computational Methods in Applied Sciences 24, DOI 10.1007/978-94-007-0317-9_12,

with one partial differential equation and boundary condition is required. In this work presented here the Trefftz method is proposed for this goal. The solution of boundary value problem is based on the method of fundamental solution for space variables [3]. By this method the general solution is found. But the von Karman equations are inhomogeneous, therefore the techniques to calculate the particular solution are applied as well. For this purpose the method of radial basis functions is chosen [4, 5]. The second proposed method is Homotopy Analysis Method (HAM). The basic ideas of HAM are presented in [6, 7]. The application for the equation of second order is presented in [8], The proposal of this work is to implement HAM for system of nonlinear coupled equations including biharmonic operator. The basic step of HAM is to solve the biharmonic boundary value problem, which is done by Method of Fundamental Solutions.

12.2 Governing Equations

12.2.1 System of Equations

The details of the derivation of equation governing the finite deflection of thin plates are given in the classical book [9]. The equations are represented here for clarity and in order to refer to them during various stages of the numerical solution. The behaviour of moderately large deflection of thin elastic plates is governed by the following equations described by von Karman:

$$\nabla^4 w = \frac{g}{D} + \frac{h}{D} NL\,(w, F) \text{ in } \Omega \tag{12.1}$$

$$\nabla^4 F = NL\,(w, w) \text{ in } \Omega \tag{12.2}$$

where $w = w\,(x, y)$ is the deflection function, $F = F\,(x, y)$ is an Airy type stress function for the membrane stress, Δ^4 is the biharmonic operator, $D = Eh^3/12\left(1 - \nu^2\right)$ is the flexural rigidity of the plate having thickness h and elastic constants E, ν, Ω denotes a 2-D arbitrarily shaped region occupied by the plate, $\partial\Omega$ is the boundary of the region Ω, $L\,(w, F)$ is an operator, known as

$$NL\,(w, F) = \frac{\partial^2 w}{\partial x^2}\frac{\partial^2 F}{\partial y^2} + \frac{\partial^2 w}{\partial y^2}\frac{\partial^2 F}{\partial x^2} - 2\frac{\partial w}{\partial x \partial y}\frac{\partial F}{\partial x \partial y}. \tag{12.3}$$

12.2.2 Boundary Conditions

At every boundary point the boundary conditions should be defined. These conditions describe the way of the plate edge supporting. For the clamped edge point s the conditions are:

$$w(s) = 0 \; \frac{\partial w}{\partial n}(s) = 0 \quad F(s) = 0 \; \frac{\partial F}{\partial n}(s) = 0. \tag{12.4}$$

Simply supported edge is defined by

$$w(s) = 0\, M(s) = 0 \quad F(s) = 0\, \frac{\partial F}{\partial n}(s) = 0. \tag{12.5}$$

For the point s on the free edge the following conditions are fulfilled

$$M(s) = 0\, V(s) = 0 \quad F(s) = 0\, \frac{\partial F}{\partial n}(s) = 0. \tag{12.6}$$

where $V(s)$ is the reactive transverse force

$$V(s) = -D\left[\frac{\partial^3 w}{\partial n^3} + (2-\nu)\frac{\partial^3 w}{\partial n \partial s^2} + \frac{1}{\rho}\left(\frac{\partial^2 w}{\partial n^2} - (3-\nu)\frac{\partial^2 w}{\partial s^2}\right) - \frac{1}{\rho}\frac{\partial w}{\partial n}\right] \tag{12.7}$$

and $M(s)$ is the bending moment

$$M(s) = -D\left[\frac{\partial^2 w}{\partial n^2} + \nu\left(\frac{\partial^2 w}{\partial s^2} + \frac{1}{\rho}\frac{\partial w}{\partial n}\right)\right]. \tag{12.8}$$

The boundary conditions (12.6), (12.7) and (12.8) are written in the general form:

$$B_1 w\,(x,y) = g_1\,(x,y), \tag{12.9}$$

$$B_2 w\,(x,y) = g_2\,(x,y), \tag{12.10}$$

$$B_3 F\,(x,y) = g_3\,(x,y), \tag{12.11}$$

$$B_4 F\,(x,y) = g_4\,(x,y). \tag{12.12}$$

The (12.1) and (12.2) have something similarly to the biharmonic equations. Therefore, the procedures to solve problem under consideration will be based on algorithms for solving boundary value problems with system of biharmonic equations.

Two methods are proposed to solve nonlinear problem of the plate large deflection.

12.3 Numerical Approach

The solution of the considered problem requires application of special techniques to treat nonlinear problems. This paper proposes two alternative procedures to solve nonlinear problems: Picard iterations, Homotopy Analysis Method (HAM).

12.3.1 Picard Iterations

The governing equations are rewritten in the iterative form

$$\nabla^4 w^{(i)} = \frac{g}{D} + \frac{h}{D} NL\left(w^{(i-1)}, F^{(i-1)}\right), \tag{12.13}$$

$$\nabla^4 F^{(i)} = NL\left(w^{(i-1)}, w^{(i-1)}\right) \tag{12.14}$$

in Ω
and the boundary conditions are

$$B_1 w^{(i)}(x,y) = g_1(x,y), \tag{12.15}$$

$$B_2 w^{(i)}(x,y) = g_2(x,y), \tag{12.16}$$

$$B_3 F^{(i)}(x,y) = g_3(x,y), \tag{12.17}$$

$$B_4 F^{(i)}(x,y) = g_4(x,y). \tag{12.18}$$

To begin the iterative procedure the initial approximations of the solutions $w^{(0)}(x,y)$, $F^{(0)}(x,y)$ are to be calculated. For this purpose the auxiliary boundary value problems are solved

$$\nabla^4 w^{(0)}(x,y) = \frac{g}{D} \tag{12.19}$$

with the conditions

$$B_1 w^{(0)}(x,y) = g_1(x,y), \tag{12.20}$$

$$B_2 w^{(0)}(x,y) = g_2(x,y) \tag{12.21}$$

and

$$\nabla^4 F^{(0)}(x,y) = 0 \tag{12.22}$$

with conditions

$$B_3 F^{(0)}(x,y) = g_3(x,y), \tag{12.23}$$

$$B_4 F^{(0)}(x,y) = g_4(x,y). \tag{12.24}$$

Once the solutions $w^{(0)}(x,y)$, $F^{(0)}(x,y)$ are obtained the iterative procedure, defined by formulas (12.13), (12.14), (12.15), (12.16), (12.17) and (12.18) can be applied.

12.3.2 Homotopy Analysis Method (HAM)

The second procedure proposed to solve problem of nonlinearity of equations is Homotopy Analysis Method. The method is based on definition of homotopy family. For the considered (12.1) and (12.2) the homotopy families are described as:

$$\begin{aligned}&(1-\lambda)\,\nabla^4\left(w\left(\mathbf{x},\lambda,\tau\right)-w_0\left(\mathbf{x}\right)\right)=\\&=\tau\lambda\left(\nabla^4 w\left(\mathbf{x},\lambda,\tau\right)-\alpha_0 NL\left(w\left(\mathbf{x},\lambda,\tau\right),F\left(\mathbf{x},\lambda,\tau\right)\right)-f_1\left(\mathbf{x}\right)\right),\end{aligned}\tag{12.25}$$

$$\begin{aligned}&(1-\lambda)\,\nabla^4\left(F\left(\mathbf{x},\lambda,\tau\right)-F_0(\mathbf{x})\right)=\\&=\tau\lambda\left(\nabla^4 F\left(\mathbf{x},\lambda,\tau\right)+\tfrac{1}{2}NL\left(w\left(\mathbf{x},\lambda,\tau\right),w\left(\mathbf{x},\lambda,\tau\right)\right)-f_2(\mathbf{x})\right).\end{aligned}\tag{12.26}$$

where $\mathbf{x}=(x,y)$, $\alpha_0=\frac{h}{D}$, $f_1(\mathbf{x})=\frac{g}{D}$, $f_2(\mathbf{x})=0$. The parameter of homotopy τ is useful to control the convergence of the method. The other parameter $\lambda\in[0,1]$ gives: $w\left(\mathbf{x},0,\tau\right)=w_0(\mathbf{x})$, $F(\mathbf{x},0,\tau)=F_0(\mathbf{x})$ which are the known solutions of auxiliary linear boundary value problems; and $w\left(\mathbf{x},1,\tau\right)=w\left(\mathbf{x}\right)$, $F(\mathbf{x},1,\tau)=F(\mathbf{x})$ which are the solutions of the considered nonlinear problem. The approximations of the solutions are given as

$$w\left(\mathbf{x}\right)=w_0(\mathbf{x})+\sum_{m=1}^{M}\frac{w_0^{[m]}(\mathbf{x})}{m!},\tag{12.27}$$

$$F\left(\mathbf{x}\right)=F_0(\mathbf{x})+\sum_{m=1}^{M}\frac{F_0^{[m]}(\mathbf{x})}{m!}.\tag{12.28}$$

The mth-order deformation derivatives $w_0^{[m]}(\mathbf{x})$, $F_0^{[m]}(\mathbf{x})$ are obtained as the solutions of the linear boundary value problems given below.

First of these problems is described by biharmonic equation

$$\nabla^4 w_0^{[m]}(\mathbf{x})=R_{m,1}\left(\mathbf{x}\right),\ \text{for}\quad m=1,2,\ldots\tag{12.29}$$

where

$$R_{1,1}\left(\mathbf{x}\right)=\tau\left(A_1\left(w_0\left(\mathbf{x}\right),\ F_0\left(\mathbf{x}\right)\right)-f_1\left(\mathbf{x}\right)\right)\tag{12.30}$$

and

$$R_{m,1}\left(\mathbf{x}\right)=m\left(\nabla^4 w_0^{[m-1]}+\tau\left.\frac{\partial^{m-1}A_1\left(w\left(\mathbf{x},\lambda,\tau\right),\bar{F}\left(\mathbf{x},\lambda,\tau\right)\right)}{\partial\lambda^{m-1}}\right|_{\lambda=0}\right)\tag{12.31}$$

$$\text{for}\quad m=2,3,\ldots.$$

with boundary conditions

$$B_1 w_0^{[m]}(\mathbf{x}) = \delta_{m,1}\left(g_1(\mathbf{x}) - B_1 w_0(\mathbf{x})\right) \text{ for } \mathbf{x} \in \Gamma, \tag{12.32}$$

$$B_2 w_0^{[m]}(\mathbf{x}) = \delta_{m,1}\left(g_2(\mathbf{x}) - B_2 w_0(\mathbf{x})\right) \text{ for } \mathbf{x} \in \Gamma \tag{12.33}$$

for $m = 1, 2, \ldots$ and $\delta_{1,1} = 1$, $\delta_{m,1} = 0$ for $m \geq 2$.

The second one of boundary value problems is defined as

$$\nabla^4 F_0^{[m]}(\mathbf{x}) = R_{m,2}(\mathbf{x}), \text{ for } \quad m = 1, 2, \ldots \tag{12.34}$$

where

$$R_{1,2}(\mathbf{x}) = \tau\left(A_2\left(w_0(\mathbf{x}), F_0(\mathbf{x})\right) - f_2(\mathbf{x})\right) \tag{12.35}$$

and

$$R_{m,2}(\mathbf{x}) = m\left(\nabla^4 F_0^{[m-1]}(\mathbf{x}) + \tau \left.\frac{\partial^{m-1} A_2\left(w(\mathbf{x},\lambda,\tau), F(\mathbf{x},\lambda,\tau)\right)}{\partial \lambda^{m-1}}\right|_{\lambda=0}\right) \tag{12.36}$$

for $m = 2, 3, \ldots$.

with the boundary conditions given in the following form:

$$B_3 F_0^{[m]}(\mathbf{x}) = \delta_{m,1}\left(g_3(\mathbf{x}) - B_3 F_0(\mathbf{x})\right) \text{ for } \mathbf{x} \in \Gamma, \tag{12.37}$$

$$B_4 F_0^{[m]}(\mathbf{x}) = \delta_{m,1}\left(g_4(\mathbf{x}) - B_4 F_0(\mathbf{x})\right) \text{ for } \mathbf{x} \in \Gamma \tag{12.38}$$

for $m = 1, 2, \ldots$ and $\delta_{1,1} = 1$, $\delta_{m,1} = 0$ for $m \geq 2$.

The auxiliary differentials of the equations operators with respect to the homotopy parameter are:

$$\begin{aligned} &\left.\frac{\partial^m A_1\left(w(\mathbf{x},\lambda,\tau), F(\mathbf{x},\lambda,\tau)\right)}{\partial \lambda^m}\right|_{\lambda=0} = \\ &= \nabla^2\nabla^2 w_0^{[m]}(\mathbf{x}) + \alpha_0 \sum_{i=0}^{m} \binom{m}{i} NL\left(w_0^{[m]}(\mathbf{x}), F_0^{[i]}(\mathbf{x})\right) \end{aligned} \tag{12.39}$$

$$\begin{aligned} &\left.\frac{\partial^m A_2\left(w(\mathbf{x},\lambda,\tau), F(\mathbf{x},\lambda,\tau)\right)}{\partial \lambda^m}\right|_{\lambda=0} = \\ &= \nabla^2\nabla^2 w_0^{[m]}(\mathbf{x}) + \frac{1}{2} \sum_{i=0}^{m} \binom{m}{i} NL\left(w_0^{[m]}(\mathbf{x}), w_0^{[i]}(\mathbf{x})\right). \end{aligned} \tag{12.40}$$

In both Picard iterations and HAM the basic step is to solve boundary value problem consists of system of linear biharmonic equation with boundary conditions. The proposal of this paper is to use the Method of Fundamental Solutions (MFS) to obtaining solutions of such problems.

12.3.3 Method of Fundamental Solutions

The linear biharmonic equation is written as

$$\nabla^4 u(x,y) = f(x,y) \tag{12.41}$$

and is considered on the region Ω.

The boundary conditions have the general form

$$B_1 u = g_1(x,y) \text{ on } \partial\Omega, \tag{12.42}$$

$$B_2 u = g_2(x,y) \text{ on } \partial\Omega, \tag{12.43}$$

where B_1, B_2 are operators imposed as boundary conditions and $\partial\Omega$ is the boundary of the region Ω.

Let us denote $\{P_i = (x_i, y_i)\}_{i=1}^{N}$ to be N collocations points in $\Omega \cup \partial\Omega$ of which $\{(x_i, y_i)\}_{i=1}^{Nl}$ are interior points; $\{(x_i, y_i)\}_{i=Nl+1}^{N}$ are boundary points.

The right-hand side function f is approximated as Radial Basis Functions (RBFs) as

$$f_N(x,y) = \sum_{j=1}^{N} a_j \varphi(r_j) + \sum_{k=1}^{l} b_k p_k(x,y) \tag{12.44}$$

where $r_j = \sqrt{(x-x_j)^2 + (y-y_j)^2}$, $\varphi(r_j) = \varphi\left(\sqrt{(x-x_j)^2 + (y-y_j)^2}\right) : R^d \to R^+$ is a RBF, $\{p_k\}_{k=1}^{l}$ is the complete basis for d-variate polynomials of degree $\leq m-1$, and C_{m+d-1}^{d} is the dimension of P_{m-1}. The coefficients $\{a_j\}$, $\{b_k\}$ can be found by solving the system

$$\sum_{j=1}^{N} a_j \varphi(r_{ji}) + \sum_{k=1}^{l} b_k p_k(x_i, y_i) = f(x_i, y_i), \quad 1 \leq i \leq N, \tag{12.45}$$

$$\sum_{j=1}^{N} a_j p_k(x_j, y_j) = 0, \quad 1 \leq k \leq l, \tag{12.46}$$

where $r_{ji} = \sqrt{(x_i-x_j)^2 + (y_i-y_j)^2}$, $\{x_i, y_i\}_{i=1}^{N}$ are the collocation points on $\Omega \cup \partial\Omega$ (see Fig. 12.1).

The approximate particular solution u_p can be obtained using the coefficients $\{a_j\}$ and $\{b_k\}$

$$u_p(x,y) = \sum_{j=1}^{N} a_j \Phi_j(r_j) + \sum_{k=1}^{N} b_k \Psi_k(x,y), \tag{12.47}$$

Fig. 12.1 Collocation points

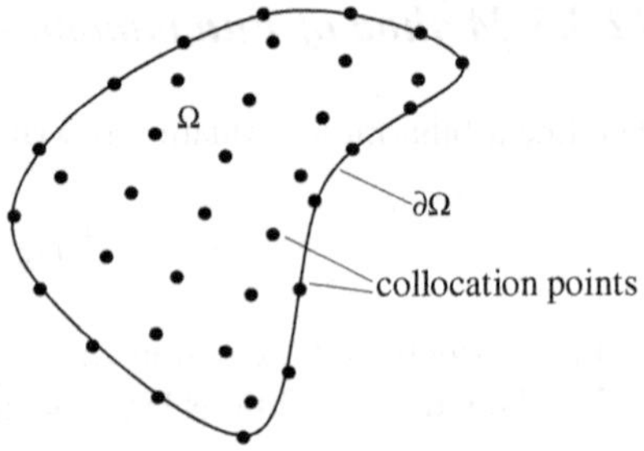

where

$$\nabla^4 \Phi_j = \varphi_j (r_j) \quad \text{for } j = 1, \ldots, N, \tag{12.48}$$

$$\nabla^4 \Psi_i = p_i (x, y) \quad \text{for } i = 1, \ldots, l. \tag{12.49}$$

Solution of the boundary value problem with the differential equation (12.41) and the boundary conditions (12.42) and (12.43) can be given as

$$u = u_p + u_h \tag{12.50}$$

where u_h is solution of boundary value problem in the form

$$\nabla^4 u_h = 0 \text{ in } \Omega, \tag{12.51}$$

$$B_1 u_h = g_1 (x, y) - B_1 u_p \text{ on } \partial\Omega, \tag{12.52}$$

$$B_2 u_h = g_2 (x, y) - B_2 u_p \text{ on } \partial\Omega. \tag{12.53}$$

The method of fundamental solution is used to solve problems (12.51), (12.52) and (12.53); this means that the approximation of the homogeneous solution is a linear superposition of the special functions, so-called fundamental solutions, as

$$u_h (x, y) = \sum_{j=1}^{Ns} c_j fs_1 \left(r_j^s\right) + \sum_{j=1}^{Ns} c_{Ns+j} fs_2 \left(r_j^s\right) \tag{12.54}$$

where $fs_1\left(r_j^s\right)$, $fs_2\left(r_j^s\right)$ are the fundamental solution functions. The set of so called source points $\left\{\left(x_j^s, y_j^s\right)\right\}_{j=1,\,Ns}$ is introduced (see Fig. 12.2). So, the distance r_j^s is defined as $r_j^s = \sqrt{\left(x - x_j^s\right)^2 + \left(y - y_j^s\right)^2}$.

Putting (12.54) into boundary conditions (12.52) and (12.53) gives the system of linear algebraic equations:

$$\sum_{j=1}^{Ns} c_j B_1 fs_1 \left(r_{ji}^s\right) + \sum_{j=1}^{Ns} c_{Ns+j} B_1 fs_2 \left(r_{ji}^s\right) = g_1 (x_i, y_i) - B_1 u_p (x_i, y_i) \tag{12.55}$$

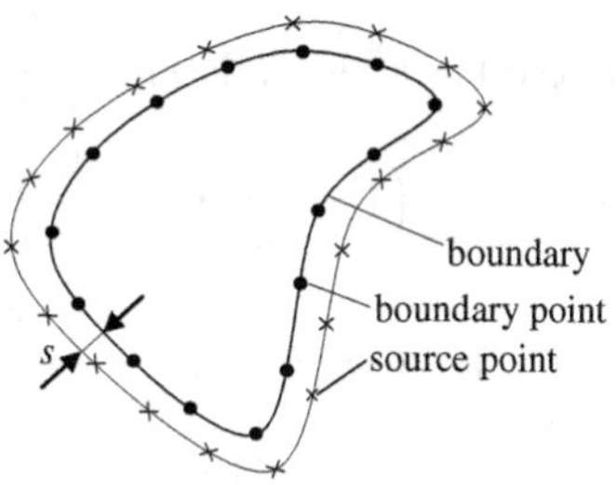

Fig. 12.2 Boundary and source points

for $Nl+1 \le i \le N$,

$$\sum_{j=1}^{Ns} c_j B_2 fs_1\left(r_{ji}^s\right) + \sum_{j=1}^{Ns} c_{Ns+j} B_2 fs_2\left(r_{ji}^s\right) = g_2\left(x_i, y_i\right) - B_2 u_p\left(x_i, y_i\right) \tag{12.56}$$

for $Nl+1 \le i \le N$,
where (x_i, y_i) for $Nl+1 \le i \le N$ are the boundary points, and $r_{ji}^s = \sqrt{\left(x_i - x_j^s\right)^2 + \left(y_i - y_j^s\right)^2}$.

The solutions of the system (12.55) and (12.56), which are the coefficients c_j give the homogeneous solution in form (12.54) of the considered boundary value problem (12.41), (12.42) and (12.43). The particular solution of the boundary problem is calculated using (12.47). Therefore the solution of the boundary value problem (12.41), (12.42) and (12.43) is calculated with the formula (12.50).

The fundamental solutions for biharmonic equation are the functions:

$$fs_1(r) = \ln(r) \tag{12.57}$$

$$fs_2(r) = r^2 \ln(r) \tag{12.58}$$

There are several proposals of the Radial Basis Functions in literature. Some of Radial Basis Functions and the particular solutions of biharmonic equation are presented below for convenience of reader:

- $\varphi(r) = 1 + r$, $\Phi(r) = r^4\left(\frac{1}{64} + \frac{r}{225}\right)$
- $\varphi(r) = r^2 \ln(r)$; $\Phi(r) = r^6\left(\frac{\ln(r)}{576} - \frac{5}{3456}\right)$
- $\varphi(r) = \sqrt{r^2 + c^2}$;

$$\Phi(r) = \frac{1}{900}\left(\sqrt{r^2+c^2}\left(-61c^4 + 48c^2r^2 + 4r^4\right) + \left(30c^2 - 75r^2\right)c^3 \ln\left[2\left(c + \sqrt{r^2+c^2}\right)\right]\right)$$

- $\varphi(r) = r^2 + r^3;\ \Phi(r) = \dfrac{r^6}{576} + \dfrac{r^7}{1225}$
- $\varphi(r) = \left(1 - \dfrac{r}{a}\right)^4 \left(1 + \dfrac{4r}{a}\right);$

$$\mathrm{r} < \mathrm{a} : \Phi(r) = r^4 \left(\frac{1}{64} - \frac{5r^2}{288a^2} + \frac{4r^3}{245a^3} - \frac{5r^4}{768a^4} + \frac{4r^5}{3969a^5}\right)$$

$$\mathrm{r} > \mathrm{a} : \Phi(r) = \frac{109r^2a^2}{23520} + \frac{r^2a^2}{56} \ln\left(\frac{r}{a}\right)$$

- $\varphi(r) = \dfrac{1}{\sqrt{r^2 + c^2}};$

$$\Phi(r) = \frac{1}{36}\left(-\left(11c^2 - 4r^2\right)\sqrt{r^2 + c^2} + \left(6c^2 - 9r^2\right) c \ln\left(2\left(c + \sqrt{r^2 + c^2}\right)\right)\right)$$

- $\varphi(r) = a_0 + a_1 r + a_2 r^2 + a_3 r^3; \quad \Phi(r) = r^4 \left(\dfrac{a_0}{64} + \dfrac{a_1 r}{225} + \dfrac{a_2 r^2}{576} + \dfrac{a_3 r^3}{1225}\right).$

12.4 Numerical Approach

12.4.1 Numerical Solution of Bihamonic Problem

To show the effectiveness of Method of Fundamental Solutions the boundary value problem with the biharmonic linear equation is solved. The problem is described by equation:

$$\nabla^4 u(x, y) = 72xy \text{ in } \Omega \tag{12.59}$$

and the boundary conditions

$$u(x, y) = x^3 y^3 \text{ on } \partial\Omega, \tag{12.60}$$

$$\nabla^2 u(x, y) = 6xy(x^2 + y^2) \text{ on } \partial\Omega \tag{12.61}$$

where $\Omega = \{(x, y)| - 1 \le x \le 1, -1 \le y \le 1\}$ and $\partial\Omega$ is the boundary of the region Ω.

The parameters for the Method of Fundamental Solutions are chosen as: $N = 400$, $N\text{-}Nl = 44$, $N_S = 36$, $s = 0.2$.

The right-hand side function of (12.59) is approximated by the Radial Basis Function, which is Thin Plate Spline of the form:

$$\varphi(r) = r^2 \ln(r) \tag{12.62}$$

The absolute error of the approximation is presented on Fig. 12.3. The error is of order 10^{-10}, which shows good quality of the approximation procedure.

The Fig. 12.4 shows the approximated solution of the considered problem.

To check the quality of calculations this solution is compared to the exact solution. The absolute error is plotted in Fig. 12.5. The error is of order 10^{-6}, which

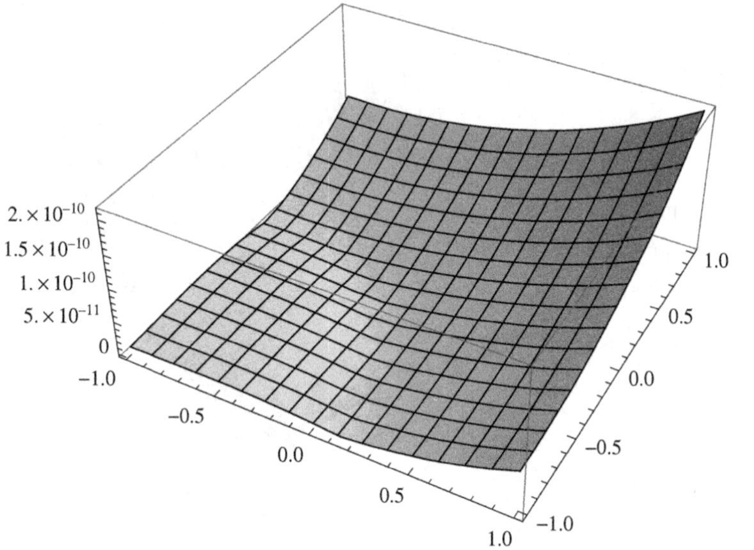

Fig. 12.3 The absolute error of the approximation

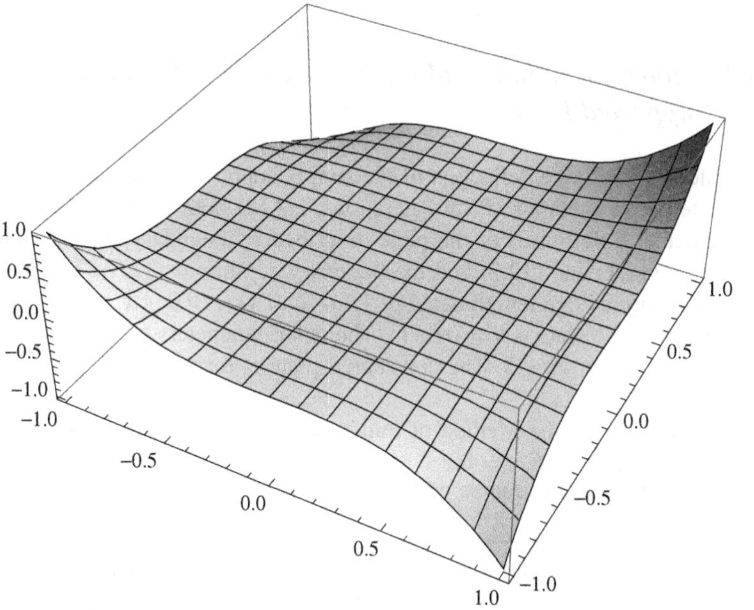

Fig. 12.4 The approximated solution

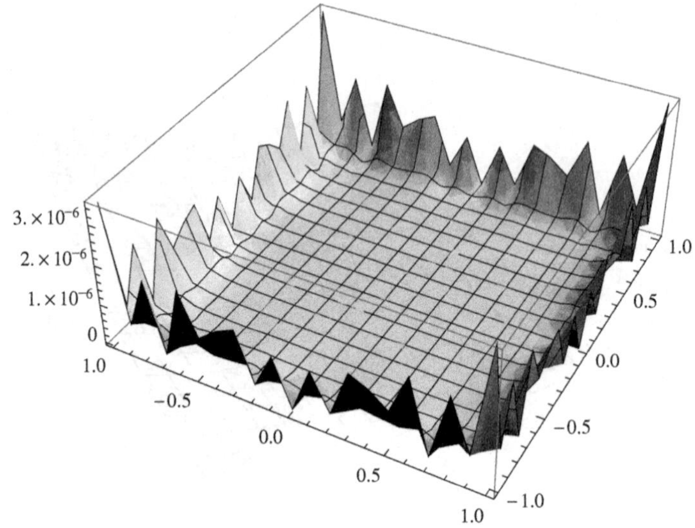

Fig. 12.5 The absolute error of the solution

shows that the Method of Fundamental Solutions is good tool to solve boundary value problem with biharmonic equation.

12.4.2 Numerical Solution of Large Deflection of Simply Supported Plate

The problem of the large deflection of the simply supported plate has been solved by two alternative methods (Picard iterations, HAM) both supported by Method of Fundamental Solutions. For the plate described by boundary value problem is described by (12.1) and (12.2) and the boundary conditions (12.5). The plate has square shape of the edge length equal to 1, so $\Omega = \{(x,y)|\, 0 \le x \le 1, 0 \le y \le 1\}$. The thickness of the plate is $h = 0.02$ m. The other parameters of the plate material are: $E = 3.10^7$ Pa, $\nu = 0.3$. The considered plate is loaded by constant external pressure $g = 2.5\,\mathrm{N/m}^2$.

The parameters for the Method of Fundamental Solutions are chosen as: $N = 400$, $N\text{-}Nl = 44$, $N_S = 36$, $s = 0.2$.

The iterative procedure can be performed until the condition

$$\min\left(\frac{\sqrt{\sum_{k=1}^{J}\left[w^{(i)}\left(X_k^t, Y_k^t\right) - w^{(i-1)}\left(X_k^t, Y_k^t\right)\right]^2}}{J}, \frac{\sqrt{\sum_{k=1}^{J}\left[F^{(i)}\left(X_k^t, Y_k^t\right) - F^{(i-1)}\left(X_k^t, Y_k^t\right)\right]^2}}{J}\right) < \varepsilon$$

$$\text{for} \quad i = 2, 3, \ldots \tag{12.63}$$

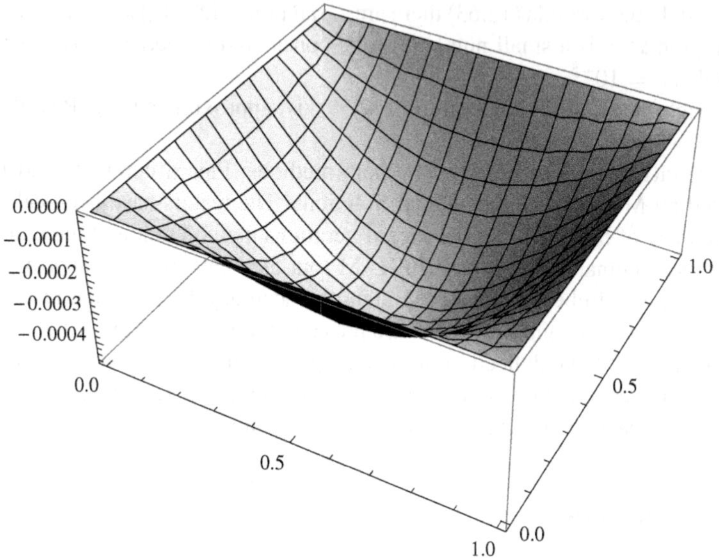

Fig. 12.6 The large deflection of the plate

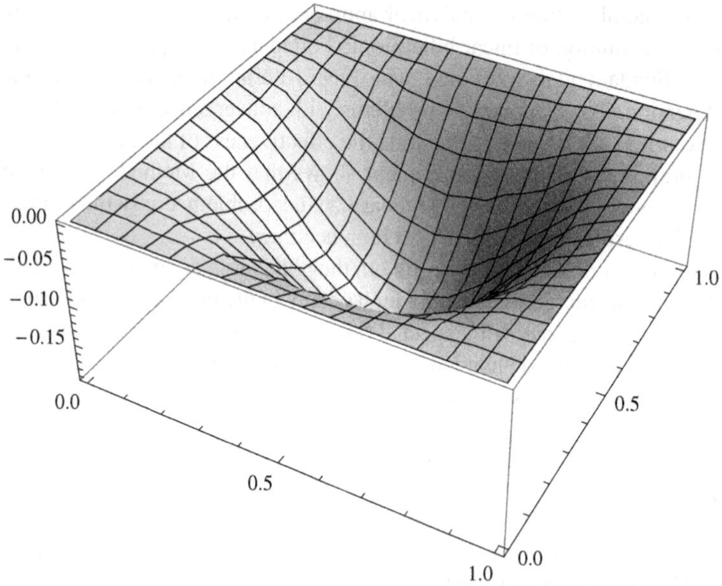

Fig. 12.7 The membrane forces in the plate

is fulfilled. In the formula (12.63) there are J trial points $\left(X_k^t, Y_k^t\right)$ $k = 1,2,..,J$, chosen in the region Ω, ε is a small number. In the considered numerical experiment it is chosen that $\varepsilon = 10^{-5}$.

The right-hand side function of (12.59) is approximated by Radial Basis Function, which is Thin Plate Spline.

The results have been obtained by both methods: Picard iterations and HAM, supported by Method of Fundamental Solutions. The results have been calculated with accuracy of order 10^{-12}. The Picard iterations method required three iteration steps to reach demanded accuracy. In HAM four elements of series approximating the solutions had to be calculated to obtain the accuracy 10^{-10} of the results. The deflection of the plate and function F are presented in Figs. 12.6 and 12.7.

On the Fig. 12.6 the deflection of the plate is presented. It is possible to observe that the boundary conditions are hold. Also, for the membrane forces, plotted in Fig. 12.7, it is easy to observe that the membrane boundary conditions hold.

12.5 Conclusions

Two methods (Picard iterations and HAM) for the analysis of thin plates undergoing large deflections are presented. The meshless method (Method of Fundamental Solutions) is proposed to support the presented methods. The main benefit of presented meshless based methods is that a novel approach to calculate the plate stressed without integral. The Homotopy Analysis Method is analytical-numerical method. It is combination of analytical approach for the differential deformations and numerical solution of linear biharmonic boundary value problems (realized by Method of Fundamental Solutions). The Picard iterations method is presented in simplest version. There appear the problems with convergence of the methods.

Two examples are applied to illustrate the effectiveness of the proposed methods. The first proposed numerical experiment shows that the Method of Fundamental Solutions is good tool to solve boundary value problem consisting biharmonic equation. The results are obtained with high accuracy.

The second example is to solve problem the plate large deflection. The deflection of the plate and the membrane force, which occurs in the plate, have been calculated. The obtained results agree with expected ones. The convergence of the both method has been achieved with demanded accuracy in very few steps. So, the proposed techniques are efficient to solve numerically problem of the plate large deflection.

References

1. Tsiatas, G.C., Katsikadelis, J.T.: Large deflection analysis of elastic space membranes. Int. J. Numer. Methods Eng. **65**, 264–294 (2006)
2. Wang, W., Ji, X., Tanaka, M.: A dual reciprocity boundary element approach for the problems of large deflection of this elastic plates. Comput. Mech. **28**, 58–65 (2000)
3. Pullikkas, A., Karageorghis, A., Georgiou, G.: Methods of fundamental solutions for harmonic and biharmonic boundary value problems. Comput. Mech. **21**, 16–423 (1998)

4. Wang, J.G., Liu, G.R.: On the optimal shape parameters of radial basis functions used for 2-D meshless methods. Comput. Methods Appl. Mech. Eng. **191**, 2611–2630 (2002)
5. Cheng, A.H.-D.: Particular solutions of Laplacian, Helmholtz-type, and polyharmonic operators involving higher order radial basis functions. Eng. Anal. Bound. Elem. **24**, 531–538 (2000)
6. Liao, S.: Homotopy Analysis Method: a new analytical technique for nonliner problems. Commun. Nonlin Sci. Numer. Simulat. **2**, 95–99 (1997)
7. Liao, S.: Boundary element method for general nonlinear differential operators. Eng. Anal. Bound. Elem. **20**, 91–99 (1997)
8. Liao, S.: General boundary element method for Poisson equation with spatially varying conductivity. Eng. Anal. Bound. Elem. **21**, 23–38 (1998)
9. Timoszenko, S.P., Woinowski-Kreiger, S.: Theory of Plates and Shells. McGraw, New York, NY (1959)

Chapter 13
Modelling and Spatial Discretization in Depletion Calculations of the Fast Reactor Cell with HELIOS 1.10

R. Zajac, P. Darilek, and V. Necas

Abstract The article presents advanced application of computational code HELIOS 1.10 and several application problems of this computer code caused by mathematical solutions of reactor theory physical methods in the program source. The first part of the article presents the HELIOS engineering application on the transmutation fuel cycle modelling where the VVER-440 and SUPERPHENIX fuel assemblies were modelled and compared. The French fast reactor SUPERPHENIX was chosen for the transmutation purposes because this fast reactor is compatible in principle with the Generation IV strategy and Slovakia uses the Russian nuclear reactor VVER-440. The code HELIOS is used mostly for the preparation of few group cross-sections for the use in 3D full core simulation codes. It is necessary to provide its testing and continual improving. The second part of the article refers to one way of sensitivity analysis connected with spatial discretization of the model cells. It shows a very high influence of the spatial discretization on the final results. The maximum effort was aimed at spatial discretization effect and its influence on infinite multiplication factor (k_{inf}) results. Multiplication factor (k_{inf}) reflects the multiplication ability of the relevant SUPERPHENIX and VVER-440 fuel assemblies situated hypothetically into the infinite lattice with no neutron leakage. All kinds of calculations were performed by a spectral code HELIOS 1.10. This study also exposes the HELIOS modelling and simulating borders.

Keywords Spatial discretization · Transmutation · Fast reactor · Fuel cycle · Plutonium · MOX

R. Zajac (✉)
VUJE, Inc., Okruzna 5, Trnava, Slovakia; Faculty of Electrical Engineering and Information Technology, Department of Nuclear Physics and Technology, Slovak University of Technology in Bratislava, Ilkovicova 3, Bratislava, Slovakia
e-mail: radoslav.zajac@vuje.sk

J. Murín et al. (eds.), *Computational Modelling and Advanced Simulations*, Computational Methods in Applied Sciences 24, DOI 10.1007/978-94-007-0317-9_13,

13.1 Introduction

Computational modelling and advanced simulations of different physical processes occur during irradiation of nuclear fuel in nuclear reactors are performed at present by various computational codes all over the world. Codes ORIGEN-S in the SCALE, HELIOS, WIMS, DYN, PERMAK and BIPR are used frequently for determination core and spent parameters of VVER reactors. Each of these codes uses certain methods and algorithms that with acceptable accuracy describe or simulate physical actions and processes happennig in the followed space, volume and time interval.

ORIGEN-S computes time-dependent concentrations and source terms of a large isotopes number, which are simultaneously generated or depleted through neutronic transmutation, fission, radioactive decay, input feed rates, and physical or chemical removal rates [1].

WIMS is a major software package containing a wide range of lattice cell and burnup methods for the design and development of all types of thermal reactor systems including experimental low power facilities as well as commercially operating power reactors [2].

The 3-dimensional neutron kinetic and dynamic models of DYN are based on a nodal expansion method for solving the two-group neutron diffusion equation in hex-z or rectangular x, y, z- geometry. The thermal-hydraulic part consists of a two phase flow model describing coolant behaviour and a fuel rod model [3].

PERMAK is a 2-dimensional fine mesh diffusion computer code for 4- or 6-group VVER core burnup calculation. This code performs 2-D pin-by-pin diffusion calculations [4].

BIPR7 is a 3-D 2-group coarse mesh diffusion code for various quasi-stationary regimes and core operation calculations [4].

Modelling results uncertainties depend significantly on chosen algorithms and also on nuclear data libraries that particularly include information about nuclear radiation interaction of different kinds with neighbouring materials, neutron and gamma cross sections depending on energy. The listed codes use US cross-section library ENDF (Evaluated Nuclear Data Files). The latest ENDF release is certificated ENDF/B-VII.0 [5] that is in the process of continual improvement. The release ENDF/B-VII.1 is supported to be available for the scientific public in 2010 and ENDF/B-VII.2 in the year 2011.

The non-stop accumulation of spent nuclear fuel initiates solution of fuel cycle back-end as spent fuel reprocessing and the high level radioactive waste deposition into the geological repository. One way how to solve the spent fuel problem is to re-use the energy of potential fissille nuclides in a closed Generation IV fuel cycle [6]. The presented calculation activities are directly connected with these ideas and an effort to apply the fast reactor under Slovak conditions.

13.2 Fundamentals of Computational Code HELIOS 1.10

HELIOS is a deterministical computational code using the transport theory. The HELIOS 1.10 is a particle – transport code that uses the physical methods to obtain fluxes, currents and number densities of the nuclear fuel. The transport method of HELIOS for particle transport is called current-coupling collision-probability method (CCCP method) [7]. His main purposes are primary neutron data calculations and preparations (cross section, diffusion coefficients and reactivity) for the 3D core simulators such as BIPR, PERMAK and DYN. The nuclear fuel irradiation is possible to simulate by creation of the critical or sub critical model in 2D lattice structure. HELIOS 1.10 uses the cross-section library prepared by computer code NJOY from ENDF/B-VI.8. The cross sections are arranged and sorted into energy groups from 10^{-5} eV to 20 MeV. The HELIOS 2 version is being prepared and it will include the cross-section library created on the ENDF/B-VII.0 base.

The HELIOS code applies for neutron current calculations the basic physical method created by the two physical models [8]:

1. Calculating fluxes and currents by current-coupling method for particle transport
2. First flight probabilities

The current-coupling and collision probability method (CCCP method) makes the HELIOS code very flexible in critical and sub critical geometric structure creation. It is possible to model each up to date known LWR fuel assemblies and also assemblies of experimental and school nuclear reactors.

HELIOS uses two independent programs named AURORA and ZENITH for the input – output data reading and processing (Fig. 13.1). The data stream between AURORA, HELIOS and ZENITH processors runs trough database called HERMES. AURORA presents an input processor that reads input file data and writes them down in HERMES. HELIOS reads recorded HERMES data, then on the their basis it runs calculation jobs and puts the results down back in HERMES. ZENITH is an output processor that reads data in HERMES and puts them down in an output file in accordance with the user requirements.

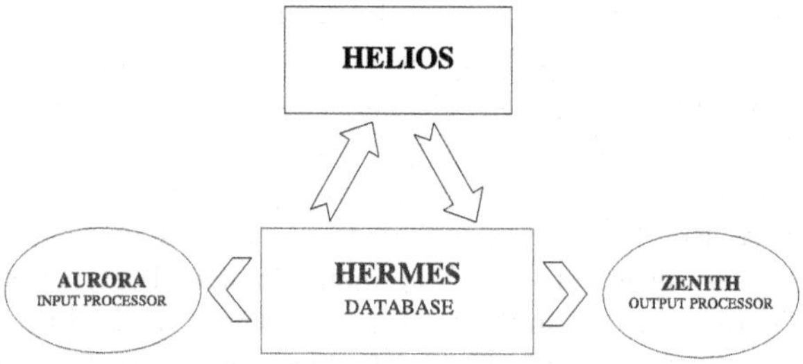

Fig. 13.1 Inside HELIOS 1.10 structure [8]

In the CCCP method the collision probability (CP) method was combined with the method of current coupling in following way. Fluxes Φ in all the regions of the calculated system are solved as linear combination of the sources Q in all the regions and in-currents j^- through all the sectors on all the interface segments of the system [7]

$$\Phi = XQ + Yj^- \tag{13.1}$$

where X and Y are diagonal block matrices of $E \times E$ (E – number of space elements) created by column vectors X_i and Y_s – vectors of response fluxes in all the regions of a space element due to a unit source in region i and a unit in-current at angular sector s respectively.

Global problem – solving the currents in the entire system (by inner iterations) – is based on equation

$$j^+ = Rj^- + PQ \tag{13.2}$$

where R and P are $E \times E$ diagonal block matrices, created by column vectors R_j and R_s – vectors of multiple flight escape and transmission probabilities through all the sectors of a space element due to a unit source in region i and a unit in-current through sector s respectively. Elements of matrices X, Y, R and P are evaluated from first flight probabilities.

Relation between j^- and j^+ is based on geometry matrix H – essential part of geometry treatment in HELIOS:

$$j^+ = Hj^- \tag{13.3}$$

The H matrix is square and has as many rows as the number of sectors in the model system.

13.3 Transmutation Fuel Cycle

At present the open fuel cycle (one-through fuel cycle) is the most spread in the nuclear energy industry. It means the complex technological process of nuclear fuel treatment from uranium ore mining to storage of spent nuclear fuel [9]. However spent fuel reprocessing offers possibilities to return energy potential nuclides (Pu-239, 241) together with the dangerous long-lived minor actinides (Np, Am, Cm, Cf) back into the fuel cycle and to reduce the geological repository cost. The reprocessing and Pu re-use close the open fuel cycle. Closed fuel cycle is generally considered to be very prospective and the methods of fuel reprocessing are being improved thereinafter. In this study the French fast reactor SUPERPHENIX was chosen hypothetically for the Slovak spent fuel transmutation. The main reason for SUPERPHENIX selection was the reactor ability to transmutate minor actinides from spent fuel. The minor actinides can be transmutated only in the high neutron flux space.

13.3.1 Inventories from NPP V1 Bohunice

The reactor unit 1 of the NPP V1 was operated for 27 core cycles (1978–2006) and the unit 2 was in operation for 28 core cycles (1979–2008). The spent fuel was transported to the Soviet Union during the first operated years and later on it was stored in the nuclear power plant (NPP) intermediate storage. This chapter shows the overview of the reactor units 1&2 inventory [10]. At the beginning of the operation both reactor units used old types of fuel assemblies with the uniform enrichment 1.6, 2.4 and 3.6% of U-235. The nonuniform fuel assemblies (radial profilation of enrichment) with average enrichment 3.82% of U-235 were loaded only during several operated cycles before NPP V1 shutdown. The number of the VVER-440 fuel assemblies stored in the intermediate storage in NPP Bohunice is listed in Table 13.1.

Table 13.1 The number of the VVER-440 spent fuel assemblies from NPP V1

Reactor unit fuel type	Unit 1 (V1)	Unit 2 (V1)	Unit1 + Unit 2 (V1)
1.6% U-235	24	18	42
2.4% U-235	491	356	847
3.6% U-235	1,788	1,824	3,612
3.82% U-235	216	426	642
Total	2,519	2,624	5,143

The major part of the total mass [10] consists of U-238. U-238 makes more than 95% of the total spent fuel amount. The weight of fissionable nuclides (U-235 + Pu-239 + Pu-241) is 10.174 t. SCALE 4.4 (ORIGEN-S) system was used for the determination of nuclide composition for the each operated and spent VVER-440 assembly. The average burn up of assembly and cycle parameters was calculated by the Russian code BIPR 7 [5].

13.3.2 VVER-440 and SUPERPHENIX

The VVER-440 core consists of 349 hexagonal fuel assemblies. At present the VVER-440 fuel assembly with enrichment 4.25% of U-235 and 3.35 wt% of Gd_2O_3 is the most advanced assembly used in Slovak nuclear reactors. This type of assembly was modelled in an infinite lattice (Fig. 13.2) and computer code HELIOS 1.10 was used.

The central part of the SUPERPHENIX core [11] consists of fuel subassemblies that include a fuel part surrounded by a fertile part. This configuration is that of a homogeneous reactor, in other words the one in which the fuel and blanket parts are distinctly separated: the fuel part in the centre, the blanket part surrounding the fuel. This is the configuration that was adopted for all the fast reactors

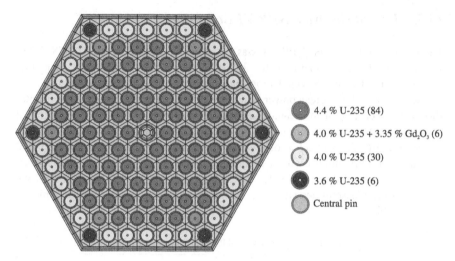

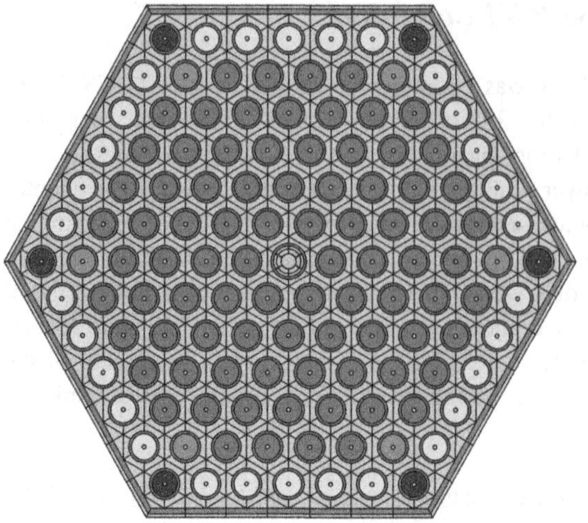

Fig. 13.2 Cross section of VVER-440 advanced fuel assembly [12]. The advanced assembly was simulated in an infinite lattice without using spatial discretisation

Table 13.2 The comparison of parameters between reactors SUPERPHENIX and VVER-440 [12]

Parameter/type of reactor	SUPERPHENIX	VVER-440
Electric power [MWe]	1,450	440
Efficiency [%]	40	30
Fuel	$(U, Pu)O_2$ – MOX	(UO_2 + 3.35% Gd_2O_3) UOX
Final burn up (fuel) [MWd/kgU]	136,000	50,000
Final burn up (fertile) [MWd/kgU]	4,000	–
Initial enrichment of Pu [%]	23.2	0.0
Initial enrichment of U-235 [%]	0.25	4.25
Initial enrichment of MA [%]	2.7	0.0
Amount of fuel [t_{HM}]	(37.1 + 76.5) (fuel + blanket)	44.1
Number of fuel assemblies in core	364	349
Number of fertile assemblies in core	233	0
Number of pins per fuel assembly	271	126
Number of pins per fertile assembly	91	0
Coolant temperature [°C]	550	280
Coolant composition	Na	H_2O

in operation up to now. The original SUPERPHENIX fuel assembly is listed in Fig. 13.4.

All of the fuel assemblies in the part 13.3 were simulated in an infinite lattice.

The differences between core parameters of SUPERPHENIX and VVER-440 are listed in Table 13.2 (Fig. 13.3).

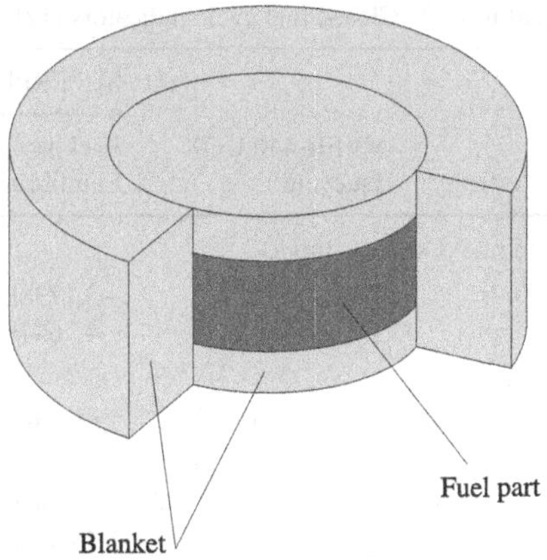

Fig. 13.3 The usual fast reactor core configuration [11]

13.3.3 Transmutation Fuel Cycle Indicators

Table 13.3 includes 14 fuel cycle indicators of VVER-440 and SUPERPHENIX fuel cycle. The values of cycle indicators of VVER-440 and SUPERPHENIX are seemingly very different but it is necessary to take in to account the big difference of electric energy production of at SUPERPHENIX and VVER-440. SUPERPHENIX is able to produce 3.295 more electric powers than VVER-440. The fuel cycle indicators are listed in the special physical units [kg/TWhe] (*nuclide weight concentration/produced electric energy unit*). This special unit optimises and rates the total different physical properties of followed rector types, so thereby it offers the real undistorted frame of the closed fuel cycle. The fuel cycle indicators are very important for engineers interested in the geological design and for decision makers.

With the best will the probability of fast reactor real application for the electricity production in Slovakia condition is low. This study only indicates the useful utilization of the fissionable nuclides included in the spent fuel. Support of fast reactor technology holders should be envisaged.

13.4 Spatial Model Discretization

Shortly described combination of two methods in introductory chapters creates the HELIOS geometric flexibility that is one of the most important aspects of HELIOS. A system consists from regions can be partitioned into a number of smaller systems by spatial discretization [13–15].

The previous listed analyse was performed without spatial model discretization. The spatial discretization of space regions leads to higher results accuracy but it takes a lot of computer time. The fuel cycle indicators are the calculation results that correspond to necessary time for fuel cycle analyse realization.

Table 13.3 Closed fuel cycle indicators [12]

Type of reactor Indicator/Cycle	VVER-440,UOX 1st cycle	SUPERPHENIX, MOX Fuel part Equilibrium	 Fertile part Equilibrium
1. Pu and MA transmutation rate (Output – Input)			
Pu [kg/TWhe]	–	−3.1738E+01	–
MA [kg/TWhe]	–	−4.1644E+00	–
2. Mass of fuel entering reactor			
Fresh fuel [t/TWhe]	2.5297E+00	7.6593E-01	7.6593E-01
3.Rate of Pu and MA in the fuel entering reactor			
Pu [%]	–	2.3236E+01	–
MA [%]	–	2.7000E+00	–
4. Rate of Pu and MA in the spent fuel (5 years of cooling)			
Pu [%]	1.2207E+00	1.9056E+01	1.6442E+00
MA [%]	1.0768E-01	2.1563E+00	5.0645E-03
5. Mass of reprocessed fuel			
Spent fuel [g/TWhe]	2.5297E+00	7.6593E-01	7.6593E-01
6. Mass of fuel to the repository			
Sum of repr. losses [kg/TWhe]	2.5297E+00	7.6593E-01	7.6341E-01
Sum of FP [kg/TWhe]	1.1558E+02	9.2940E+01	2.2648E+00
Total loss. + FP [kg/TWhe]	1.1811E+03	9.3599E+01	3.0282E+00
Irradiated U[kg/TWhe]	2.3263E+00	4.9609E-01	7.5003E-01
7. Energy recovery per t of natural U			
Fission actinides [kg/TWhe]	4.0658E+01	7.5905E+01	1.3872E+01
Equivalent weight [tU_{nat}]	3.4788E+00	2.1450E+01	3.9202E+00
8. Average quantity of Pu separated in fuel cycle			
Pu [kg/TWhe]	2.9279E+01	1.9037E+02	1.6425E+01
9. Average content of Pu-239 and Pu-241 in Pu in the moment of separation			
Pu-239 [%]	5.2558E+01	4.7417E+01	9.7215E+01
Pu-241 [%]	1.4726E+01	4.0487E+00	4.3805E-02
10. Average content of Pu-239 and Pu-241 in Pu after separation			
Pu-239 [%]	5.2505E+01	4.7370E+01	9.7118E+01
Pu-241 [%]	1.4711E+01	4.0446E+00	4.3761E-02
11. Amount of finally disposed Pu			
Pu [kg/TWhe]	2.9306E+01	1.4596E+02	1.2593E+01
12. Inventory of the enriched U in front end			
U-235 in fresh fuel [kg/TWhe]	1.0751E+02	1.9148E+00	1.9148E+00
U-235 in spent fuel [kg/TWhe]	2.0938E+01	7.8585E-01	1.6241E+00
13. Mass of separated U			
Separated U [t/TWhe]	2.3625E+00	4.9558E-01	7.4905E-01
14. Spent natural uranium per one cycle			
Spent U_{nat} [t/TWhe]	2.3271E+01	7.0964E-02	1.5150E-02

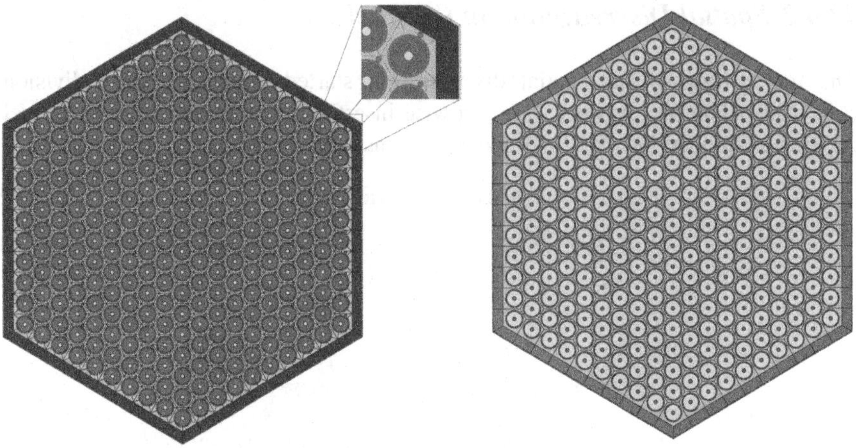

Fig. 13.4 The original SUPERPHENIX cross section fuel assembly (*left*) with distance wire and cross section of pseudo SUPERPHENIX fuel assembly (*right*) modelled in an infinite lattice

The followed problem is closely connected with a spatial discretization of the cells and regions in the fuel and coolant of the fast reactor. The discretization influence on infinite multiplication factor (k_{inf}) calculations was evaluated by use of the methods – neutron current coupling and collision probability, especially the first flight probability.

Figure 13.4 includes the full geometry of the SUPERPHENIX fuel assembly. On detailed view of the pin is demonstrated the distance wire. For the spatial discretization analysis was the distance wire integrated into the fuel cladding of pseudo-fuel assembly.

13.4.1 Strategies for Optimal Discretization

The starting calculations point was to model the SUPERPHENIX pseudo-assembly (Fig. 13.4). The pseudo-assembly cells were divided with rings into the regions. The 112-neutron groups library was used in all of calculation cases (Fig. 13.5).

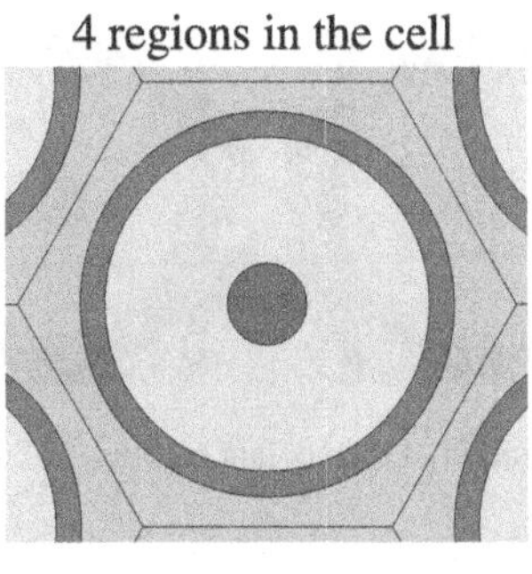

Fig. 13.5 The basis cell in discretization strategy

13.4.2 Spatial Discretization in the Fuel

The real testing of the spatial discretization started with fuel space division (Fig. 13.6). The fuel regions were stepwise inserted into forms of rings. The fuel region was divided into 5, 11, 21 and 31 regions with rings.

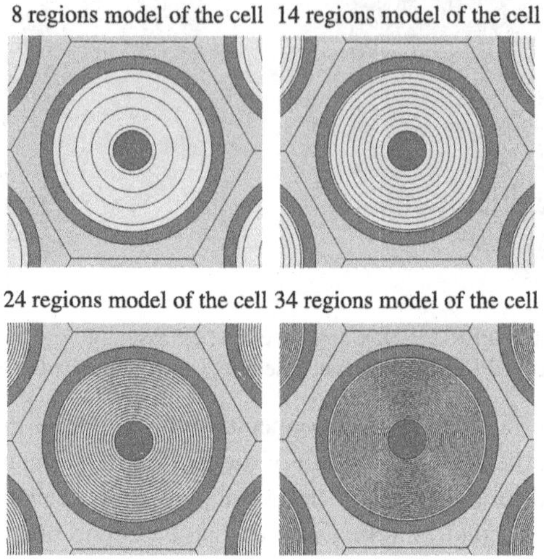

Fig. 13.6 The stepwise discretization of the fuel regions

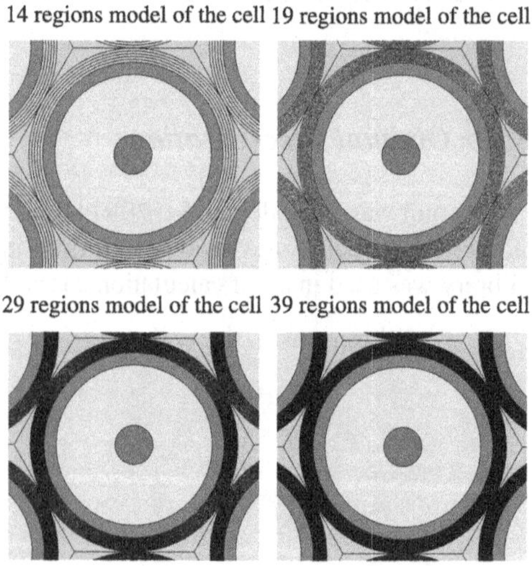

Fig. 13.7 The stepwise discretization of the coolant regions

13.4.3 Spatial Discretization in the Coolant

The next step in the analysis of the spatial discretization effect was the stepwise partition of the coolant (Fig. 13.7). Coolant parts were divided also with rings. The corners of hexagonal cells weren't partitioned because Na does not influence the neutron fluxes and currents significantly.

13.4.4 Optimized Strategy

It was necessary to calculate the credible infinite multiplication factor k_{inf} value by combination of the coolant and fuel discretization (Fig. 13.8). The model consists of the finest cell discretization that was used as reference model. But the high cost of this calculation caused the limits in the worth neutron flux and current calculations. The model can include only fine discretization cells that consist of 69 regions. The 69 region cell is the maximum cell which is possible to use without other model reductions. The main limit is the operation memory allocation.

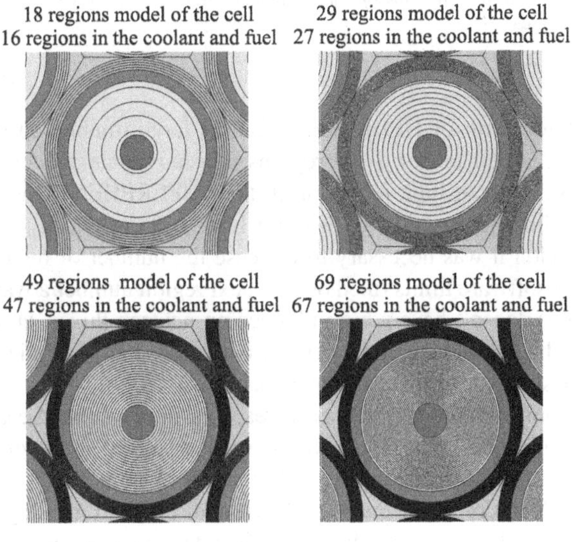

Fig. 13.8 The combination of the coolant and fuel discretization

13.5 Results and Discussion

The Fig. 13.9 shows the dependence of k_{inf} on fuel region number. The infinite multiplication factor (k_{inf}) curves for fuel and combined discretization exhibits strong linear rise with growing number of regions. The flux distribution is very strongly dependent on the spatial discretization as well. It is nearly impossible to get a reliable result for the k_{inf} without performing a real very fine mesh calculation. The

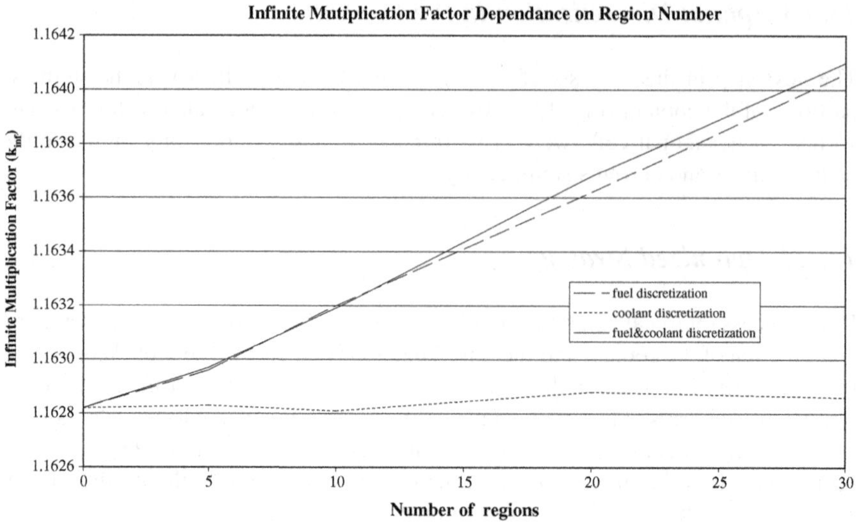

Fig. 13.9 Infinite multiplication factor k_{inf} depends on the number of regions for different discretization strategies

memory allocation limit of HELIOS 1.10 causes large troubles in the worth k_{inf} determination. The cell model of 69 regions is a maximum model that was possible to calculate without high reduction in the SUPERPHENIX assembly. In the Fig. 13.9 the solid and dashed curves don't lead to the stable k_{inf} value. For the system stabilization it was necessary to increase the number of regions in the cell but the HELIOS software limits ruled out the next region number advancement. The coolant (natrium) doesn't have a big influence on neutron fluxes, currents and k_{inf} for that reason the k_{inf} values in the case of coolant spatial discretization are very similar to the basic model value (Table 13.4).

Table 13.4 compares the k_{inf} values for each of the calculated cases. The optimized model divided into 69 regions with rings was designed as the reference model. The highest deviations were achieved in all of the cases of fuel spatial discretization model. The k_{inf} deviation in the entire fuel spatial discretization is higher than 0.1% with the respect to reference k_{inf} value. The listed deviations are very hardly acceptable because the tolerable deviations for k_{inf} comparison are at the level of persent thousandths. The calculation time of the 69 regions cell model is app. 34 more times longer than calculation of the basic model.

13.6 Conclusions

Non-negligible dependence of infinite multiplication factor k_{inf} on space discretization was shown at calculations of fast reactor by HELIOS 1.10 code. As no acceptable solution of this problem was found, another analyses with even more fine

Table 13.4 Comparison of criticality and computation time for different discretization schemes, final burn up 136,000 MWd/t_{HM} (Intel (R) Core(TM)2 CPU, 2.67 GHz, 3.00 GB RAM)

Relevant quantities	Basis model	Fuel discretization, number of fuel regions			
	Fuel – 1 region *Coolant – 1 region*	5	11	21	31
k_{inf}	1.16282	1.16296	1.16320	1.16362	1.16406
Deviation [%]	−0.11	−0.098	−0.077	−0.041	−0.0034
CPU (min) time	9.04	43.19	92.87	185.23	306.35
CPU time factor	1.0	4.77	10.27	20.49	33.89
Relevant quantities	Basis model	Coolant discretization, number of coolant regions			
	Fuel – 1 region *Coolant – 1 region*	11	16	26	36
k_{inf}	1.16282	1.16283	1.16281	1.16288	1.16286
Deviation [%]	−0.11	−0.109	−0.111	−0.105	−0.107
CPU (min) time	9.04	10.54	12.46	14.93	19.71
CPU time factor	1.0	1.17	1.38	1.65	2.18
Relevant quantities	Basis model	Optimized strategy, number of fuel and coolant regions			
	Fuel – 1 region *Coolant – 1 region*	18	29	49	69
k_{inf}	1.16282	1.16297	1.16319	1.16368	1.16410
Deviation [%]	−0.11	−0.097	−0.078	−0.036	Reference
CPU (min) time	9.04	52.50	101.02	197.75	306.66
CPU time factor	1.0	5.81	11.17	21.88	33.92

discretization and with other values of angular discretization factor k are needed. Broad enough analyses can be enabled by improved memory management (dynamic memory allocation) and faster performance (multiprocessing) in future versions of the code HELIOS.

Acknowledgements The thanks for this project realization belong to the engineering and research organization VUJE, Inc. that sponsored this project. One of the authors (V. Necas) expresses acknowledgments to the Slovak Grant Agency for Science for support through the grant No. VEGA 1/0685/09.

References

1. Herman, W.O., Westfall, M.R.: ORIGEN-S: SCALE system module to calculation fuel depletion actinide transmutaion, fission product buildup and decay, and associated radiation source terms. ORNL, ORNL/NUREG/CSD-2/V2/R6, vol. 2, section F7 (1998)
2. Newton, D.T., Hutton, L.J.: The Next Generation WIMS Lattice Code: WIMS9. PHYSOR, Seoul (2002)

3. Grundmann, U., et al.: DYN3D Version 3.2. Code for Calculation of Transients in Light Water Reactors with Hexagonal or Quadratic Fuel Elements. Code Manual and Input Data Description for Release, Forschungszentrum, Dresden (2007)
4. Sidorenko, D.V., et al.: Verification of 3-D generation code package for neutronic calculations of VVERs. Proceedings of the 10th Symposium of AER. KFKI, Budapest (2000)
5. Chadwick, M.B., Obložinský, P., Herman, M.: ENDF/B-VII.0: Next generation evaluated nuclear data library fo nuclear science and technology. Nucl. Data Sheets **107**(12), 2931–3060, ISSN 0090-3752 (2006)
6. US Department of Energy: Office of Nuclear Energy: Science and Technology Advanced Fuel Cycle Initiative (AFCI) Comparison Report, FY 2005. http://afci.sandia.gov/reading_room/reading_list.htm (2005). Accessed 25 Nov 2009
7. Giust, F.D., Sobieska, M.M., Stamm'ler, R.J.J.: HELIOS: benchmarking against hexagonal TIC lattices. Proceedings of the 7th Symposium of AER. KFKI, Budapest (1997)
8. Studsvik Scandpower: HELIOS Methods (2008)
9. JAVYS: Spent Fuel Management: final Solution: Fuel Reprocessing. www.javys.sk/en/index.php?page=manazment-vjp/konecne-riesenie/ prepracovanie-paliva (2009). Accessed 30 Nov 2009
10. Chrapciak, V., Malatin, P.: Inventory from NPP V-1 Jaslovske Bohunice. Proc. 19th Working AER Group E. **19**, 1–4 (2009)
11. Bailly, H. et al.: The Nuclear Fuel of Pressurized Water Reactors and Fast Reactors. Lavoisier Publishing, Paris (1999)
12. Zajac, R., Darilek, P., Necas, V.: Transmutation of the WWER-440 spent fuel in fast reactor. Proceedings of the 18th Symposium of AER. KFKI, Budapest (2008)
13. Merk, B., Koch, R.: The effect of spatial discretization in LWR cell calculations with HELIOS 1.9. Proceedings of the 18th Symposium of AER. KFKI, Budapest (2008)
14. Darilek, P., Zajac, R., Breza, J., Necas, V.: Comparison of PWR-IMF and FR Fuel Cycles. GLOBAL 2007. Advanced Nuclear Fuel Cycles and Systems, Boise (2007)
15. Darilek, P., Zajac, R.: Sustainable fuel cycle alternatives for small nuclear economies. Proceedings of the 16th Symposium of AER. KFKI, Budapest (2006)

Chapter 14
Linear Algebra Issues in a Family of Advanced Hybrid Finite Elements

N.A. Dumont and C.A. Aguilar Marón

Abstract The hybrid finite element method, proposed more than 40 years ago on the basis of the Hellinger-Reissner potential, was a conceptual breakthrough among the discretization formulations, as extensively explored since then both academically and in commercial codes, including an independent branch of more recent developments called Trefftz methods. More recently, the method was extended – on the basis of Przemieniecki's proposition of frequency-dependent stiffness and mass matrices – to the analysis of time-dependent problems in terms of an advanced mode superposition procedure that may take into account general initial conditions as well as general body actions. It was also shown that the whole concept can be used with non-singular fundamental solutions, thus leading to general families of hybrid, macro finite elements of 2D and 3D, static and transient problems of elasticity and potential. They are the subject of the present paper, both in the frame of the advanced mode superposition procedure and in terms of numerical Laplace and Fourier transforms. The possibilities of application of the formulation are illustrated by means of a numerical example.

Keywords Hybrid finite elements · Advanced mode superposition procedure · Laplace and Fourier transformation

14.1 Introduction

The modern variational developments in continuum mechanics were inaugurated with a now historical paper by Hellinger [1], which motivated Reissner [2] to establish the Hellinger-Reissner potential, apparently the first of a series of remarkable achievements in the field [3, 4]. Pian [5] proposed the first systematic, computer-oriented, implementation of the new methodology, which opened up a new area

N.A. Dumont (✉)
Civil Engineering Department, Pontifical Catholic University of Rio de Janeiro, Rio de Janeiro, Brazil
e-mail: dumont@puc-rio.br

J. Murín et al. (eds.), *Computational Modelling and Advanced Simulations*, Computational Methods in Applied Sciences 24, DOI 10.1007/978-94-007-0317-9_14,

of applications in the then young finite element method. The name "hybrid finite element method" was coined by Pian in 1967"... to signify elements which maintain either equilibrium or compatibility in the element and then to satisfy compatibility or equilibrium respectively along the interelement boundary" [6]. Theodore Pian's work has deserved worldwide recognition, as in the special conference organized by Atluri et al. [7]. The German Society for Applied Mathematics and Mechanics (ZAMM) also honoured Eric Reissner posthumously with a minisymposium [8], not only for his achievements in the variational methods, but also for his not lesser work in plate theory.

A completely independent chain of developments was also inaugurated by Trefftz [9], who, apparently unaware of Hellinger's [1] paper, wrote with many decades of advance a classical article covering exactly the same subject of Reissner's and proposing basically the same methodology that was eventually baptized as the "Hellinger-Reissner potential". Trefftz' paper remained half forgotten for many decades, until Jirousek brought it to the front line of computer applications with a series of papers [10]. Jirousek's works gave rise to an incredibly large number of developments in a field that is now known as "Trefftz methods", although many of these formulations are completely unrelated to Trefftz' original proposition.

It is remarkable that Hellinger-Reissner's and Trefftz' achievements have given rise to two independent communities of researchers, which seem to behave as if unaware of each other. Jirousek's papers and some books on Trefftz' method(s) (see [11], for instance) don't mention Hellinger, Reissner, Washizu or Pian. Conferences in honour to Trefftz have also been organized Maunder [12], but with no word on a possible controversy. It is also worth mentioning one of Reddy's highly qualified books [13], which makes no reference to Trefftz in the introduction of the variational methods, but gets hold of the historical paper of 1926 in the chapter that deals with torsion of a prismatic shaft.

With respect to time-dependent problems, the developments due to Pian have been applied much in the sense of the traditional displacement finite element method, with just adaptations of concepts inherited from the theory of the classical structural dynamics [7]. Ad hoc formulations of an uncountable number of specific finite elements are available in the technical literature, which however still lack an underlying, unified conceptual pattern. By the way, it is worth remarking that the most recent book co-authored by Pian on hybrid finite elements simply ignores time-dependent problems [14].

As occurred in the above outlined counter position of the concepts by Hellinger-Reissner and Trefftz, two parallel, independent developments with regard to time-dependent problems also took place recently. Przemieniecki [15] is undeniably the original author of the proposition that mass and stiffness matrices may be obtained as functions of a structure's vibration frequency, as he developed for truss and beam elements on the basis of a displacement formulation. This was the confessed inspiration of a series of papers on displacement-based "finite dynamic elements", which should be better called "finite harmonic elements", as they have been only applied to frequency analyses [16–18]. Independently from these developments, a proposition made by Beskos and Narayanan [19] and generalized by Doyle and collaborators

[20] culminated in the "spectral finite element method". Gopalakrishnan et al. [21] outline this method thoroughly, with a series of conceptual justifications and an interesting account of its historical developments as well as a motivating exposition of the many application possibilities. Although Gopalakrishnan et al. mention a "dynamic stiffness method" in their outlines, they never refer to Przemieniecki's work or followers.

An established technique to solve time-dependent problems is the formulation of a complete frequency-domain analysis via Laplace or Fourier transforms, with subsequent ad hoc expression of results by numerical inversion. Although usually easy to implement, such a transform inversion is computationally intensive, if accurate results are desired, and is not void of numerical instabilities. Moreover, the inclusion of non-homogeneous initial conditions and general body forces may become troublesome. The Gaver-Stehfest algorithm [22, 23] is suited to diffusion-type problems [24]. General dynamics problems demand more robust algorithms usually based on Fourier series expansions, which descend from a proposition firstly made by Dubner and Abate [25]. Algorithms of either kind are already implemented in mathematical languages such as Matlab and Mathematica. The Fortran IMSL library has in the subroutine INLAP a Fourier-series algorithm proposed by Crump [26] and improved by de Hoog et al. [27]. The spectral finite element method is based on transforms such as the fast Fourier transform [21].

The present paper has its roots in Reissner's and Pian's works [28, 29]. A generalized Hellinger-Reissner potential was proposed to model time-dependent problems in a frequency-domain framework that uses the singular fundamental solutions of the conventional, collocation boundary element method as interpolation functions [30, 31]. A brief outline of this formulation is given in the next Section, as applied to non-singular fundamental solutions, thus resulting in a generalized hybrid finite element method [32–34].

One has firstly proposed to solve transient problems of potential and elasticity using an advanced mode superposition technique that applies to equilibrium-based finite element and boundary element models [31, 35]. The developments combine and extend Pian's [5] hybrid finite element formulation and Przemieniecki's [15] suggestion of displacement-based, frequency-dependent elements, thus arriving at a hybrid finite/boundary element method for the general analysis of transient problems. Actually, one is dealing with generalized finite element families, as the interpolation functions that satisfy domain equilibrium may be singular (free-field Green's functions) or not (non-singular transcendental functions) [33]. Transient problems of potential and elasticity are modelled, also including viscous damping. Starting from a frequency-domain formulation, one shows that there is an underlying complex-symmetric (if viscous damping is included), non-linear eigenvalue problem related to the λ-matrices of a free-vibration analysis, with an effective stiffness matrix expressed as the frequency power series $\mathbf{K}_0 - i\lambda\mathbf{C}_1 - \lambda^2\mathbf{M}_1 - i\lambda^3\mathbf{C}_2 - \lambda^4\mathbf{M}_2 - i\lambda^5\mathbf{C}_3 - \lambda^6\mathbf{M}_3 - \cdots$, where $\mathbf{K}_0, \mathbf{C}_j, \mathbf{M}_j \in \Re^{n\times n}$ are generalized stiffness, damping and mass matrices [35]. The eigenvectors of this problem fulfill generalized orthogonality properties [36] that enable the implementation of an advanced mode superposition technique. This leads to the solution in the time domain and the immediate expression of all results of interest. Owing to the equilibrium-based formulation, general domain actions (including body forces and moving loads) as well as boundary and initial conditions are dealt with in a straightforward way.

More recently, it was decided to implement the hybrid finite/boundary element method for time-dependent problems also in the frame of Laplace and Fourier transforms and compare it with the advanced mode superposition technique [24] in terms of accuracy of results and total computational time. It is not a simple task, as the performance is problem dependent and – not less important – also dependent on the code writer's skills.

The present contribution addresses some issues that may have prevented the development of the hybrid finite element method as a unified methodology. Experience is borrowed from the concepts developed in the frame of the variationally-based, hybrid boundary element method mainly in conjunction with the theory of generalized inverses. Zero energy modes are an issue that has haunted the developers of incompatible finite elements from the beginning [14]. Although a strict mathematical assessment of the present proposition still lacks – and space restrictions prevents a thorough outline –, one shows how to correctly and systematically address the present subject.

This paper is a sequel of a contribution presented recently [32], in which one proposes an alternative variational principle with respect to the inertia term. However, this alternative does not bring any improvement in the formulation – in fact, it poses more conceptual difficulties – and had to be disregarded. The original title is kept, as justified in Section 14.5. The basic variational formulation is presented in Section 14.2. Non-singular fundamental solutions for 2D and 3D problems of potential and elasticity are displayed in Section 14.3, where some relevant spectral properties are also discussed. The basic features of the advanced modal analysis are briefly outlined in Section 14.4 and a simple numerical example is presented in Section 14.6.

14.2 Generalized Variational Formulation for Time-Dependent Problems

The time effect is due to the inertia of an elastic body. The formulation for (diffusion-type) potential problems may be obtained as a particular case. Viscous damping is not included only for the sake of simplicity [35]. Starting point of the formulation is a generalized version of Hamilton's principle [30, 31]:

$$\begin{aligned}\int_{t_1}^{t_2}\Big(&-\int_\Omega\left(\delta\sigma^s_{ij,j}-\rho\delta\ddot{u}^s_i\right)\left(u^s_i-u^d_i\right)\mathrm{d}\Omega+\int_\Gamma\delta\sigma^s_{ij}\eta_j\left(u^s_i-u^d_i\right)\mathrm{d}\Gamma\\&+\int_\Omega\delta u^d_i\left(\sigma^s_{ij,j}+\bar{f}_i-\rho\ddot{u}^s_i\right)\mathrm{d}\Omega-\int_\Gamma\delta u^d_i\left(\sigma^s_{ij}\eta_j-\bar{t}_i\right)\mathrm{d}\Gamma\Big)\,dt=0\end{aligned}\tag{14.1}$$

The elastic body has domain Ω, boundary Γ with outward unit normal η_j, and is observed along the time interval $[t_1,\ t_2]$. The subscripts i and j can assume values 1,

2 and 3, as referred to global coordinates x, y and z, respectively. A subscript after a comma denotes derivative with respect to the corresponding coordinate direction. Repeated subscripts indicate a three-term summation, in the general case of three-dimensional problems. The body is subjected both to body forces $\bar{f}_i$ in the domain Ω and to traction forces $\bar{t}_i$ along part Γ_σ of the boundary.

This is a two-field formulation, in terms of a *stress* field σ_{ij}^s, which a priori fulfils all domain equilibrium and kinematic equations (but no boundary conditions), and of an independent boundary *displacement* field u_i^d such that $u_i^d = \bar{u}_i$ along Γ_u for prescribed displacements $\bar{u}_i$. For convenience, the last integral in the above equation is carried out along the whole boundary $\Gamma = \Gamma_\sigma \cup \Gamma_u$, since $\delta u_i^d = 0$ along Γ_u.

The stress field σ_{ij}^s satisfies the differential equilibrium equation

$$\sigma_{ij,j}^s + \bar{f}_i - \rho \ddot{u}_i^s = 0 \text{ in } \Omega \tag{14.2}$$

This is achieved by expressing σ_{ij}^s as a sum of two terms, also for the corresponding displacement field u_i^s,

$$\sigma_{ij}^s = \sigma_{ij}^* + \sigma_{ij}^p, \quad u_i^s = u_i^* + u_i^p, \tag{14.3}$$

where σ_{ij}^p and u_i^p characterize an arbitrary particular solution, and σ_{ij}^* and u_i^* are the homogeneous solution of (14.2).

Since no boundary conditions are specified for the solution of the partial differential Equation (14.2), an infinite number of homogeneous solutions is possible – the so-called fundamental solutions. A convenient set of such solutions may be expressed as a series of non-singular functions, as proposed to arrive at a hybrid finite element formulation,

$$\sigma_{ij}^* = \sigma_{ijr}^* p_r^*, \qquad u_i^* = u_{ir}^* p_r^*, \tag{14.4}$$

in terms of a set of n^* yet unknown parameters p_r^*. The expression of u_i^* above, if particularized to the static solution, is valid except for arbitrary amounts of rigid body displacements.

Owing to the equilibrium requirement of (14.2), the independent displacement field u_i^d of (14.1) needs to be defined exclusively along the boundary Γ in terms of (polynomial) interpolation functions u_{is} and a set d_s of $n^d \leq n^*$ yet unknown boundary nodal displacements:

$$u_i^d = u_{is} d_s \text{ along } \Gamma \tag{14.5}$$

As a consequence of these assumptions, (14.1) becomes ultimately expressed in matrix notation, for a given time instant t, as

$$\delta\, \mathbf{p}^{*\mathrm{T}} \left(\mathbf{F}^* \mathbf{p}^* - \mathbf{H}\mathbf{d} + \mathbf{b}\right) - \delta\, \mathbf{d}^{\mathrm{T}} \left(\mathbf{H}^{\mathrm{T}} \mathbf{p}^* + \mathbf{p}^p - \mathbf{p}\right) = \mathbf{0} \tag{14.6}$$

in which $\mathbf{p}^*$ and $\mathbf{d}$ are vectors containing the unknown parameters p_r^* and d_s, respectively. The $n^* \times n^*$ symmetric flexibility matrix $\mathbf{F}^*$, the $n^* \times n^d$ kinematic transformation matrix $\mathbf{H}$ and the vector $\mathbf{b}$ of n^* nodal displacements equivalent to the body forces are defined in terms of boundary integrals as

$$[\mathbf{F}^* \; \mathbf{H} \; \mathbf{b}] := [F^*_{rs} \; H_{rs} \; b_r] = \int_\Gamma \left\{\sigma^*_{ijr}\eta_j\right\} \langle u^*_{is} \; u_{is} \; u^p_i \rangle \, \mathrm{d}\Gamma \tag{14.7}$$

If a singular fundamental solution σ^*_{ijr} is used, $\mathbf{H}$ and $\mathbf{b}$ must be obtained in terms of finite-part integrals plus the local evaluation of discontinuous terms where $\sigma^*_{ijr} \rightarrow \infty$. This is a standard, mathematically consolidated procedure. In such a case, however, $\mathbf{F}^*$ cannot be obtained in (14.7) when both σ^*_{ijr} and u^*_{is} refer to the same nodal point. The evaluation of the coefficients of $\mathbf{F}^*$ about its main diagonal is carried out in a linear algebra framework related to the mechanical meanings of the matrix nodal expressions [28, 29, 31, 37]. For non-singular fundamental solutions – the present case -, $\mathbf{F}^*$ is completely evaluated by (14.7), although still consistent with the linear algebra features of the singular case, if one is dealing with a homogeneous material [38].

The vectors $\mathbf{p}^p$ and $\mathbf{p}$ in (14.6) of nodal forces equivalent to the body forces $\bar{f}_i$ and the traction forces $\bar{t}_i$, respectively, are defined as

$$\left[\mathbf{p}^p \; \mathbf{p}\right] := \left[p^p_r \; p_r\right] = \int_\Gamma \{u_{ir}\} \left\langle \sigma^p_{ij}\eta_j \; \bar{t}_i \right\rangle \mathrm{d}\Gamma \tag{14.8}$$

For arbitrary values of $\delta\,\mathbf{p}^*$ and $\delta\,\mathbf{d}$, (14.6) leads to the sets of nodal compatibility and equilibrium equations:

$$\begin{aligned} \mathbf{F}^*\mathbf{p}^* - \mathbf{H}\mathbf{d} + \mathbf{b} &= \mathbf{0} \\ \mathbf{H}^{\mathbf{T}}\mathbf{p}^* + \mathbf{p}^p - \mathbf{p} &= \mathbf{0} \end{aligned} \tag{14.9}$$

Solving for $\mathbf{p}^*$ in the above equations, one obtains

$$\mathbf{H}^{\mathbf{T}}\mathbf{F}^{*(-1)}\mathbf{H}\mathbf{d} = \mathbf{p} - \mathbf{H}^{\mathbf{T}}\mathbf{F}^{*(-1)}\mathbf{b} \tag{14.10}$$

For general time-dependent problems whether in the frequency or in the time domain, $\mathbf{F}^*$ is non-singular (although ill conditioned for very small values of time or frequency) and can be inverted. For the static case, $\mathbf{F}^* \equiv \mathbf{F}^*_0$ is singular and can only be inverted in the frame of the theory of generalized inverses, which is actually a quite straightforward procedure [28]. In the above equation,

$$\mathbf{H}^{\mathbf{T}}\mathbf{F}^{*(-1)}\mathbf{H} = \mathbf{K} \tag{14.11}$$

constitutes a symmetric, positive definite (semi-definite, in the static case) stiffness matrix that transforms nodal displacements into forces in equilibrium with the set of equivalent nodal forces defined at the right-hand side of (14.10).

Equations (14.6), (14.7), (14.8), (14.9), (14.10) and (14.11) are formally the same ones obtained for static problems [28, 31]. As occurs in the conventional boundary element formulation [28, 39], the numerical implementation of the equations above in either the time or the frequency domain involves no further difficulties than the ones that have already been dealt with in the frame of a static formulation. However, one restricts the developments to the analysis in the frequency domain.

In the frequency domain, the fundamental solution of (14.4) may be adequately expressed as $u^*_{is}(\omega) \leftarrow u^*_{is}(0) + u^*_{is}(\omega)$ and $\sigma^*_{ijs}(\omega) \leftarrow \sigma^*_{ijs}(0) + \sigma^*_{ijs}(\omega)$, where $u^*_{is}(0)$ and $\sigma^*_{ijs}(0)$ correspond to the static case. As a result, the matrices $\mathbf{F}^*$ and $\mathbf{H}$ of (14.6) may be formally represented as

$$\mathbf{F}^*(\omega) \equiv \mathbf{F}^*_0 + \mathbf{F}^*_\omega, \quad \mathbf{H}(\omega) \equiv \mathbf{H}_0 + \mathbf{H}_\omega \tag{14.12}$$

In the case of a finite domain, rigid body displacements, represented by an orthogonal matrix $\mathbf{W}$, cannot be transformed by the static part of the kinematic matrix, $\mathbf{H}_0$, that is,

$$\mathbf{H}_0\mathbf{W} = \mathbf{0} \quad \Rightarrow \quad \mathbf{H}_0^{\mathrm{T}}\mathbf{V} = \mathbf{0}, \tag{14.13}$$

where one already introduces an orthogonal basis $\mathbf{V}$ of the null space of $\mathbf{H}_0$. Then, for consistency of (14.9), one must have

$$\mathbf{F}^*_0\mathbf{V} = \mathbf{0}, \tag{14.14}$$

which is exactly checked for homogeneous materials and only approximately otherwise [38]. For singular fundamental solutions, the above equation is a means of evaluating the coefficients about the main diagonal of $\mathbf{F}^*_0$ [28], which cannot be obtained by (14.7) – see, however, important conceptual developments in [38] and [40]. Whether singular or non-singular fundamental solutions are used, $\mathbf{F}_0^{*(-1)}$, as needed in (14.10) and (14.11) in the case of a static problem, is obtained according to the theory of generalized inverses [41]. For the static case, one simply replaces $\mathbf{F}_0^{*(-1)}$ with $\left(\mathbf{F}^*_0 + \mathbf{V}\mathbf{V}^{\mathrm{T}}\right)^{-1}$ [29].

14.3 Non-singular Fundamental Solutions in the Frequency Domain

This Section briefly presents the general families of 2D and 3D non-singular fundamental solutions of a homogeneous, isotropic continuum, that is, sets of field interpolation functions in balance with fluxes or stresses, for potential and elasticity problems, respectively [33].

For 2D problems, the general family of solutions u^0_r of the Laplace equation in polar co-ordinates (r, θ), such that $x = r\cos\theta$ and $y = r\sin\theta$, is

$$u^0_1 = 1, \quad u^0_m = r^m \cos m\theta, \quad u^0_{2m+1} = r^m \sin m\theta, \qquad m = 1, \ldots, n \tag{14.15}$$

which correspond to $n^* = 2n + 1$ polynomial solutions comprised by a complete polynomial of degree n in Cartesian co-ordinates. The index r in u^0_r corresponds to a parameter p_r^* in (14.4). For illustration, the set of such polynomial terms for $n = 3$ is given by

$$\begin{aligned} u_r^0 &= \langle 1 \mid r\cos\theta \; r\sin \mid r^2\cos 2\theta \quad r^2 \sin 2\theta \mid r^3\cos 3\theta \; r^3\sin 3\theta \rangle \\ &\equiv \langle 1 \mid xy \mid x^2 - y^2 \quad 2xy \mid -3xy^2 + x^3 - y^3 + 3yx^2 \rangle \end{aligned} \tag{14.16}$$

For 3D problems, the general family of solutions u_r^0 of the Laplace equation is most adequately expressed in spherical co-ordinates (r, θ, φ), such that $z = r\cos\theta$, $x = r\sin\theta\cos\varphi$, $y = r\sin\theta\sin\varphi$, given as

$$\left.\begin{aligned} u^0_{m^2+\ell+1} &= r^m L_\ell^m(\cos\theta)\cos\ell\varphi \\ u^0_{m^2+\ell+2} &= r^m L_\ell^m(\cos\theta)\sin\ell\varphi, \quad \ell \neq 0 \end{aligned}\right\}, \quad \ell = 0,\ldots,m; \quad m = 0,\ldots,n \tag{14.17}$$

where $L_\ell^m(\cos\theta)$ is the associate Legendre function of the first kind with argument $\cos\theta$, in fact a trigonometric function of θ evaluated in the frame of *spherical harmonics*, which exists only for $\ell \leq m$. Then, for every $m \geq 0$ there are $2m + 1$ solutions comprised by (14.17) and, as a result, the set of possible polynomials in Cartesian coordinates up to degree n is $n^* = (n + 1)^2$. For $n = 3$, for example, one obtains a vector u_r^0 with 16 fundamental solutions:

$$\begin{aligned} u_r^0 = \langle 1 \mid & rL_0^1 \;\; rL_1^1\sin\varphi \;\; rL_1^1\cos\varphi \mid r^2L_0^2 \;\; r^2L_1^2\sin\varphi \;\; r^2L_1^2\cos\varphi \;\; r^2L_2^2\sin 2\varphi \;\; r^2L_2^2\cos 2\varphi \mid \\ & r^3L_0^3 \;\; r^3L_1^3\sin\varphi \;\; r^3L_1^3\cos\varphi \;\; r^3L_2^3\sin 2\varphi \;\; r^3L_2^3\cos 2\varphi \;\; r^3L_3^3\sin 3\varphi \;\; r^3L_3^3\cos 3\varphi \rangle \\ \equiv \langle 1 \mid & x\,y\,z \mid xy \;\; -x^2+y^2 \;\; xz \;\; yz \;\; -x^2+z^2 \mid \\ & x^3-3xy^2-3x^2y+y^3 \;\; xyz \;\; y^2z-x^2z \;\; xz^2-xy^2-x^2y+yz^2 \;\; -3x^2z+z^3 \rangle \end{aligned} \tag{14.18}$$

The fundamental solution u_r^* of the Helmholtz equation $u_{,ii} + k^2u = 0$ needed as in (14.4) for the frequency-domain formulation of a transient problem is obtained simply by replacing r^m in either (14.15) or (14.17) with $(2/k)^m m! J_m(kr)$ or $(2m + 1)!\sqrt{\pi}/m!/(2\,k)^m J_{m+1/2}(kr)$, for 2D or 3D problems, respectively. In these expressions, $J_m(kr)$ is the Bessel function of first kind and order m in the argument kr. The Bessel functions of order $m + 1/2$ are in fact trigonometric functions. The terms that multiply $J_m(kr)$ and $J_{m+1/2}(kr)$ are chosen in such a way that $\lim_{k\to 0} u_r^* = u_r^0$.

For the general elasticity problem in a homogeneous, isotropic medium, the displacement fundamental solution of (14.2) is given, in a frequency-domain formulation, from

$$\left(k_1^2/k_2^2 - 1\right) u^*_{jr,ji} + u^*_{ir,jj} + k_2^2 u^*_{ir} = 0, \tag{14.19}$$

where $k_1 = \dfrac{\omega}{c_1}$ and $k_2 = \dfrac{\omega}{c_2}$, for $c_1 = \sqrt{\dfrac{2G(1-\nu)}{\rho(1-2\nu)}}$ and $c_2 = \sqrt{\dfrac{G}{\rho}}$ the propagation speed of irrotational and shear waves in the elastic medium, respectively. These expressions are given in terms of the transverse elasticity modulus G, the Poisson's ratio ν, the specific mass ρ and the natural frequency ω. The solution of (14.19) may be obtained in terms of a potential function Φ_r:

$$\Phi_r = \frac{\left(J_d(k_1 r)/k_1^d - J_d(k_2 r)/k_2^d\right) 2^{d+2}(d+1)!}{r^d \left(k_2^2 - k_1^2\right)} u_r^0 \text{ for 2D} \tag{14.20}$$

$$\Phi_r = \frac{\left(J_{d+1/2}(k_1 r)\Big/k_1^{d+1/2} - J_{d+1/2}(k_2 r)\Big/k_2^{d+1/2}\right)\sqrt{\pi}(2d+3)!}{r^{d+1/2}\left(k_2^2 - k_1^2\right) 2^{d+1/2}(d+1)!} u_r^0 \text{ for 3D} \tag{14.21}$$

where d is the degree of the polynomial u_r^0 of either (14.15) or (14.17) in Cartesian coordinates. The multiplying factors in the above equations are chosen in such a way that, for the static case, $\lim_{\omega\to 0} \Phi_r = r^2 u_r^0$.

Then the solution of (14.19) is simply

$$u_{ir}^* = \left(\Phi_{\ell,jj} + k_1^2\Phi_\ell\right)\delta_{ik} + \left(k_1^2\Big/k_2^2 - 1\right)\Phi_{\ell,ik} \quad \ell = 1,\ldots,\dim(u_r^0) \tag{14.22}$$

in Cartesian coordinates with directions i, j, k, where $r \equiv D\ell + k - D$ and D is equal to 2 or 3, for 2D or 3D problems, respectively. The number of solutions comprised by the above equation, corresponding to a complete polynomial of order n in the static case, is $2(2n+1)$ for 2D problems and $3(n+1)^2$ for 3D problems. For the sake of illustration, the development of u_{ir}^*in (14.22) for the static case is given for $n = 1$ as

$$u_{ir}^0 = \left[\begin{array}{cc|cccc|c} 3-4\nu & 0 & (5-8\nu)x & -y & (7-8\nu)y & -x & \cdots \\ 0 & 3-4\nu & -y & (7-8\nu)x & -x & (5-8\nu)y & \cdots \end{array}\right] \tag{14.23}$$

$$u_{ir}^0 = \left[\begin{array}{ccc|cccc} 3-4\nu & 0 & 0 & (7-10\nu)x & -y & -z & (9-10\nu)y \\ 0 & 3-4\nu & 0 & -y & (9-10\nu)x & 0 & -x \\ 0 & 0 & 3-4\nu & -z & 0 & (9-10\nu)x & 0 \end{array}\right.$$

$$\left.\begin{array}{cccccc|c} -x & 0 & (9-10\nu)z & 0 & -x & \cdots \\ (7-10\nu)y & -z & 0 & (9-10\nu)z & -y & \cdots \\ -z & (9-10\nu)y & -x & -y & (7-10\nu)z & \cdots \end{array}\right] \tag{14.24}$$

The frequency-domain implementation of the fundamental solutions derived above gives rise to full rank matrices $\mathbf{F}^*$ and $\mathbf{H}$, as outlined in Section 14.2. However, the matrices $\mathbf{F}_0^*$ and $\mathbf{H}_0$ of the static case are not of full rank, with null-spaces given in (14.13) and (14.14). The non-normalized matrix $\tilde{\mathbf{W}}$ of either constant potentials or rigid body displacements is straightforward to construct, as given below for a discrete model with nodal points of co-ordinates (x_s, y_s, z_s):

$$\tilde{\mathbf{W}} = \left[\,1\ 1\ \cdots\ 1\,\right]^{\mathbf{T}} \text{ for 2D and 3D potential problems} \tag{14.25}$$

$$\tilde{\mathbf{W}} = \left[\begin{array}{cc|cc|c|cc} 1 & 0 & 1 & 0 & \cdots & 1 & 0 \\ 0 & 1 & 0 & 1 & \cdots & 0 & 1 \\ \hline y_1 & -x_1 & y_2 & -x_2 & \cdots & y_{nnp} & -x_{nnp} \end{array}\right]^{\mathbf{T}} \quad \text{2D elasticity} \tag{14.26}$$

$$\tilde{\mathbf{W}} = \left[\begin{array}{ccc|ccc|c|ccc} 1 & 0 & 0 & 1 & 0 & 0 & \cdots & 1 & 0 & 0 \\ 0 & 1 & 0 & 0 & 1 & 0 & \cdots & 0 & 1 & 0 \\ 0 & 0 & 1 & 0 & 0 & 1 & \cdots & 0 & 0 & 1 \\ \hline 0 & -z_1 & y_1 & 0 & -z_2 & y_2 & \cdots & 0 & -z_{nnp} & y_{nnp} \\ z_1 & 0 & -x_1 & z_2 & 0 & -x_2 & \cdots & z_{nnp} & 0 & -x_{nnp} \\ -y_1 & x_1 & 0 & -y_2 & x_2 & 0 & \cdots & -y_{nnp} & x_{nnp} & 0 \end{array}\right]^{\mathbf{T}} \quad \text{3D elasticity} \tag{14.27}$$

$\mathbf{W}$ is obtained by orthogonalization of $\tilde{\mathbf{W}}$, such that $\mathbf{W}^{\mathbf{T}}\mathbf{W} = \mathbf{I}$. Equations (14.25), (14.26) and (14.27) are also valid for non-homogeneous materials.

Non-orthogonal bases represented by $\tilde{\mathbf{V}}$ in (14.13) and (14.14) are

$$\tilde{\mathbf{V}} \equiv \mathbf{V} = \left[\, 1\; 0 \cdots 0 \,\right]^{\mathbf{T}} \text{ for 2D and 3D potential problems} \tag{14.28}$$

$$\mathbf{V} = \left[\begin{array}{cc|cc|cc|c} 1 & 0 & 0 & 0 & 0 & 0 & \cdots 0 \\ 0 & 1 & 0 & 0 & 0 & 0 & \cdots 0 \\ \hline 0 & 0 & 0 & 1 & -1 & 0 & \cdots 0 \end{array}\right]^{\mathbf{T}} \quad \text{2D elasticity} \tag{14.29}$$

$$\mathbf{V} = \left[\begin{array}{ccc|ccc|ccc|ccc|c} 1 & 0 & 0 & 0 & 0 & 0 & 0 & 0 & 0 & 0 & 0 & 0 & \cdots 0 \\ 0 & 1 & 0 & 0 & 0 & 0 & 0 & 0 & 0 & 0 & 0 & 0 & \cdots 0 \\ 0 & 0 & 1 & 0 & 0 & 0 & 0 & 0 & 0 & 0 & 0 & 0 & \cdots 0 \\ \hline 0 & 0 & 0 & 0 & 0 & 0 & 0 & 0 & -1 & 0 & 1 & 0 & \cdots 0 \\ 0 & 0 & 0 & 0 & 0 & 1 & 0 & 0 & 0 & -1 & 0 & 0 & \cdots 0 \\ 0 & 0 & 0 & 0 & -1 & 0 & 1 & 0 & 0 & 0 & 0 & 0 & \cdots 0 \end{array}\right]^{\mathbf{T}} \quad \text{3D elasticity} \tag{14.30}$$

Table 14.1 summarizes some possibilities of finite element construction in the present hybrid formulation using non-singular fundamental solutions [34]. For 2D problems, examples are given for 3-node and 6-node triangles (CST and T6) as well as for 4-node and 8-node quadrilaterals (Q4 and Q8). For 3D problems, examples are given for 4-node and 10-node tetrahedrals (Te4 and Te10) as well as for 8-node and 20-node hexahedrals (H8 and H20). The number of fundamental solutions must be at least equal to the number of external degrees of freedom, $n^* \geq n^d$. A row with the required polynomial orders corresponding to the static case is given, from which follows the number n^* of functions of the least complete set of solutions. Except for the CST and the Te4 elements, $\mathbf{H}$ is a rectangular matrix, as shown, with ranks of

Table 14.1 Some data for hybrid 2D and 3D elements for elasticity problems

	2D				3D			
Polynomial order n	1	2	3	4	1	2	3	4
Number n^* of internal d.o.f.	$n^* = 2(2n+1)$				$n^* = 3(n+1)^2$			
	6	10	14	18	12	27	48	75
Element type (n^d)	CST (6)	Q4 (8)	T6 (14)	Q8 (16)	Te4 (12)	H8 (24)	Te10 (30)	H20 (60)
Shape of **H**	6×6	10× 8	14× 12	18× 16	12× 12	27× 24	48× 30	75× 60
Rank of $\mathbf{H}_0$	3	5	9	13	6	18	24	54
Order of $\mathbf{F}^*$	6	10	14	18	12	27	48	75
Rank of $\mathbf{F}_0^*$	3	7	11	15	6	21	42	69

$\mathbf{H}_0$ given by $n^d - n^r$, where n^r is the number of rigid body displacements. The order of $\mathbf{F}^*$ is n^* and the rank of $\mathbf{F}_0^*$ is $n^* - n^r$.

For a formulation using Laplace or Fourier transforms, $\mathbf{F}^*(\omega)$ is non-singular and its inverse, as required in (14.10), may in principle be found by any standard way. For the static case, or for the modal advanced analysis of Section 14.4, $\mathbf{F}_0^*$ is singular, according to (14.14), and one must resort to the theory of generalized inverses [41, 42], which is well consolidated.

14.4 Advanced Modal Analysis of the Time-Dependent Problem

An established technique to solve time-dependent problems is the formulation of a complete frequency-domain analysis via Laplace or Fourier transforms, with subsequent ad hoc expression of results by numerical inversion, as already mentioned in Section 14.1. In such a case, $\mathbf{F}^*$ (ω), as given in (14.7), is always non-singular. Although already implemented in the frame of the hybrid finite/boundary element method, as used in the numerical example of Section 14.6, this formulation is not further developed in the present paper [24].

One may express the fundamental solution of (14.4) as a power series of frequencies with an arbitrary number m of terms [31]:

$$u_j^* = \sum_{i=0}^{m} \omega^{2i} u_{ijr}^* p_r^* = \left(u_{0jr}^* + \omega^2 \, u_{1jr}^* + \omega^4 \, u_{2jr}^* + \cdots + \omega^{2m} \, u_{mjr}^* \right) p_r^* \quad (14.31)$$

In this Section, summation carried out over terms of matrix polynomials is explicitly indicated. Observe that u_{0jr}^* of the above equation is the same matrix u_{jr}^0 of

Section 14.2. The matrices $\mathbf{F}^*$ and $\mathbf{H}$, defined in (14.7), as well as $\mathbf{K}$, introduced in (14.11), also become power series of frequencies:

$$\mathbf{F}^* = \sum_{i=0}^{m} \omega^{2i} \mathbf{F}_i^*, \qquad \mathbf{H} = \sum_{i=0}^{m} \omega^{2i} \mathbf{H}_i, \qquad \mathbf{K} = \sum_{i=0}^{m} \omega^{2i} \mathbf{K}_i \tag{14.32}$$

In such a case, the evaluation of the inverse $\mathbf{F}^{*-1}$ as a power series is not a trivial task in terms of linear algebra, since, although $\mathbf{F}^*$ is non-singular, the leading term $\mathbf{F}_0^*$ of the power series indicated in (14.32) is singular. However, since the product $\mathbf{V}^{\mathrm{T}}\mathbf{F}_1^*\mathbf{V}$ is non-singular, the inverse of $\mathbf{F}^*$ may be obtained by an algorithm whose implementation is straightforward [42].

This matrix power series can be used to transform the frequency-domain Equation (14.10) into the time-domain expression

$$\left(\mathbf{K}_0 - \sum_{i=1}^{m} (-1)^i \mathbf{M}_i \frac{\partial^{2i}}{\partial t^{2i}} \right) (\mathbf{d} - \mathbf{d}^p) = \mathbf{p}(t) - \mathbf{p}^p(t), \tag{14.33}$$

which is a coupled set of higher-order time derivatives that makes use of the matrices obtained in the frequency formulation. In this equation, one expresses $\mathbf{K}_0$ explicitly as the stiffness matrix of the static discrete-element formulation and renames the remaining terms of the power series of $\mathbf{K}$ in (14.32) as $-\mathbf{M}_i$, as generalized mass matrices, although they are actually a blending of mass and stiffness matrices. The vectors $\mathbf{d}$ of displacements are the problem unknowns, to be determined for applied equivalent nodal forces $\mathbf{p}(t)$ as well as initial conditions, given the boundary conditions and some particular solution in the time domain. The number m of frequency-related matrices can be in principle arbitrarily large, although increasing round-off errors may become a concern. The advantage of such a formulation – as compared with the developments in the classical literature – is that one guarantees a more accurate fulfillment of the dynamic differential equilibrium equations in the domain $\mathbf{\Omega}$ as the number m of terms increases.

Let the non-linear eigenvalue problem related to (14.33) be formulated as

$$\mathbf{K}_0 \mathbf{\Phi} - \sum_{i=1}^{m} \mathbf{M}_i \mathbf{\Phi} \mathbf{\Omega}^{2i} = \mathbf{0}, \tag{14.34}$$

where $\mathbf{\Omega}^2$ is a diagonal matrix with as many eigenvalues ω^2 as required to model the structural behaviour and $\mathbf{\Phi}$ is the matrix of the corresponding eigenvectors. The normality conditions of $\mathbf{\Phi}$ are expressed as [36]:

$$\sum_{i=1}^{m} \sum_{j=i}^{m} \mathbf{\Omega}^{2j-2i} \mathbf{\Phi}^{\mathrm{T}} \mathbf{M}_j \mathbf{\Phi} \mathbf{\Omega}^{2i-2} = \mathbf{I}, \quad \left(\mathbf{\Phi}^{\mathrm{T}} \mathbf{K}_0 \mathbf{\Phi} + \sum_{i=1}^{m-1} \sum_{j=1}^{m-i} \mathbf{\Omega}^{2i} \mathbf{\Phi}^{\mathrm{T}} \mathbf{M}_{j+i} \mathbf{\Phi} \mathbf{\Omega}^{2j} \right) = \mathbf{\Omega} \tag{14.35}$$

In the context of a mode-superposition procedure, one may approximate the time-dependent potentials $\mathbf{d}(t)$ as a finite sum of contributions of the normalized eigenvectors $\mathbf{\Phi}$ (according to the equations above) multiplied by a vector of amplitudes $\boldsymbol{\eta} \equiv \boldsymbol{\eta}(t)$, which are the new unknowns of the problem:

$$\mathbf{d} = \Phi\eta \tag{14.36}$$

After some manipulation [35], one transforms (14.33) into

$$\mathbf{\Omega}^2(\boldsymbol{\eta} - \boldsymbol{\eta}^p) + (\ddot{\boldsymbol{\eta}} - \ddot{\boldsymbol{\eta}}^p) = \mathbf{\Phi}^{\mathbf{T}}(\mathbf{p} - \mathbf{p}^p), \tag{14.37}$$

an uncoupled system of first order differential equations of time, which may be integrated either analytically or by means of standard numerical algorithms.

The initial conditions for the solution of (14.37) are given in terms of subsets $\mathbf{\Phi}_\epsilon$ and $\mathbf{\Omega}_\epsilon$ of modes and frequencies related to a state of pure deformation, as well as subsets $\mathbf{\Phi}_r$ and $\mathbf{\Omega}_r \equiv 0$ related to rigid body displacements:

$$\boldsymbol{\eta}_\varepsilon = \left[\mathbf{\Phi}_\varepsilon^{\mathbf{T}}\mathbf{K}_0\mathbf{\Phi}_\varepsilon\right]^{-1}\mathbf{\Phi}_\varepsilon^{\mathbf{T}}\mathbf{K}_0\mathbf{d}, \quad \boldsymbol{\eta}_r = \mathbf{\Phi}_r^{\mathbf{T}}\mathbf{M}_1\mathbf{d} \tag{14.38}$$

The set of uncoupled first order differential Equations (14.37), together with (14.38) for inclusion of the initial conditions, is an amenable form of (14.33) for the solution of a wide range of time-dependent problems on the basis of a frequency formulation. This formulation has been generalized to problems with viscous damping, in which case the non-linear eigenvalue problem of (14.34) comprises a set of complex-symmetric matrices [35, 36]. All mass matrices $\mathbf{M}_j$ must be positive definite, which can only be ensured if the whole formulation is flawless conceived and implemented.

14.5 Accuracy and Linear Algebra Issues

The hybrid boundary element method – as developed on the basis of the same singular fundamental solutions of the conventional boundary element method – has been tested for a wide range of problems ever since its proposition more than 2 decades ago and has evolved to a mathematically grounded analysis tool [28–31]. Since this method relies on fundamental solutions as the domain trial functions, according to (14.3), – and not just as test functions, as in the conventional boundary element method, – one decided to apply back to finite elements the concepts already developed. The implementations for the transient analysis of structures made out of truss and beam elements followed successfully [24, 33, 34]. In fact, the boundaries of truss and beam elements coalesce to just two points and there is no displacement approximation implied by (14.5). As a result, a displacement finite element formulation coincides with the present hybrid formulation and also with the spectral finite element implementation. Conceptually, truss and beam elements are at the same time finite and boundary elements, with the particularity that the fundamental solutions are the actual solutions of the problem one is attempting to solve. Then, any numerical method presents the same convergence pattern for structures made out of truss and beam elements whether mode superposition or Laplace transform is being used.

In the hybrid boundary element method, which is based on singular fundamental solutions, as well as in hybrid implementations of truss and beam elements, the domain displacements u_i^s and the boundary displacements u_i^d, as introduced in Section 14.2, give, when evaluated at the nodal points, the same results, which may be expressed by the equation

$$\mathbf{U}^*\mathbf{p}^* = \mathbf{d} - \mathbf{d}^p, \tag{14.39}$$

where $\mathbf{U}^* := U^*_{rs}$ are the values of u^*_{ir} in (14.4) measured at the nodal points and the the nodal values $\mathbf{d}^p$ come from the second of (14.3). When applied to a static problem, the above equation holds except for an arbitrary amount of rigid body displacements, as observed after (14.4), and should be replaced with $(\mathbf{I} - \mathbf{W}\mathbf{W}^{\mathrm{T}})\mathbf{U}_0^*\mathbf{p}^* = (\mathbf{I} - \mathbf{W}\mathbf{W}^{\mathrm{T}})(\mathbf{d} - \mathbf{d}^p)$ [29, 31].

For a finite element implementation – which excludes truss and beam elements –, (14.39) does not hold and there is an error $\varepsilon = \left|\mathbf{U}^*\mathbf{p}^* - (\mathbf{d} - \mathbf{d}^p)\right| > 0$. In fact, one obtains from the energy potential Equation (14.1) that the inertia term $\rho\ddot{u}_i^s$ is actually disconnected from the displacement field u_i^d (which is defined only along the boundary). Nevertheless, the energy principle makes sure that the error ϵ is bounded and tends to decrease with mesh refinement.

There was an attempt by Dumont and Aguilar [32] to improve the energy potential of (14.1) by using $\rho\ddot{u}_i^d$ to take the inertia term into account. However, some linear algebra issues immediately showed that the problem is mechanically unfeasible, a fact that was not completely understood at that time. The present formulation is conceptually flawless and leads to excellent results provided that some precautions are taken, as outlined in the following.

One must make sure that the fundamental solutions proposed in Section 14.3 are invariant with respect to coordinate translation of the finite element. This is particularly important for 2D elements, as the formulation turns out to be based on Bessel functions, which, for large arguments, are poorly represented by power series. Then, for a modal analysis, it is important that the Cartesian coordinates be always referred to the element's middle point, so that the radius r present in the equations throughout Section 14.3 be kept as small as possible. Moreover, it is expected that higher eigenvalues can only be accurately obtained for a sufficiently refined mesh. This problem should not occur when Laplace or Fourier transforms are used. However, the matrix $\mathbf{F}^*$ may become ill conditioned for too small wave numbers if one does not take the precaution suggested for the modal analysis.

One remarkable feature of the present formulation – whether for static or time-dependent problems – is that no reference is needed to *zero energy modes*, since one is always working with a number $n^* \geq n^d$ of solutions that correspond to complete polynomials of the static solution. Most important, the adequate inversion of $\mathbf{F}^*$

accounts automatically for the static or dynamic condensation of the overabundant internal degrees of freedom.

14.6 Numerical Example

This problem was proposed by Bruch and Zyvoloski [43] and consists in the homogeneous heat conduction in the square domain of dimensions 1×1, completely insulated along the edges $x = 0$ and $y = 0$ and with prescribed temperature $u(t) = 1$ along the edges $x = 1$ and $y = 1$. Initial temperature condition is $u(x, y, 0) = 0$ in the whole domain. Isotropic thermal conductivity $k = 1$ and specific heat $c = 1$ are assumed.

In the present analysis, the plate is discretized with a 4×4 mesh with distorted quadratic elements (Q8), as shown on the left of Fig. 14.1. The graphics on the right of the figure show results along the edge $x = 0$ (which are the same along the edge $y = 0$, due to symmetry) for several time instants (in seconds), using either one or three generalized mass matrices, according to (14.34), as compared with the

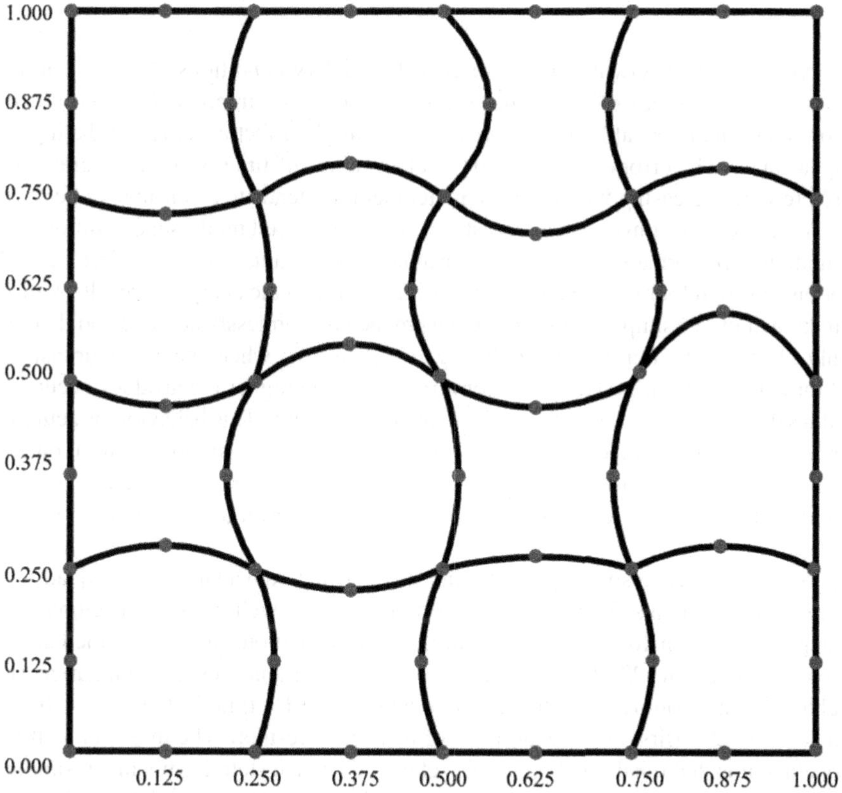

Fig. 14.1 (continued)

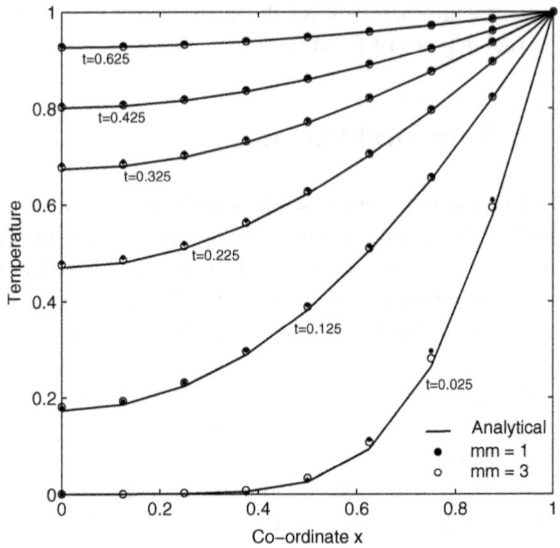

Fig. 14.1 Distorted 4×4 mesh and numerical results along x = 0 for different time instants (s)

analytical solution. Owing to the prescribed boundary conditions, the problem has a total of 48 degrees of freedom. One sees that results improve with the use of more mass matrices, although round-off errors might affect accuracy of the higher eigenvalues. The errors are larger for smaller values of time, when high gradients are present, decreasing with time, as the temperature tends to a constant value.

An interesting convergence assessment of the advanced mode-superposition formulation is performed in terms of the evaluated eigenvalues, as shown in Fig. 14.2 for one through four mass matrices and confronted with the analytical results, which are known in this simple study case. Convergence is seamlessly attained and there is only a very small improvement in the higher eigenvalues when one uses four instead of three mass matrices. However, convergence of the higher eigenvalues occurs to values that are only close – not equal – to the target results. This behaviour is actually expected, as the spatial discretization – a fixed 4×4 mesh – imposes a limit on the achievable accuracy. To compare with, implementations using truss and beam elements, whose boundaries coalesce to points, always converge to the correct values [24, 34].

The results were also compared against a classical modal analysis using quadratic Serendipity elements for the same distorted mesh, as well as with developments using Laplace transform and numerical inverses carried out in terms of the Gaver-Stehfest algorithm [22, 23]. A total of seven different analyses are summarized in Table 14.2 and compared with the analytical solution for time instants $t = 0.025$ s and $t = 0.625$ s (first and second rows of respective results). The most compelling conclusion is that quadratic Serendipity elements, which stem from a much simpler formulation, perform quite well, at least for the present numerical example. The results with Serendipity elements are comparable in accuracy with the ones of the

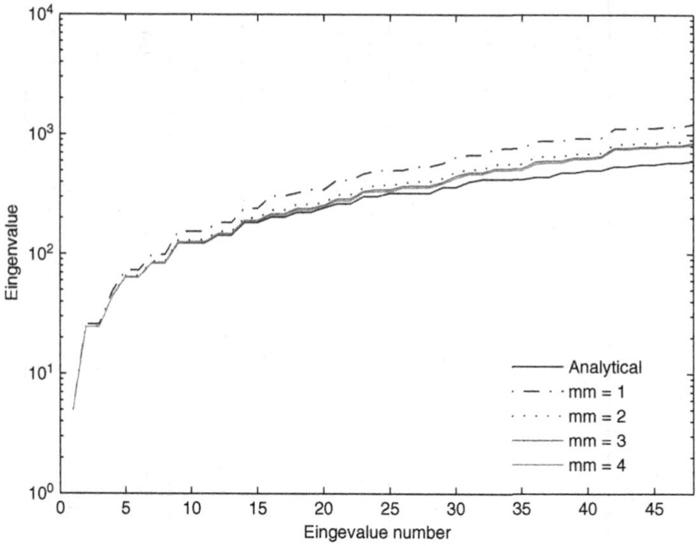

Fig. 14.2 Eigenvalues for the numerical example modelled with one through four mass matrices

hybrid formulation with three mass matrices. On the other hand, the results using the Gaver-Stehfest algorithm show excellent accuracy of the hybrid finite elements, definitely better than the ones with Serendipity elements. This example is just a flavour of how difficult it is to carry out numerical comparisons. A battery of examples, in which the meshes are also refined, must be run before one is able to draw any definitive conclusions in terms of accuracy, computational effort and ease of code development.

14.7 Conclusions

This paper shows that a variationally-based boundary element formulation, obtained as a generalization of the Hellinger-Reissner potential to take the time effect into account, may be applied to finite elements. (Conceptually, this is a journey back home, as the hybrid boundary element method has been conceived as a generalization of Pian's achievements for finite elements.) In such a generalized formulation, the only difference between a boundary element and a finite element is the use of singular or non-singular fundamental solutions as domain interpolation functions. For non-singular solutions, one proposes that the number of internal degrees of freedom be always given by the whole set of solutions comprised by a complete polynomial, as applied to the static problem. Then, the number of internal degrees of freedom is overabundant. Inversion of a flexibility matrix is, if needed, carried out in the frame of the theory of generalized inverses. The formulation is proposed in the frequency domain, with implementation initially carried out in the frame of and

Table 14.2 Numerical results along x = 0 the time instants $t = 0.025$ s and $t = 0.625$ s

Co-ordinate x		0.000	0.125	0.250	0.375	0.500	0.625	0.750	0.875
Analytical solution		0.0000 0.9600	0.0001 0.9607	0.0008 0.9630	0.0052 0.9667	0.0254 0.9717	0.0936 0.9778	0.2636 0.9847	0.5762 0.9922
HFEM: advanced mode superposition	m=1	0.0005 0.9621	−0.0005 0.9629	−0.0003 0.9650	0.0020 0.9685	0.0286 0.9732	0.1141 0.9790	0.2968 0.9855	0.6099 0.9926
	m=2	0.0008 0.9605	0.0006 0.9613	0.0024 0.9635	0.0068 0.9672	0.0313 0.9720	0.1080 0.9781	0.2828 0.9849	0.5968 0.9923
	m=3	0.0005 0.9605	0.0002 0.9612	0.0024 0.9635	0.0077 0.9671	0.0331 0.9720	0.1086 0.9780	0.2811 0.9849	0.5941 0.9923
	m=4	0.0005 0.9605	0.0001 0.9612	0.0024 0.9635	0.0080 0.9671	0.0339 0.9720	0.1092 0.9780	0.2806 0.9849	0.5931 0.9923
HFEM: laplace transform		0.0006 0.9598	0.0001 0.9605	0.0016 0.9628	0.0051 0.9665	0.0266 0.9714	0.0941 0.9775	0.2592 0.9845	0.5770 0.9921
Serendipity FEM: mode superposition		0.0015 0.9608	−0.0014 0.9611	−0.0037 0.9634	0.0086 0.9674	0.0310 0.9720	0.0778 0.9779	0.2734 0.9851	0.5924 0.9924
Serendipity FEM: laplace transform		0.0019 0.9605	−0.0010 0.9608	−0.0028 0.9631	0.0103 0.9671	0.0335 0.9717	0.0789 0.9777	0.2715 0.9849	0.5902 0.9923

advanced modal analysis. However, more recent implementations have also been done in terms of numerical Laplace or Fourier inverse transforms. For problems of structural dynamics, algorithms such as the one proposed by Crump [26] have led to good results. Diffusion-type problems are best dealt with in the frame of algorithms as proposed by Stehfest [22, 23]. Wave propagation – thus including scattering – is a problem that still deserves a numerical investigation in the frame of the present formulation, as the adequate representation of higher eigenmodes may become a challenge for the advanced modal analysis. The main task ahead is the adequate assessment of the computational performance of the proposed method for the analysis of large scale problems – whether using advanced modal analysis or numerical Laplace transforms.

Acknowledgements This work was supported by the Brazilian agencies CAPES, CNPq and FAPERJ.

References

1. Hellinger, E.: Die allgemeinen Ansätze der Mechanik der Kontinua. Enz. math. Wis. **4**, 602–694 (1914)
2. Reissner, E.: On a variational theorem in elasticity. J. Math. Phys. **29**, 90–95 (1950)
3. Hu, H.C.: On some variational principles in the theory of elasticity and the theory of plasticity. Sci. Sin. **4**, 33–54 (1955)
4. Washizu, K.: On the variational principles of elasticity and plasticity. Techn Report 25–18, MIT, Cambridge (1955)
5. Pian, T.H.H.: Derivation of element stiffness matrices by assumed stress distribution. AIAA J. **2**, 1333–1336 (1964)
6. Pian, T.H.H.: Reflections and remarks on hybrid and mixed finite element methods. In: Atluri, S.N., Gallagher, R.H., Zienkiewicz, O.C. (eds.) Hybrid and Mixed Finite Element Methods. Wiley, New York, NY, pp. 565–570 (1983)
7. Atluri, S.N., Gallagher, R.H., Zienkiewicz, O.C.: Hybrid and Mixed Finite Element Methods. Wiley, Chichester (1983)
8. Stein, E.: Eröffnung des minisymposiums "in memoriam Eric Reissner". ZAMM **80**(1), 49–52 (2000)
9. Trefftz, E.: Ein gegenstück zum ritzschen verfahren. Proceedings of the 2nd International Congress of Applied Mechanics, Zurich (1926)
10. Jirousek, J., Leon, N.: A powerful finite element for plate bending. Comput. Methods Appl. Mech. Eng. **12**, 77–96 (1977)
11. Qin, Q.H.: The Trefftz Finite and Boundary Element Method. WIT Press, Southampton (2003)
12. Maunder, E. (org.): Trefftz method – development and applications in computational mechanics (IACM special interest conference), 16–18 September 2002, Exeter (2002)
13. Reddy, J.N.: Energy Principles and Variational Methods in Applied Mechanics, 2nd edn. Wiley, New York, NY (2002)
14. Pian, T.H.H., Wu, C.C.: Hybrid and Incompatible Finite Element Methods. Chapman & Hall/CRC, New York, NY (2006)
15. Przemieniecki, J.S.: Theory of Matrix Structural Analysis. McGraw-Hill, New York. (1968)
16. Gupta, K.K.: On a finite dynamic element method for free vibration analysis of structures. Comput. Methods Appl. Mech. Eng. **9**, 105–120 (1976)
17. Paz, M., Dung, L.: Power series expansion of the general stiffness matrix for beam elements. Int. J. Numer. Methods Eng. **9**, 449–459 (1975)

18. Voss, H.A.: New justification of finite dynamic element methods. Int. Ser. Numer. Math. **83**, 232–242 (1987)
19. Beskos, D., Narayanan, G.: Dynamic response o frameworks by numerical Laplace transforms. Comput. Methods Appl. Mech. Eng. **37**, 289–307 (1983)
20. Doyle, J.: Wave Propagation in Structures. Springer, New York, NY (1997)
21. Gopalakrishnan, S., Chakraborty, A., Mahapatra, D.R.: Spectral Finite Element Method. Springer, New York, NY (2008)
22. Stehfest, H.: Algorithm 368: numerical inversion of Laplace transforms [D5]. Commun. ACM **13**(1), 47–49 (1970)
23. Stehfest, H.: Remark on algorithm 368 numerical inversion of Laplace transforms [D5]. Commun. ACM **13**(10), 624 (1970)
24. Aguilar, C.A.: Efficiency assessment of advanced modal analysis as compared to techniques based on numerical inverse transforms. M.Sc. Thesis (in Portuguese), PUC-Rio, Rio de Janeiro (2008)
25. Dubner, H., Abate, J.: Numerical inversion of Laplace transforms by relating them to the finite fourier cosine transform. J. Assoc. Comput. Mach. **15**(1), 115–123 (1968)
26. Crump, K.S.: Numerical inversion of Laplace transforms using a Fourier series approximation. J. A. Comput. Mach. **23**, 89–96 (1976)
27. De Hoog, F.R., Knight, J.H., Stokes, A.N.: An improved method for numerical inversion of Laplace transforms. SIAM J. Sci. Stat. Comput. **3**, 357–366 (1982)
28. Dumont, N.A.: The hybrid boundary element method: an alliance between mechanical consistency and simplicity. Appl. Mech. Rev. **42**(11) Part 2, S54–S63 (1989)
29. Dumont, N.A.: Variationally-based, hybrid boundary element methods. Comput. Assist. Mech. Eng. Sci. **10**, 407–430 (2003)
30. Dumont, N.A., Oliveira, R.: The exact dynamic formulation of the hybrid boundary element method. Proceedings of the 18th CILAMCE, Brasília, 29–31 October, pp. 357–364 (1997)
31. Dumont, N.A., Oliveira, R.: From frequency-dependent mass and stiffness matrices to the dynamic response of elastic systems. Int. J. Solids Struct. **38**, 10–13, 1813–1830 (2001)
32. Dumont, N.A., Aguilar, C.A.: Linear algebra issues in a family of advanced hybrid finite elements. In: Murín, J., Kutis, V., Duris, R. (eds.) CMAS2009 – International Conference on Computational Modelling and Advanced Simulations, 15 pp on CD, Bratislava (2009)
33. Dumont, N.A., Prazeres, P.G.C.: Hybrid dynamic finite element families for the general analysis of time-dependent problems. In: Proceedings of the ICSSD 2005 – 3rd International Conference on Structural Stability and Dynamics, Kissimmee, 10 pp on CD (2005)
34. Prazeres, P.G.C.: Development of hybrid finite elements for the analysis of dynamics problems using advanced modal superposition. M.Sc. Thesis (in Portuguese), PUC-Rio, Rio de Janeiro (2005)
35. Dumont, N.A.: An advanced mode superposition technique for the general analysis of time-dependent problems. In: Selvadurai, A.P.S., Tan, C.L., Aliabadi, M.H. (eds.) Advances in Boundary Element Techniques VI, pp. 333–344. CL Ltd, London (2005)
36. Dumont, N.A.: On the solution of generalized non-linear complex-symmetric eigenvalue problems. Int. J. Numer. Methods Eng. **71**, 1534–1568 (2007)
37. Dumont, N.A., Chaves, R.A.P.: General time-dependent analysis with the frequency-domain hybrid boundary element method. Comput. Assist. Mech. Eng. Sci. **10**, 431–452 (2003)
38. Dumont, N.A.: Topological aspects in the equilibrium-based implementation of finite/boundary element methods for FGMs. In: FGM 2006 – Multiscale and Functionally Graded Materials Conference 2006, Proceedings of the International Conference FGM IX, Honolulu – O'ahu, Hawai, 16–18 October 2006, American Institute of Physics, pp. 658–663 (2008)
39. Dominguez, J.: Boundary Elements in Dynamics. Computational Mechanics Publications, Southampton (1993)

40. Oliveira, M.F.F., Dumont, N.A.: Conceptual completion of the simplified hybrid boundary element method. In: Sapountzakis, E.J., Aliabadi, M.H. (eds.) Advances in Boundary Element Techniques X, Proceedings of the 10th International Conference, pp. 49–54. EC. Ltd., UK
41. Ben-Israel, A., Greville, T.N.E.: Generalized Inverses: Theory and Applications. Krieger, New York, NY (1980)
42. Dumont, N.A.: On the inverse of generalized λ-matrices with singular leading term. Int. J. Numer. Methods Eng. **66**(4), 571–603 (2006)
43. Bruch, J.C., Zyvoloski, G.: Transient two dimensional heat conduction problems solved by the finite element method. Int. J. Numer. Meth. Eng. **8**, 481–494 (1974)

Chapter 15
On Drilling Degrees of Freedom

Stephan Kugler, Peter A. Fotiu, and Justín Murín

Abstract In this paper the development of a new quadrilateral membrane finite element with drilling degrees of freedom is discussed. A variational principle employing an independent rotation field around the normal of a plane continuum element is derived. This potential is based on the Cosserat continuum theory where skew symmetric stress and strain tensors are introduced in connection with the rotation of a point. From this higher continuum theory a formulation that incorporates rotational degrees of freedom is extracted, while the stress tensor is symmetric in a weak form. The resulting potential is found to be similar to that obtained by the procedure of Hughes and Brezzi. However, Hughes and Brezzi derived their potential in terms of pure mathematical investigations of Reissner's potential, while the present procedure is based on physical considerations. This framework can be enhanced in terms of assumed stress and strain interpolations, if the numerical model is based on a modified Hu-Washizu functional with symmetric and asymmetric terms. The resulting variational statement enables the development of a new finite element that is very efficient since all parts of the stiffness matrix can be obtained analytically even in terms of arbitrary element distortions. Without the addition of any internal degrees of freedom the element shows excellent performance in bending dominated problems for rectangular element configurations.

Keywords Drilling degrees of freedom · Quadrilateral membrane element · Cosserat continuum · Assumed strain method

15.1 Introduction

In-plane rotational degrees of freedom Φ_z^I are referred to as "drilling degrees of freedom" in membrane elements (see Fig. 15.1). The subscript z indicates that the rotation vector $\mathbf{\Phi}$ is orientated in the positive z direction of a Cartesian

S. Kugler (✉)
Department of Applied and Numerical Mechanics, University of Applied Sciences Wiener Neustadt, Wiener Neustadt, Austria
e-mail: kugler@fhwn.ac.at

J. Murín et al. (eds.), *Computational Modelling and Advanced Simulations*, Computational Methods in Applied Sciences 24, DOI 10.1007/978-94-007-0317-9_15,

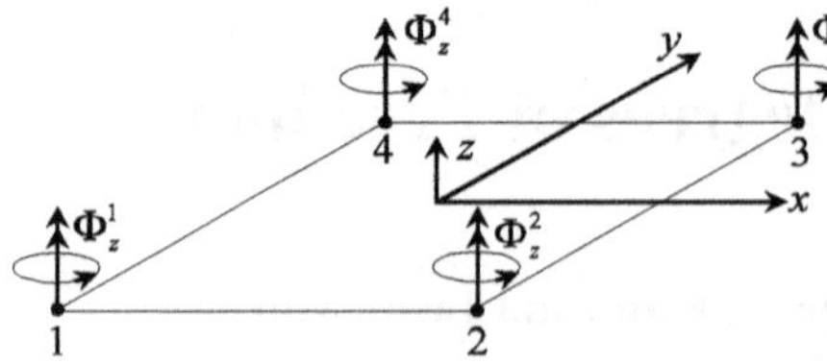

Fig. 15.1 Membrane element with drilling degrees of freedom

coordinate system laying in the plane of the membrane, while the superscript I indicates the corresponding node number (i.e. $I = 1, 2, 3, 4$ in an arbitrarily shaped quadrilateral).

The need for membrane elements with drilling degrees of freedom arises if shell elements are derived in terms of a plane plate bending element that is coupled with a plane stress membrane continuum element. Since an arbitrary three dimensional shell element is assumed to have six degrees of freedom at each node, we find that there is no stiffness associated with the drilling degree of freedom. Clearly, the corresponding stiffness matrix is singular owing to the zero column/row which corresponds to the in-plane rotation Φ_z. In principle it is always possible to work in a locally defined five degrees of freedom system at each node but in practice one encounters numerous difficulties in terms of model construction, programming, numerical sensitivity, etc. Thus the presence of this "sixth" degree of freedom based on a convergent numerical framework seems to be very attractive. Attempts to resolve this problem can be categorized in three methods:

- Incorporating an artificial stiffness for the plane rotational degree of freedom.
- Design of shape functions for rotations that are constructed out of higher order displacement based shape functions.
- Derivation of a potential that consists of an in-plane rotational degree of freedom.

The incorporation of an artificial stiffness which can be interpreted as a fictitious tensorial field seems to be inconsistent and produces numerical difficulties. Obviously, if a shell element is combined with a beam that is orientated orthogonal to the shell plane and loaded by a torsion moment the obtained stiffness of this assembly appears rather "accidentally" than stemming from mechanical considerations.

The derivation of shape functions for the in-plane rotations becomes tedious since the necessary criteria for convergence – namely the rigid body motion and the constant strain criteria – have to be fulfilled simultaneously. In the paper by MacNeal and Harder [1] and in the references therein a compilation of some early efforts is presented. However, Allman [2] and independently Bergan and Felippa [3] derived useful elements within this approach. Further refinement and the development of a corresponding quadrilateral was published by MacNeal and Harder [1].

They use quadratic displacement functions rather than cubic ones as in the early attempts to introduce a vertex rotation in terms of the midpoint normal displacement components of adjacent edges (see Fig. 15.2). However, as noted in [1] such a

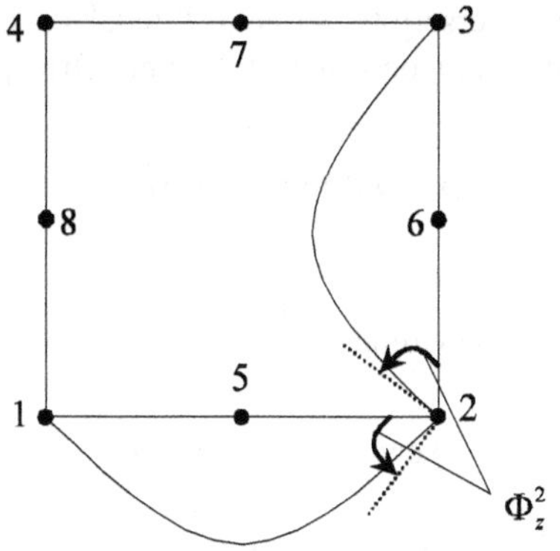

Fig. 15.2 Quadratic displacement shape functions at node 5 and 6 introduce the vertex rotation at node 2

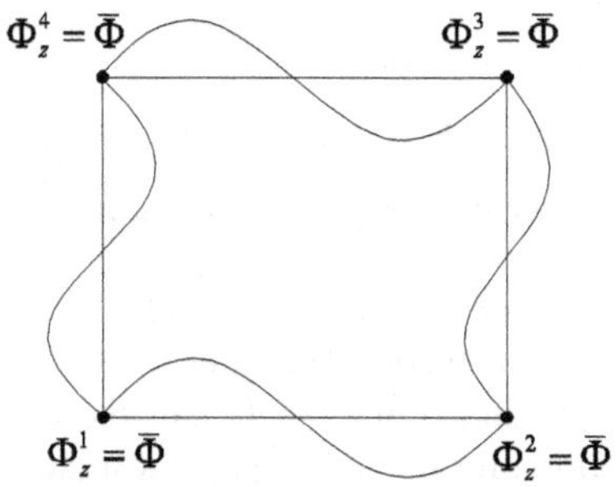

Fig. 15.3 Spurious mode of MacNeal and Harder element

procedure leads to two zero energy modes,[1] where the relevant one is depicted in Fig. 15.3 with all vertex rotations being equal to $\bar{\Phi}$. Nevertheless, possible stabilization procedures are available in [1]. The main ingredient for this stabilization is an appropriate penalty function which elastically ties the Gauss weighted average of corner rotations to the in-plane rotation at the center of the element.

For the third procedure different attempts can be found in literature. One possibility originates from Reissner's potential [4] which contains the in-plane rotation as a separate degree of freedom

$$\Pi = -\frac{1}{2}\int_{\Omega} \text{symm}\boldsymbol{\sigma}\cdot\mathbf{c}^{-1}\cdot\text{symm}\boldsymbol{\sigma}\,\mathrm{d}\Omega + \int_{\Omega} \boldsymbol{\sigma}^T\cdot(\text{grad}\mathbf{u} - \boldsymbol{\Phi})\,\mathrm{d}\Omega - \int_{\Omega} \mathbf{u}\cdot\mathbf{f}\mathrm{d}\Omega. \quad (15.1)$$

In (15.1) we temporarily use the tensorial notation as in [5], whereby $\boldsymbol{\sigma}$, $\mathbf{u}$ and $\boldsymbol{\Phi}$ represent the stress tensor, the displacement vector and the in-plane rotations, respectively. The variables $\mathbf{c}$ and $\mathbf{f}$ denote the elasticity tensor and the volumetric

[1] One spurious mode is elastically restrained for non rectangular element shapes.

force vector, while the superscripts -1 and T label an inverse and a transpose operation. The operator *symm* describes the symmetric part of a second order tensor, i.e.

$$\boldsymbol{\sigma} = \text{symm}\boldsymbol{\sigma} + \text{skew}\boldsymbol{\sigma}, \tag{15.2}$$

where

$$\text{symm}\boldsymbol{\sigma} = \frac{1}{2}\left(\boldsymbol{\sigma} + \boldsymbol{\sigma}^T\right), \tag{15.3}$$

$$\text{skew}\boldsymbol{\sigma} = \frac{1}{2}\left(\boldsymbol{\sigma} - \boldsymbol{\sigma}^T\right), \tag{15.4}$$

while the operator *grad* denotes the gradient. It is shown in [5] that (15.1) represents a boundary value problem of the form

$$\text{div}\boldsymbol{\sigma} + \mathbf{f} = 0, \tag{15.5}$$

$$\text{skew}\boldsymbol{\sigma} = 0, \tag{15.6}$$

$$\boldsymbol{\Phi} = \text{skew grad}\mathbf{u}, \tag{15.7}$$

$$\text{symm}\boldsymbol{\sigma} = \mathbf{c} \cdot \text{symm grad}\mathbf{u}. \tag{15.8}$$

Since (15.6) enforces a symmetric stress state (15.5), (15.6), (15.7) and (15.8) describe a classical continuum theory. However, if the variables are interpolated in a bilinear way it is shown in [5] that (15.1) fails the LBB conditions as well as the constraint count condition for the mixed patch test (see e.g. [6]). Consequently, Hughes and Brezzi enriched Reissner's potential (15.1) by

$$\Pi^{HB} = \Pi - \frac{1}{2}\gamma^{-1} \int_{\Omega} |\text{skew}\boldsymbol{\sigma}|^2 \, d\Omega, \tag{15.9}$$

where γ is a material dependent factor. It has to be noted that the motivation for this additive functional term arises from purely mathematical considerations. Nevertheless, a class of very useful finite elements can be developed from this assumption (see [7, 8]).

A different possibility to derive a potential that consists of in-plane rotational degrees of freedom originates from [9–11]. There, a functional for large strain rate problems consisting of stretching, spin and unsymmetric Piola stress rate tensors is reduced to linear isotropic plane stress elasticity, where the spin is set to be the in-plane rotational degree of freedom. Although some promising elements have been derived according to the theory [10, 11], the basic idea to derive this potential does not seem to stem from physical considerations, i.e. a classical large strain potential without rotational parameters is used to artificially incorporate drilling degrees of freedom within a linear theory.

Many different formulations have been proposed to maximize accuracy and efficiency of membrane elements. For implicit analysis the bottleneck for computational throughput tends to be the solution of simultaneous equations and consequently the focus has been on achieving coarse mesh accuracy. Several membrane or shell finite elements for linear and geometrically nonlinear elasticity have been developed in recent years [12–16]. Usually the formulations rely on the regularized variational statement according to (15.9). All proposed elements introduce the incompatible mode technique or somehow equivalently an enhanced strain method to improve the performance of lower order elements in bending dominated deformation patterns. The elements in the above mentioned references behave similar to the well known assumed stress element proposed by Pian and Sumihara [17] (which contains no drilling degrees of freedom), i.e. they give accurate results under pure bending loads in rectangular element configurations and their trapezoidal locking (the sensitivity to mesh distortions) is moderate. The deficiency of those strategies is that computational efficiency is sacrificed to provide coarse mesh accuracy, since the internal degrees of freedom have to be condensed at element level and at least a four point (sometimes even a nine point) Gaussian quadrature is needed.

In explicit analysis the bottleneck tends to be the element computations (constitutive relations, strain calculations, etc.). Consequently the focus has been on cheaper element formulations using single point reduced integration schemes with hourglass control. Elements based on physical hourglass control rely on Simo and Hughes' B-bar method [18] and perform very well in rectangular element configurations [19, 20] (they, however, lack of any drilling degrees of freedom). Even though these elements show more sensitivity to mesh distortions they become increasingly popular in explicit simulations. Consequently, an element that contains drilling degrees of freedom within a highly efficient formulation would be useful.

The remainder of the paper is as follows: In Section 15.2 we derive a potential similar to (15.1) and (15.9), where the in-plane rotations can be interpolated separately. However, we take a higher continuum theory (Cosserat continuum) – introducing mechanical consideration – as a starting point. The resulting potential is further enlarged by ideas similar to the Hu-Washizu variational principle for mixed elements. In Section 15.3 we propose a useful and very efficient element according to our procedure which is tested in Section 15.4. Further test cases all indicating suitable predictive quality can be found in [21].

Remark. The field of application of our derived element tends to be within explicit simulations. An extension of the applied theory to large deformations/ rotations and finite strain is inexpensive if the problem is formulated in an updated Lagrangian scheme in connection with a corotational coordinate system approach.

15.2 The Cosserat Continuum as a Basis for Membrane Elements with Drilling Degrees of Freedom

In order to avoid confusion we use an index notation within the remainder of the text and a summation of repeated indices is understood. The classical continuum theory assumes that external loads cause only internal forces (and no moments) within a continuous body. Owing to this assumption the stress and strain tensors (σ_{ij} and ε_{ij}, respectively) are symmetric, i.e. $\sigma_{ij} = \sigma_{ji}$ and $\varepsilon_{ij} = \varepsilon_{ji}$. This supposition is in good agreement with experimental verifications as far as we are not concerned with high stress gradients at holes or cracks and dynamic experiments with high frequencies that produce wavelengths in the dimension of the micro structure of the medium. On the other hand, in a body with couple stresses μ_{ij}, we get equilibrium equations with an asymmetric stress tensor. If the deformation of a body is described only by displacements and their gradients the continuum theory is called Cosserat pseudo continuum. In contrast to the Cosserat pseudo continuum the complete theory of asymmetric elasticity assumes that each molecule does not only translate but also rotates independently. Thus, every point of such a continuum has six degrees of freedom in space and three degrees of freedom in plane problems. This theory of elasticity was pioneered by Cosserat and Cosserat [22] and comprehensively summarized e.g. by Nowacki [23] and Eringen [24]. Clearly, this kind of continuum theory is well suited to formulate elements with drilling degrees of freedom. The key step of this work is to reformulate the potential of this higher continuum theory in terms of a classical continuum model which contains point rotations but shows a symmetric stress state.

15.2.1 Derivation of the Field Problem

Let us imagine a Cartesian infinitesimal volume $dx\,dy\,dz$ where force stresses σ_{ij} and couple stresses μ_{ij} are acting according to Fig. 15.4 which depicts the plane behavior for sake of clarity. Under such conditions we obtain equilibrium equations of the form

$$\sigma_{ji,j} + f_i = 0, \tag{15.10}$$

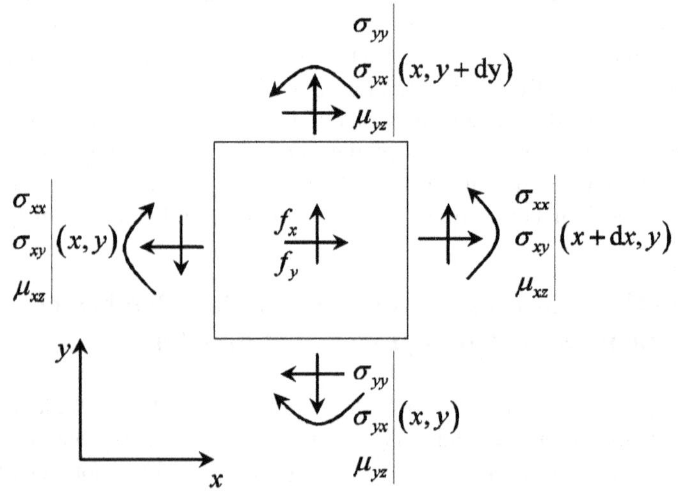

Fig. 15.4 Plane infinitesimal area with force stresses and couple stresses acting on it

$$\sigma_{ji}n_j = t_i \text{ for } \mathbf{x} \in \Gamma_t, \tag{15.11}$$

$$e_{ijk}\sigma_{jk} + \mu_{ji,j} = 0, \tag{15.12}$$

$$\mu_{ji}n_j = m_i \text{ for } \mathbf{x} \in \Gamma_t. \tag{15.13}$$

Equation (15.10) represents the force equilibrium with the associated boundary condition (15.11), where f_i is a volumetric force vector and t_i depicts the vector of surface loads acting on the boundary region Γ_t. Note that a comma in an index indicates a partial derivative. The variable n_j denotes the outward normal vector of the surface. Without any body couples the equilibrium of moments is given by (15.12) with the corresponding boundary condition (15.13). Here m_i denotes a surface couple loading on the boundary Γ_t, while e_{ijk} is the alternator tensor defined by

$$e_{123} = e_{231} = e_{312} = 1, \tag{15.14}$$

$$e_{321} = e_{213} = e_{132} = -1, \quad \text{else,} \ e_{ijk} = 0. \tag{15.15}$$

It is crucial to keep the order of indices of the stress tensors since they are no longer symmetric.

The change of virtual work done by external forces is now given by

$$\delta\Pi^{\text{ext}} = \int\limits_{\Omega} f_i\delta u_i \mathrm{d}\Omega + \int\limits_{\Gamma_t} (t_i\delta u_i + m_i\delta\phi_i)\,\mathrm{d}\Gamma, \tag{15.16}$$

where δu_i and $\delta\phi_i$ are the associated virtual displacements and rotations fields. By inserting the boundary conditions (15.11) and (15.13) into (15.16) we get after employing the Gaussian theorem

$$\begin{aligned}
\delta\Pi^{\text{ext}} &= \int\limits_{\Omega} f_i\delta u_i \mathrm{d}\Omega + \int\limits_{\Omega} \left[\left(\sigma_{ji}\delta u_i\right)_{,j} + \left(\mu_{ji}\delta\phi_i\right)_{,j}\right]\mathrm{d}\Omega = \\
&= \int\limits_{\Omega} [\underbrace{\left(\sigma_{ji,j} + f_i\right)}_{=0}\delta u_i + \underbrace{\left(\mu_{ji,j} + e_{ijk}\sigma_{jk}\right)}_{=0}\delta\phi_i]\mathrm{d}\Omega \\
&+ \int\limits_{\Omega} \left[\sigma_{ji}\delta u_{i,j} + \mu_{ji}\delta\phi_{i,j} - e_{ijk}\sigma_{jk}\delta\phi_i\right]\mathrm{d}\Omega = \\
&= \int\limits_{\Omega} \left[e_{ijk}\sigma_{ji}\delta\phi_k + \sigma_{ji}\delta u_{i,j} + \mu_{ji}\delta\phi_{i,j}\right]\mathrm{d}\Omega = \\
&= \int\limits_{\Omega} [\sigma_{ji}\underbrace{\left(\delta u_{i,j} + e_{ijk}\delta\phi_k\right)}_{\delta\varepsilon_{ji}} + \mu_{ji}\underbrace{\delta\phi_{i,j}}_{\delta\kappa_{ji}}]\mathrm{d}\Omega.
\end{aligned} \tag{15.17}$$

According to the outcome (15.17) we define the micro polar strains

$$\varepsilon_{ji} = u_{i,j} + e_{ijk}\phi_k, \tag{15.18}$$

$$\kappa_{ji} = \phi_{i,j}. \tag{15.19}$$

A constitutive relation can be obtained by expanding the free energy function. In an isotropic homogeneous and centrosymmetric[2] body one finds

$$\sigma^s_{ji} = 2\mu\varepsilon^s_{ji} + \lambda\varepsilon^s_{kk}\delta_{ji}, \tag{15.20}$$

$$\sigma^a_{ji} = 2\alpha\varepsilon^a_{ji}, \tag{15.21}$$

$$\mu^s_{ji} = 2\gamma\kappa^s_{ji} + \beta\kappa^s_{kk}\delta_{ji}, \tag{15.22}$$

$$\mu^a_{ji} = 2\varepsilon\kappa^a_{ji}. \tag{15.23}$$

The superscripts s and a indicate symmetric and asymmetric components of the micro polar strains and stresses. The symmetric and asymmetric parts of a second order tensor are found according to the Euclidean decomposition theorem, i.e.

$$\varepsilon_{ji} = \varepsilon^s_{ji} + \varepsilon^a_{ji} \tag{15.24}$$

with

$$\varepsilon^s_{ji} = \frac{1}{2}\left(u_{i,j} + u_{j,i}\right), \tag{15.25}$$

$$\varepsilon^a_{ji} = \frac{1}{2}\left(u_{i,j} - u_{j,i}\right) + e_{ijk}\phi_k. \tag{15.26}$$

In (15.20), (15.21), (15.22) and (15.23) the variables μ and λ are the Lamé constants and α, β, γ and ε are new elastic constants.

15.2.2 Reduction of the Field Problem and Its Weak Form

In the previous sections we derived a boundary value problem for the asymmetric Cosserat theory of elasticity. The governing equations are the partial differential equations of equilibrium with boundary conditions (15.10), (15.11), (15.12) and (15.13), the kinematic relations (15.25) and (15.26) and the constitutive laws (15.20), (15.21), (15.22) and (15.23). Since classical approximation methods are based on weighted residuals (e.g. the Galerkin procedure) we state in the absence of

[2] Invariant with respect to rotations.

surface couple loads m_i[3].

$$\int_{\Omega} \left(\sigma_{ji,j} + f_i\right) \delta u_i \mathrm{d}\Omega + \int_{\Gamma_t} \left(t_i - \sigma_{ji} n_j\right) \delta u_i \mathrm{d}\Gamma + \int_{\Omega} \left(e_{ijk}\sigma_{jk} + \mu_{ji,j}\right) \delta\phi_i \mathrm{d}\Omega = 0, \tag{15.27}$$

where δu_i and $\delta\phi_i$ are virtual displacements and rotations, respectively. Note that (15.27) approximates the equilibrium equations while the kinematic and constitutive relations will be fulfilled exactly. We have to point out that we intend to consider a classical elasticity theory with a symmetric stress tensor and use the Cosserat elasticity only as an auxiliary model to introduce in-plane rotations. Therefore, we assume that the couple stresses μ_{ij} are vanishing (see footnote 3) giving

$$\int_{\Omega} \left(\sigma_{ji,j} + f_i\right) \delta u_i \mathrm{d}\Omega + \int_{\Gamma_t} \left(t_i - \sigma_{ji} n_j\right) \delta u_i \mathrm{d}\Gamma + \int_{\Omega} e_{ijk}\sigma_{jk}\delta\phi_i \mathrm{d}\Omega = 0. \tag{15.28}$$

After the application of Gauss's theorem and slight rearrangements (15.28) can be rewritten as

$$\delta\Pi^{\mathrm{int}} = \delta\Pi^{\mathrm{ext}} \tag{15.29}$$

with

$$\delta\Pi^{\mathrm{int}} = \int_{\Omega} \sigma_{ji}\delta u_{i,j} \mathrm{d}\Omega - \int_{\Omega} e_{ijk}\sigma_{jk}\delta\phi_i \mathrm{d}\Omega \tag{15.30}$$

$$\delta\Pi^{\mathrm{ext}} = \int_{\Omega} f_i \delta u_i \mathrm{d}\Omega + \int_{\Gamma_t} t_i \delta u_i \mathrm{d}\Gamma. \tag{15.31}$$

The Euclidean decomposition of (15.30) gives

$$\begin{aligned} \delta\Pi^{\mathrm{int}} &= \int_{\Omega} \sigma^s_{ji}\delta u^s_{i,j} \mathrm{d}\Omega + \int_{\Omega} \sigma^a_{ji}\delta u^a_{i,j} \mathrm{d}\Omega + \int_{\Omega} e_{ijk}\sigma^a_{ji}\delta\phi_k \mathrm{d}\Omega = \\ &= \int_{\Omega} \sigma^s_{ji} \underbrace{\delta u^s_{i,j}}_{\delta\varepsilon^s_{ji}} \mathrm{d}\Omega + \int_{\Omega} \sigma^a_{ji} \underbrace{\left(\delta u^a_{i,j} + e_{ijk}\delta\phi_k\right)}_{\delta\varepsilon^a_{ji}} \mathrm{d}\Omega, \end{aligned} \tag{15.32}$$

since the double contractive product of a symmetric and an asymmetric tensor vanishes. Incorporating the constitutive relations we get

$$\int_{\Omega} C_{ijkl}\varepsilon^s_{kl}\delta\varepsilon^s_{ij} \mathrm{d}\Omega + 2\alpha \int_{\Omega} \varepsilon^a_{ij}\delta\varepsilon^a_{ij} \mathrm{d}\Omega = \int_{\Omega} f_i \delta u_i \mathrm{d}\Omega + \int_{\Gamma_t} t_i \delta u_i \mathrm{d}\Gamma \tag{15.33}$$

[3]This represents the key step of our procedure. We are not involved with a continuum theory where boundary moments are acting since we focus on classical elasticity formulations.

with

$$c_{ijkl} = \lambda \delta_{ij} \delta_{kl} + \mu \left(\delta_{ik} \delta_{jl} + \delta_{il} \delta_{jk} \right) \tag{15.34}$$

representing the isotropic, positive definite fourth order elasticity tensor with major and minor symmetry. The integrated counterpart of (15.33) gives the internal potential of a modified Cosserat continuum, where the symmetry of the stress tensor is enforced in a weak sense (see the third summand of (15.28)), i.e.

$$\Pi^{\text{int}} = \frac{1}{2} \int\limits_{\Omega} c_{ijkl} \varepsilon^{s}_{kl} \varepsilon^{s}_{ij} \mathrm{d}\Omega + \alpha \int\limits_{\Omega} \varepsilon^{a}_{ij} \varepsilon^{a}_{ij} \mathrm{d}\Omega. \tag{15.35}$$

Observe that (15.35) is similar to the modified potential derived in [5] which has been used to deduce a number of useful elements (see [7] or [8]). However, our derivation is based on physical considerations while the enhancements of Reissner's potential to achieve convergence [5] are only justified in a mathematical sense. The elastic constant α is usually[4] assumed to be the shear modulus μ [5] and as noted in [5] the theory does not anticipate much sensitivity to its particular value. Nevertheless, parameter studies over a wide range of α are carried out in Section 15.4.

15.2.3 Hu-Washizu Enhancements and the B-Bar Form

The unified derivation of (15.35) is based on approximating the equilibrium equations and the corresponding boundary conditions in a weak sense. This framework can even be generalized by using Hu-Washizu multi-field variational principles, where all equations of the boundary value problem are only fulfilled in a weak sense (see e.g. [6]). We start with the following potential

$$\begin{aligned} \bar{\Pi} = & \frac{1}{2} \int\limits_{\Omega} \bar{\sigma}^{s}_{ij} \bar{\varepsilon}^{s}_{ij} \mathrm{d}\Omega + \alpha \int\limits_{\Omega} \bar{\sigma}^{a}_{ij} \bar{\varepsilon}^{a}_{ij} \mathrm{d}\Omega + \int\limits_{\Omega} \left(\bar{\sigma}^{s}_{ij} - c_{ijkl} \bar{\varepsilon}^{s}_{kl} \right) \lambda^{(1)}_{ij} \mathrm{d}\Omega + \\ & + \int\limits_{\Omega} \left[\bar{\varepsilon}^{s}_{ij} - \frac{1}{2} \left(u_{i,j} + u_{j,i} \right) \right] \lambda^{(2)}_{ij} \mathrm{d}\Omega + \int\limits_{\Omega} \left(\bar{\sigma}^{a}_{ij} - 2\alpha \bar{\varepsilon}^{a}_{ij} \right) \lambda^{(3)}_{ij} \mathrm{d}\Omega + \\ & + \int\limits_{\Omega} \left[\bar{\varepsilon}^{a}_{ij} - \frac{1}{2} \left(u_{j,i} - u_{i,j} \right) - e_{jik} \phi_k \right] \lambda^{(4)}_{ij} \mathrm{d}\Omega - \int\limits_{\Omega} f_i u_i \mathrm{d}\Omega - \int\limits_{\Gamma_t} t_i u_i \mathrm{d}\Gamma, \end{aligned} \tag{15.36}$$

whereby an overbar indicates that the corresponding field is assumed and can be interpolated independently. The variable $\lambda^{(I)}_{ij}$ with $I = 1, 2, 3, 4$ depicts the I-th second order Lagrangian multiplicator, which has to be identified in a way that the

[4]The choice $\alpha = \mu$ is derived straightforward in terms of a convergence proof in [5].

enlarged potential (15.36) models the boundary value problem of Section 15.2.1 in connection with symmetry of the stress tensor. Variation of (15.36) gives finally

$$\lambda_{ij}^{(1)} = -\frac{1}{2}\bar{\varepsilon}_{ij}^{s}, \quad \lambda_{ij}^{(2)} = -\bar{\sigma}_{ij}^{s}, \tag{15.37}$$

$$\lambda_{ij}^{(3)} = -\frac{1}{2}\bar{\varepsilon}_{ij}^{a}, \quad \lambda_{ij}^{(4)} = -\bar{\sigma}_{ij}^{a}. \tag{15.38}$$

In principle the interpolation of the variables with an over bar can be carried out without any restrictions, so we may postulate certain orthogonalities between the particular fields. Hence, we suppose that the assumed stress tensors $\bar{\sigma}_{ij}^{s}$ and $\bar{\sigma}_{ij}^{a}$ are orthogonal to the corresponding constraint equations, i.e.

$$\int_{\Omega} \left[\bar{\varepsilon}_{ij}^{s} - \frac{1}{2}\left(u_{i,j} + u_{j,i}\right)\right] \bar{\sigma}_{ij}^{s}\, \mathrm{d}\Omega = 0 \tag{15.39}$$

$$\int_{\Omega} \left[\bar{\varepsilon}_{ij}^{a} - \frac{1}{2}\left(u_{j,i} - u_{i,j}\right) - e_{jik}\phi_k\right] \bar{\sigma}_{ij}^{a}\, \mathrm{d}\Omega = 0. \tag{15.40}$$

The key step (15.39) and (15.40) was pioneered by Simo and Hughes [18] and Hughes [25] and is often called B-bar form. Though, we do not consider the detailed elaboration of stress interpolations satisfying (15.39) and (15.40) we remark that such functions can always be found since the entire space of continuous functions is admissible. Consequently, the assumed stress fields do not appear in any of the final equations. If we rewrite (15.36) in terms of (15.39) and (15.40) we get after some manipulations

$$\bar{\Pi} = \frac{1}{2}\int_{\Omega} c_{ijkl}\,\bar{\varepsilon}_{kl}^{s}\,\bar{\varepsilon}_{ij}^{s}\,\mathrm{d}\Omega + \alpha\int_{\Omega} \bar{\varepsilon}_{ij}^{a}\,\bar{\varepsilon}_{ij}^{a}\,\mathrm{d}\Omega - \int_{\Omega} f_i u_i \mathrm{d}a - \int_{\Gamma_t} t_i u_i \mathrm{d}\Gamma. \tag{15.41}$$

Equation (15.41) gives the variational basis to assume strain fields in a way that improves the element's performance and computational efficiency. If the assumed strain field $\bar{\varepsilon}_{ij}^{s}$ and $\bar{\varepsilon}_{ij}^{a}$ are interpolated only in terms nodal displacements and rotations and (15.39) and (15.40) hold, the formulation does not consist of any internal discontinous fields that have to be interpolated separately. Since no internal degrees of freedom have to be condensed at element level the formulation (15.41) is well suited to derive highly efficient elements. Since the equilibrium state requires the potential (15.41) to be stationary we finally write

$$\int_{\Omega} c_{ijkl}\delta\,\bar{\varepsilon}_{kl}^{s}\,\bar{\varepsilon}_{ij}^{s}\,\mathrm{d}\Omega + 2\alpha\int_{\Omega} \delta\,\bar{\varepsilon}_{ij}^{a}\,\bar{\varepsilon}_{ij}^{a}\,\mathrm{d}\Omega = \int_{\Omega} f_i\delta u_i \mathrm{d}\Omega + \int_{\Gamma_t} t_i\delta u_i \mathrm{d}\Gamma. \tag{15.42}$$

15.3 Numerical Procedure – Development of Membrane Elements

In this section we use the enlarged potential (15.41) to derive two plane quadrilateral elements with drilling degrees of freedom. The stiffness matrix **K** of such an element has to satisfy the following requirements for convergence and stability:

- The stiffness matrix has to be of correct rank i.e. it has to be able to represent the zero state under three independent rigid body motions (including rotations). No other zero eigenvalues are allowed.
- **K** has to fulfill the patch test. If an arbitrary patch of elements is loaded by tractions that cause a linear displacement field the corresponding strain has to be constant.

Apart from these necessary features it is desirable that the element is not based on numerically adjusted factors. We want to derive elements that are accurate and have good predictive capabilities. The computational effectiveness of finite elements has taken a backseat in recent years because of the enlarged capacity of modern computer systems. Usually the effectiveness and the accuracy of a numerical procedure are contradictory within the development of finite elements. Clearly, the order of Gaussian quadrature influences the speed of calculations remarkably and we restrict ourselves to a 2×2 point integration and aspire to single point integration. Compared to common procedures (see e.g. [7, 8, 12–16]) our element is at least twice as fast because main ingredients of those element's stiffness matrices have to be integrated using a 3×3 Gaussian quadrature. The accuracy of the elements developed in the above mentioned elements is related to an incompatible mode technique or relies equivalently on enhanced strain fields. Clearly, such a procedure requires the implementation of static condensation to eliminate the corresponding generalized displacements, i.e. the internal degrees of freedom associated with the additional modes. The B-bar method (15.42) circumvents these internal degrees of freedom and consequently the efficiency is raised considerably.

It will be shown that the variational statement (15.42) can be used to derive elements, where the stiffness matrix can be evaluated analytically, even for arbitrary element distortions.[5] The corresponding procedure can also be interpreted as a reduced integrated one-point element with a physical hourglass stabilization.[6]

The basis of all derivations is a procedure where the strain tensors (both symmetric and asymmetric) are found from displacement and rotation interpolations such that the kinematic relations (15.25) and (15.26) hold. These strain interpolations can

[5] This procedure also leads to the derivation of elements without drilling degrees of freedom, i.e. classical plane continuum elements where all parts of the stiffness matrix are evaluated analytically.

[6] Admirably, even the hourglass stiffness matrices avoid any numerical quadrature which considerably furnishes the element's effectiveness.

be modified according to the assumed strain method of Section 15.2.3 to achieve more effectiveness and accuracy.

In terms of implementation it is useful to change from indicial notation to a matrix notation (Voigt notation). Apart from this we observe that the (assumed) asymmetric strain tensor $\bar{\varepsilon}^a_{ij}$ can be reformulated as a vector, i.e.

$$\bar{\varepsilon}^a_k = \frac{1}{2} e_{ijk}\, \bar{\varepsilon}^a_{ji} \,. \tag{15.43}$$

Thus, for a plane problem ($k = z$) we obtain from (15.26)

$$\varepsilon^a_z = \frac{1}{2}\left(u_{x,y} - u_{y,x}\right) + \phi_z, \tag{15.44}$$

which is rewritten by omitting the dispensable subscript z according to

$$\varepsilon_a = \frac{1}{2}\left(u_{x,y} - u_{y,x}\right) + \phi. \tag{15.45}$$

15.3.1 Elements that Fulfill the Kinematic Relations in a Strong Form

If we find strain interpolations which fulfill the kinematic relations in terms of bilinear interpolations of displacements and rotations, respectively the variational statements for equilibrium (15.33) and (15.42) are identical and we call the resulting element QM4D4. Equation (15.33) is reformulated for two dimensional problems with a plane stress constraint in matrix notation, i.e.,

$$\int_\Omega \delta \boldsymbol{\varepsilon}_s^T \mathbf{C} \boldsymbol{\varepsilon}_s \mathrm{d}\Omega + 2\alpha \int_\Omega \delta\varepsilon_a \varepsilon_a \mathrm{d}\Omega = \int_\Omega \delta \mathbf{u}^T \mathbf{f} \mathrm{d}a + \int_{\Gamma_t} \delta \mathbf{u}^T \mathbf{t} \mathrm{d}\Gamma, \tag{15.46}$$

with

$$\boldsymbol{\varepsilon}_s = \begin{bmatrix} u_{x,x} \\ u_{y,y} \\ u_{x,y} + u_{y,x} \end{bmatrix}, \quad \varepsilon_a = \frac{1}{2}\left(u_{x,y} - u_{y,x}\right) + \phi \tag{15.47}$$

and

$$\mathbf{C} = \frac{E}{1-\nu^2}\begin{bmatrix} 1 & \nu & 0 \\ \nu & 1 & 0 \\ 0 & 0 & (1-\nu)/2 \end{bmatrix}. \tag{15.48}$$

In (15.48) E depicts the elastic modulus while ν denotes the Poisson's ratio. It is convenient to use two different representations of bilinear shape functions for

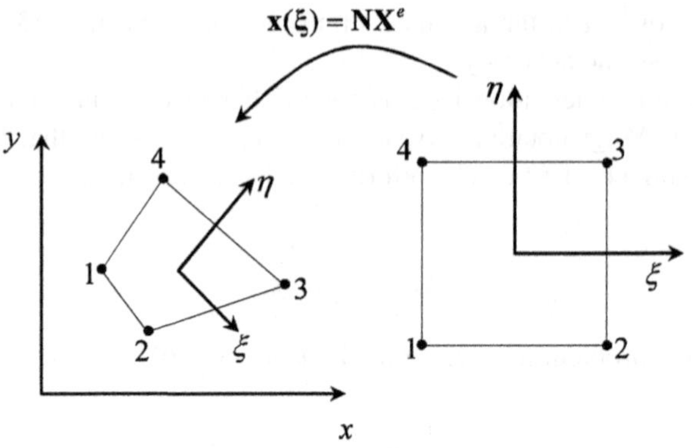

Fig. 15.5 Plane quadrilateral with skew angled coordinates ξ and η

the interpolations of displacements and rotations (see Fig. 15.5). The standard representation in terms of elemental coordinates ξ and η is

$$N_I = \frac{1}{4}\left(1+\xi_I\xi\right)\left(1+\eta_I\eta\right) \text{ (no sum on } I) \tag{15.49}$$

with

$$\begin{bmatrix}\xi_1\\ \eta_1\end{bmatrix}=\begin{bmatrix}-1\\ -1\end{bmatrix},\ \begin{bmatrix}\xi_2\\ \eta_2\end{bmatrix}=\begin{bmatrix}1\\ -1\end{bmatrix},\ \begin{bmatrix}\xi_3\\ \eta_3\end{bmatrix}=\begin{bmatrix}1\\ 1\end{bmatrix},\ \begin{bmatrix}\xi_4\\ \eta_4\end{bmatrix}=\begin{bmatrix}-1\\ 1\end{bmatrix}. \tag{15.50}$$

Note that capital indices $I = 1, 2, 3, 4$ refer to node numbers and the variables ξ_I and η_I indicate no tensorial quantities. Hence, no sum of repeated indices is understood. An alternative form of the shape functions (15.49) is given by

$$\mathbf{N} = \left[N_1\ N_2\ N_3\ N_4\right], \tag{15.51}$$

$$\mathbf{N} = \boldsymbol{\Delta} + \mathbf{b}_x x + \mathbf{b}_y y + \boldsymbol{\gamma}\xi\eta, \tag{15.52}$$

with

$$\mathbf{b}_x = \frac{1}{2A_{\text{elem}}}\left[y_{24}\ y_{31}\ y_{42}\ y_{13}\right], \tag{15.53}$$

$$\mathbf{b}_y = \frac{1}{2A_{\text{elem}}}\left[x_{42}\ x_{13}\ x_{24}\ x_{31}\right], \tag{15.54}$$

$$\boldsymbol{\gamma} = \frac{1}{4A_{\text{elem}}}\begin{bmatrix}X_2y_{34}+X_3y_{42}+X_4y_{23}\\ X_1y_{43}+X_3y_{14}+X_4y_{31}\\ X_1y_{24}+X_2y_{41}+X_4y_{12}\\ X_1y_{32}+X_2y_{13}+X_3y_{21}\end{bmatrix}^T, \tag{15.55}$$

$$\boldsymbol{\Delta} = \frac{1}{4}\left[\mathbf{t} - (X_1 + X_2 + X_3 + X_4)\,\mathbf{b}_x - (Y_1 + Y_2 + Y_3 + Y_4)\,\mathbf{b}_y\right], \tag{15.56}$$

$$\mathbf{t} = \left[\,1\;1\;1\;1\,\right], \tag{15.57}$$

$$x_{IJ} = X_I - X_J\,,\; y_{IJ} = Y_I - Y_J, \tag{15.58}$$

$$A_{\text{elem}} = \frac{1}{2}\,(x_{24}y_{31} + x_{31}y_{42})\,. \tag{15.59}$$

A similar representation of shape functions is e.g. given in [19]. Note that all upper case letters refer to nodal variables i.e. U_{iJ} represents the displacement in i-th direction of node J, X_J and Y_J depict the nodal x and y coordinates while Φ_J is the in-plane rotation at node J.

According to Fig. 15.5 the element kinematics of an isoparametric element is given by

$$\mathbf{x}(\xi) = \begin{bmatrix} x \\ y \end{bmatrix} = \mathbf{N}_u \begin{bmatrix} \mathbf{X} \\ \mathbf{Y} \end{bmatrix}, \tag{15.60}$$

$$\mathbf{N}_u = \begin{bmatrix} \mathbf{N} & \mathbf{0} \\ \mathbf{0} & \mathbf{N} \end{bmatrix}, \tag{15.61}$$

$$\mathbf{X} = \left[\,X_1\;X_2\;X_3\;X_4\,\right]^T, \tag{15.62}$$

$$\mathbf{Y} = \left[\,Y_1\;Y_2\;Y_3\;Y_4\,\right]^T. \tag{15.63}$$

From the interpolations of displacements and rotations according to

$$\mathbf{u}(\boldsymbol{\xi}) = \begin{bmatrix} u_x \\ u_y \end{bmatrix} = \mathbf{N}_u\mathbf{U}, \tag{15.64}$$

$$\mathbf{U} = \begin{bmatrix} \mathbf{U}_x \\ \mathbf{U}_y \end{bmatrix}, \tag{15.65}$$

$$\mathbf{U}_x = \left[\,U_{x1}\;U_{x2}\;U_{x3}\;U_{x4}\,\right]^T, \tag{15.66}$$

$$\mathbf{U}_y = \left[\,U_{y1}\;U_{y2}\;U_{y3}\;U_{y4}\,\right]^T \tag{15.67}$$

and

$$\phi = \mathbf{N}_\phi \boldsymbol{\Phi}, \tag{15.68}$$

$$\mathbf{N}_\phi = \mathbf{N} = \left[\,N_1\;N_2\;N_3\;N_4\,\right], \tag{15.69}$$

$$\boldsymbol{\Phi} = \left[\,\Phi_1\;\Phi_2\;\Phi_3\;\Phi_4\,\right]^T, \tag{15.70}$$

the symmetric and asymmetric strain interpolations can be evaluated from the definitions (15.47) as

$$\boldsymbol{\varepsilon}_s = \mathbf{B}_{su0}\begin{bmatrix}\mathbf{U}_x\\ \mathbf{U}_y\end{bmatrix} + \mathbf{B}_{suH}\begin{bmatrix}\boldsymbol{\gamma} & \mathbf{0}\\ \mathbf{0} & \boldsymbol{\gamma}\end{bmatrix}\begin{bmatrix}\mathbf{U}_x\\ \mathbf{U}_y\end{bmatrix}, \tag{15.71}$$

$$\varepsilon_a = \frac{1}{2}\left(\mathbf{B}_{au0}\begin{bmatrix}\mathbf{U}_x\\ \mathbf{U}_y\end{bmatrix} + \mathbf{B}_{auH}\begin{bmatrix}\boldsymbol{\gamma} & \mathbf{0}\\ \mathbf{0} & \boldsymbol{\gamma}\end{bmatrix}\begin{bmatrix}\mathbf{U}_x\\ \mathbf{U}_y\end{bmatrix}\right) + \mathbf{N}_\phi\boldsymbol{\Phi}. \tag{15.72}$$

The strain interpolation functions appearing in (15.71) and (15.72) are given by

$$\mathbf{B}_{su0} = \begin{bmatrix}\mathbf{b}_x & \mathbf{0}\\ \mathbf{0} & \mathbf{b}_y\\ \mathbf{b}_y & \mathbf{b}_x\end{bmatrix}, \tag{15.73}$$

$$\mathbf{B}_{suH} = \begin{bmatrix}(\xi\eta)_{,x} & 0\\ 0 & (\xi\eta)_{,y}\\ (\xi\eta)_{,y} & (\xi\eta)_{,x}\end{bmatrix}, \tag{15.74}$$

$$\mathbf{B}_{au0} = \begin{bmatrix}\mathbf{b}_y & -\mathbf{b}_x\end{bmatrix}, \tag{15.75}$$

$$\mathbf{B}_{auH} = \begin{bmatrix}(\xi\eta)_{,y} & -(\xi\eta)_{,x}\end{bmatrix}. \tag{15.76}$$

The components of $\mathbf{B}_{suH}$ and $\mathbf{B}_{auH}$ are evaluated as

$$(\xi\eta)_{,x} = \frac{1}{4det\mathbf{J}}\left(\eta\begin{bmatrix}-1 & -1 & 1 & 1\end{bmatrix}\mathbf{X} - \xi\begin{bmatrix}-1 & 1 & 1 & -1\end{bmatrix}\mathbf{Y}\right) \tag{15.77}$$

$$(\xi\eta)_{,y} = \frac{1}{4det\mathbf{J}}\left(\xi\begin{bmatrix}-1 & 1 & 1 & -1\end{bmatrix}\mathbf{X} - \eta\begin{bmatrix}-1 & -1 & 1 & 1\end{bmatrix}\mathbf{Y}\right), \tag{15.78}$$

respectively where $\mathbf{J}$ denotes the Jacobian of the elements kinematics, i.e.

$$\mathbf{J} = \begin{bmatrix}x_{,\xi} & y_{,\xi}\\ x_{,\eta} & y_{,\eta}\end{bmatrix}. \tag{15.79}$$

The orthogonalities between the constant and hourglass parts of strain interpolations are obvious, i.e.

$$\int_\Omega \mathbf{B}_{su0}^T\mathbf{C}\mathbf{B}_{suH}\mathrm{d}\Omega = 0, \tag{15.80}$$

$$\int_\Omega \mathbf{B}_{au0}^T\mathbf{B}_{auH}\mathrm{d}\Omega = 0, \tag{15.81}$$

and we can derive the terms of the stiffness matrix of the QM4D4 element according to (15.46)

$$\begin{bmatrix} \mathbf{K}_{11} & \mathbf{K}_{21}^T \\ \mathbf{K}_{21} & \mathbf{K}_{22} \end{bmatrix} \begin{bmatrix} \mathbf{U} \\ \boldsymbol{\Phi} \end{bmatrix} = \begin{bmatrix} \mathbf{F}_u \\ \mathbf{0} \end{bmatrix}, \tag{15.82}$$

$$\mathbf{K}_{11} = \mathbf{K}_{suu0} + \mathbf{K}_{suuH} + \mathbf{K}_{auu0} + \mathbf{K}_{auuH}, \tag{15.83}$$

$$\mathbf{K}_{21} = \mathbf{K}_{a\phi u0} + \mathbf{K}_{a\phi uH}, \tag{15.84}$$

$$\mathbf{K}_{22} = \mathbf{K}_{a\phi\phi}, \tag{15.85}$$

$$\mathbf{F}_u = \int_{\Omega} \mathbf{N}_u^T \mathbf{f} d\Omega + \int_{\Gamma_t} \mathbf{N}_u^T \mathbf{t} d\Gamma. \tag{15.86}$$

The submatrices in (15.83), (15.84) and (15.85) are given below

$$\mathbf{K}_{suu0} = \int_{\Omega} \mathbf{B}_{su0}^T \mathbf{C} \mathbf{B}_{su0} d\Omega, \tag{15.87}$$

$$\mathbf{K}_{suuH} = \boldsymbol{\Gamma}^T \left[\int_{\Omega} \mathbf{B}_{suH}^T \mathbf{C} \mathbf{B}_{suH} \mathrm{d}\Omega \right] \boldsymbol{\Gamma}, \tag{15.88}$$

$$\mathbf{K}_{auu0} = \frac{\alpha}{2} \int_{\Omega} \mathbf{B}_{au0}^T \mathbf{B}_{au0} \mathrm{d}\Omega, \tag{15.89}$$

$$\mathbf{K}_{auuH} = \frac{\alpha}{2} \boldsymbol{\Gamma}^T \left[\int_{\Omega} \mathbf{B}_{auH}^T \mathbf{B}_{auH} \mathrm{d}\Omega \right] \boldsymbol{\Gamma}, \tag{15.90}$$

$$\mathbf{K}_{a\phi u0} = \alpha \int_{\Omega} \mathbf{N}_{\phi}^T \mathbf{B}_{au0} \mathrm{d}\Omega, \tag{15.91}$$

$$\mathbf{K}_{a\phi uH} = \alpha \left[\int_{\Omega} \mathbf{N}_{\phi}^T \mathbf{B}_{auH} \mathrm{d}\Omega \right] \boldsymbol{\Gamma}, \tag{15.92}$$

$$\mathbf{K}_{a\phi\phi} = 2\alpha \int_{\Omega} \mathbf{N}_{\phi}^T \mathbf{N}_{\phi} \mathrm{d}\Omega \tag{15.93}$$

with

$$\boldsymbol{\Gamma} = \begin{bmatrix} \boldsymbol{\gamma} & \mathbf{0} \\ \mathbf{0} & \boldsymbol{\gamma} \end{bmatrix}. \tag{15.94}$$

15.3.2 Modifications for Improved Accuracy and Efficiency

Within this section we introduce two modifications of the strain interpolations which are justified by the assumed strain method. Considering the derivatives of

Section 15.2.3 (15.46) can be rewritten with assumed strain fields $\bar{\boldsymbol{\varepsilon}}_s$, $\bar{\varepsilon}_a$,

$$\int_{\Omega} \delta\bar{\boldsymbol{\varepsilon}}_s^T \mathbf{C}\bar{\boldsymbol{\varepsilon}}_s \mathrm{d}\Omega + 2\alpha \int_{\Omega} \delta\bar{\varepsilon}_a \bar{\varepsilon}_a \mathrm{d}\Omega = \int_{\Omega} \delta\mathbf{u}^T \mathbf{f} \mathrm{d}\Omega + \int_{\Gamma_t} \delta\mathbf{u}^T \mathbf{t} \mathrm{d}\Gamma. \tag{15.95}$$

The implementation of the considerations of Section 15.2.3 is striking in its elegance and simplicity: Equations (15.46) and (15.95) are identical but the kinematic fields that are indicated by an overbar can be designed in order to avoid any locking phenomena and to improve the efficiency. We state

$$\bar{\boldsymbol{\varepsilon}}_s = \bar{\mathbf{B}}_{su0} \begin{bmatrix} \mathbf{U}_x \\ \mathbf{U}_y \end{bmatrix} + \bar{\mathbf{B}}_{suH} \begin{bmatrix} \boldsymbol{\gamma} & \mathbf{0} \\ \mathbf{0} & \boldsymbol{\gamma} \end{bmatrix} \begin{bmatrix} \mathbf{U}_x \\ \mathbf{U}_y \end{bmatrix}, \tag{15.96}$$

$$\bar{\varepsilon}_a = \frac{1}{2} \left(\bar{\mathbf{B}}_{au0} \begin{bmatrix} \mathbf{U}_x \\ \mathbf{U}_y \end{bmatrix} + \bar{\mathbf{B}}_{suH} \begin{bmatrix} \boldsymbol{\gamma} & \mathbf{0} \\ \mathbf{0} & \boldsymbol{\gamma} \end{bmatrix} \begin{bmatrix} \mathbf{U}_x \\ \mathbf{U}_y \end{bmatrix} \right) + \bar{\mathbf{N}}_{\varphi} \boldsymbol{\Phi}. \tag{15.97}$$

with the strain interpolation functions given by

$$\bar{\mathbf{B}}_{su0} = \sqrt{\frac{j_0}{j}} \begin{bmatrix} \mathbf{b}_x & \mathbf{0} \\ \mathbf{0} & \mathbf{b}_y \\ \mathbf{b}_y & \mathbf{b}_x \end{bmatrix}, \tag{15.98}$$

$$\bar{\mathbf{B}}_{suH} = \sqrt{\frac{j}{j_0}} \begin{bmatrix} (\xi\eta)_{,x} & -e\,(\xi\eta)_{,y} \\ -e\,(\xi\eta)_{,x} & (\xi\eta)_{,y} \\ 0 & 0 \end{bmatrix}, \tag{15.99}$$

$$\bar{\mathbf{B}}_{au0} = \sqrt{\frac{j_0}{j}} \left[\mathbf{b}_y \ -\mathbf{b}_x \right], \tag{15.100}$$

$$\bar{\mathbf{B}}_{auH} = \aleph \sqrt{\frac{j}{j_0}} \left[(\xi\eta)_{,y} - (\xi\eta)_{,x} \right], \tag{15.101}$$

$$\bar{\mathbf{N}}_{\phi} = \sqrt{\frac{j_0}{j}} \mathbf{N}_{\phi} = \sqrt{\frac{j_0}{j}} \left[N_1 \ N_2 \ N_3 \ N_4 \right]. \tag{15.102}$$

At first we have modified the hourglass parts. In (15.99) and (15.101) the variables e and $\aleph$ indicate arbitrary real scalar constants. The modification (15.99) stems from considerations of Belytschko and Bachrach [19] and is summarized in e.g. [26]. Then, we introduced two different, i.e. $\sqrt{j_0/j}$ and $\sqrt{j/j_0}$, where $j = det\mathbf{J}(\xi,\eta)$, $j_0 = det\mathbf{J}(\xi = 0; \eta = 0)$. These factors ensure the orthogonality conditions (15.80) and (15.81) to hold, while all parts of the stiffness matrix can be integrated analytically, i.e. without any numerical quadrature.

The element related to these modifications is called **A**nalytically **I**ntegrated **Q**uintessential **B**ending element with **D**rilling degrees of freedom (AIQBD) and the stiffness matrix is given according to (15.82)–(15.93), i.e.

$$\bar{\mathbf{K}} = \begin{bmatrix} \bar{\mathbf{K}}_{11} & \bar{\mathbf{K}}_{21}^T \\ \bar{\mathbf{K}}_{21} & \bar{\mathbf{K}}_{22} \end{bmatrix}, \tag{15.103}$$

with

$$\bar{\mathbf{K}}_{11} = \bar{\mathbf{K}}_{suu0} + \bar{\mathbf{K}}_{suuH} + \bar{\mathbf{K}}_{auu0} + \bar{\mathbf{K}}_{auuH}, \tag{15.104}$$

$$\bar{\mathbf{K}}_{21} = \bar{\mathbf{K}}_{a\varphi u0} + \bar{\mathbf{K}}_{a\varphi uH}, \tag{15.105}$$

$$\bar{\mathbf{K}}_{22} = \bar{\mathbf{K}}_{a\varphi\varphi}, \tag{15.106}$$

and

$$\bar{\mathbf{K}}_{suu0} = \mathbf{K}_{suu0} = \int_{\Omega} \bar{\mathbf{B}}_{su0}^T \mathbf{C} \bar{\mathbf{B}}_{su0} \mathrm{d}\Omega, \tag{15.107}$$

$$\bar{\mathbf{K}}_{suuH} = \mathbf{\Gamma}^T \left[\int_{\Omega} \bar{\mathbf{B}}_{suH}^T \mathbf{C} \bar{\mathbf{B}}_{suH}^T \mathrm{d}\Omega \right] \mathbf{\Gamma}, \tag{15.108}$$

$$\bar{\mathbf{K}}_{auu0} = \mathbf{K}_{auu0} = \frac{\alpha}{2} \int_{\Omega} \bar{\mathbf{B}}_{au0}^T \bar{\mathbf{B}}_{au0} \mathrm{d}\Omega, \tag{15.109}$$

$$\bar{\mathbf{K}}_{auuH} = \frac{\alpha}{2} \mathbf{\Gamma}^T \left[\int_{\Omega} \bar{\mathbf{B}}_{auH}^T \bar{\mathbf{B}}_{auH} \mathrm{d}\Omega \right] \mathbf{\Gamma}, \tag{15.110}$$

$$\bar{\mathbf{K}}_{a\varphi u0} = \alpha \int_{\Omega} \bar{\mathbf{N}}_{\varphi}^T \bar{\mathbf{B}}_{au0} \mathrm{d}\Omega, \tag{15.111}$$

$$\bar{\mathbf{K}}_{a\varphi uH} = \alpha \left[\int_{\Omega} \bar{\mathbf{N}}_{\varphi}^T \bar{\mathbf{B}}_{auH} \mathrm{d}\Omega \right] \mathbf{\Gamma}, \tag{15.112}$$

$$\bar{\mathbf{K}}_{a\varphi\varphi} = 2\alpha \int_{\Omega} \bar{\mathbf{N}}_{\varphi}^T \bar{\mathbf{N}}_{\varphi} \mathrm{d}\Omega, \tag{15.113}$$

where $\mathbf{\Gamma}$ is given in (15.94).

Now, we claim that the enhanced stiffness matrix according to (15.103) has to reproduce the exact internal energy in terms of pure bending, if the element is in a rectangular configuration (see Fig. 15.6).

The differential equation that describes the deformation of a beam under a constant moment M reads

$$EI \frac{\mathrm{d}^2 u_y}{\mathrm{d}x^2} = M \tag{15.114}$$

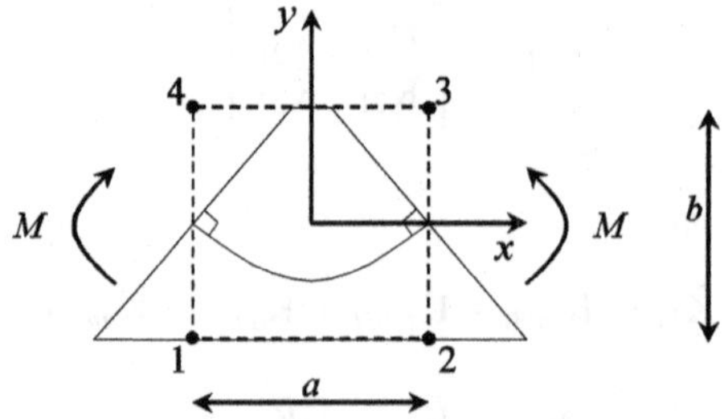

Fig. 15.6 Deformation of a rectangular element in terms of pure bending

with $I = b^3t/12$ being the moment of inertia. Integration of (15.114) and considering the boundary conditions

$$u_y\left(-\frac{a}{2}\right) = u_y\left(\frac{a}{2}\right) = 0 \tag{15.115}$$

yields

$$EI\frac{\mathrm{d}u_y}{\mathrm{d}x} = EI\phi = Mx \tag{15.116}$$

$$EIu_y = \frac{M}{8}\left(4x^2 - a^2\right). \tag{15.117}$$

Equation (15.116) can be used to calculate the nodal horizontal displacement $u_x = -y\phi$ giving the nodal rotations and translations

$$U_x^I = \frac{-MX^IY^I}{EI}, \qquad U_y^I = 0, \qquad \Phi^I = \frac{MX^I}{EI}. \tag{15.118}$$

The corresponding exact internal energy is given by

$$\Pi^{\text{exact}} = \frac{1}{2}\int_{-b/2}^{b/2}\int_{-a/2}^{a/2}\sigma_{xx}\varepsilon_{xx}\mathrm{d}x\mathrm{d}y = \frac{1}{2}\int_{-b/2}^{b/2}\int_{-a/2}^{a/2}\frac{M^2y^2}{E\,I^2}\mathrm{d}x\mathrm{d}y = \frac{6M^2a}{Eb^3t}. \tag{15.119}$$

If we determine the internal energy of an element that is deformed according to (15.118) we get, taking into account (15.98)–(15.113)

$$\Pi^{\text{Fem}} = \frac{1}{2}\left[\mathbf{U}^T\ \Phi^T\right]\bar{\mathbf{K}}\begin{bmatrix}\mathbf{U}\\ \Phi\end{bmatrix} =$$

$$= \frac{aM^2\left[b^2\left(12\nu e - 6 - 6e^2\right)\right]}{b^5Et\left(\nu^2 - 1\right)} + \frac{aM^2\left[\frac{3a^2}{2}\left(\nu - 1\right)\left(\aleph - 2\right)^2\right]}{b^5Et\left(\nu^2 - 1\right)}. \tag{15.120}$$

Requiring that

$$\Pi^{\mathrm{Fem}} = \Pi^{\mathrm{exact}}, \tag{15.121}$$

we obtain a value for

$$\aleph = 2 \pm i \left(\frac{8b^2 \nu e - 4b^2 \left(e^2 + \nu^2\right)}{a^2 \left(\nu - 1\right)} \right)^{12}, \tag{15.122}$$

where $i = \sqrt{-1}$. Since meaningful results for $\aleph$ can only be expected for real numbers we require the imaginary part of $\aleph$ to vanish which results in

$$e = \nu, \quad \aleph = 2. \tag{15.123}$$

Under such conditions all parts of the stiffness matrix (15.107), (15.108), (15.109), (15.110), (15.111), (15.112) and (15.113) can be calculated analytically, i.e.

$$\bar{\mathbf{K}}_{su0} = 4tj_0 \mathbf{B}_{su0}^T \mathbf{C} \mathbf{B}_{su0}, \tag{15.124}$$

$$\bar{\mathbf{K}}_{suH} = \mathbf{\Gamma}^T \frac{Et}{6j_0} \begin{bmatrix} \left(y_{13}^2 + y_{24}^2\right) & \nu(x_{13}y_{13} + x_{24}y_{24}) \\ \nu\left(x_{13}y_{13} + x_{24}y_{24}\right) & \left(x_{13}^2 + x_{24}^2\right) \end{bmatrix} \mathbf{\Gamma}, \tag{15.125}$$

$$\bar{\mathbf{K}}_{auu0} = 2\alpha tj_0 \mathbf{B}_{auo}^T \mathbf{B}_{au0}, \tag{15.126}$$

$$\bar{\mathbf{K}}_{auuH} = \mathbf{\Gamma}^T \frac{\alpha t}{3j_0} \begin{bmatrix} \left(x_{13}^2 + x_{24}^2\right) & (x_{13}y_{13} + x_{24}y_{24}) \\ (x_{13}y_{13} + x_{24}y_{24}) & \left(y_{13}^2 + y_{24}^2\right) \end{bmatrix} \mathbf{\Gamma}, \tag{15.127}$$

$$\bar{\mathbf{K}}_{a\phi u0} = \frac{\alpha t}{8} \begin{bmatrix} x_{42} & x_{13} & x_{24} & x_{31} & y_{42} & y_{13} & y_{24} & y_{31} \\ x_{42} & x_{13} & x_{24} & x_{31} & y_{42} & y_{13} & y_{24} & y_{31} \\ x_{42} & x_{13} & x_{24} & x_{31} & y_{42} & y_{13} & y_{24} & y_{31} \\ x_{42} & x_{13} & x_{24} & x_{31} & y_{42} & y_{13} & y_{24} & y_{31} \end{bmatrix}, \tag{15.128}$$

$$\bar{\mathbf{K}}_{a\phi u0} = \frac{\alpha t}{3} \begin{bmatrix} x_{42} & y_{42} \\ x_{31} & y_{31} \\ x_{24} & y_{24} \\ x_{13} & y_{13} \end{bmatrix} \mathbf{\Gamma}, \tag{15.129}$$

$$\bar{\mathbf{K}}_{a\phi\phi} = \frac{2\alpha tj_0}{9} \begin{bmatrix} 4 & 2 & 1 & 2 \\ 2 & 4 & 2 & 1 \\ 1 & 2 & 4 & 2 \\ 2 & 1 & 2 & 4 \end{bmatrix}. \tag{15.130}$$

Note that $\mathbf{B}_{su0}$ and $\mathbf{B}_{au0}$ of (15.124) and (15.126) are given in (15.73) and (15.75). An element with a stiffness matrix according to (15.133), (15.134),

(15.135), (15.136), (15.137), (15.138) and (15.139) is called Analytically Integrated Quintessential Bending element with Drilling degrees of freedom (AIQBD).

15.4 Results

A cantilever beam (see Fig. 15.7) consisting of two elements is loaded by an end moment. A detailed analysis of the vertical displacements and end rotations is performed, and the achieved outcome is related to the exact results according to the Euler-Bernoulli theory, i.e.

$$w(l) = \frac{Ml^2}{2EI} = 0.54,$$

$$\phi(l) = \frac{Ml}{EI} = 0.36.$$

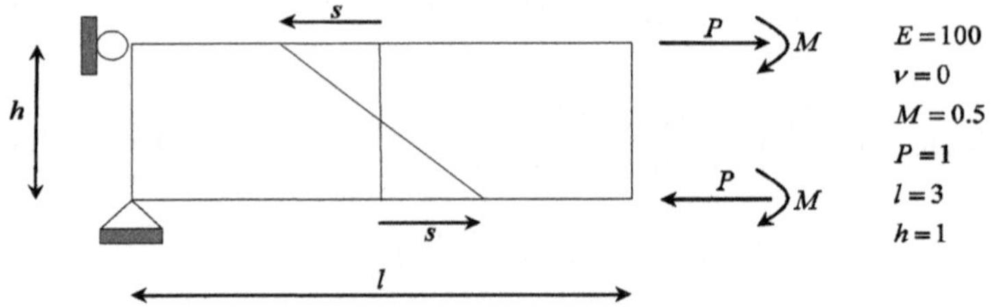

Fig. 15.7 A cantilever beam loaded by an end moment (either applied by forces or by concentrated couples)

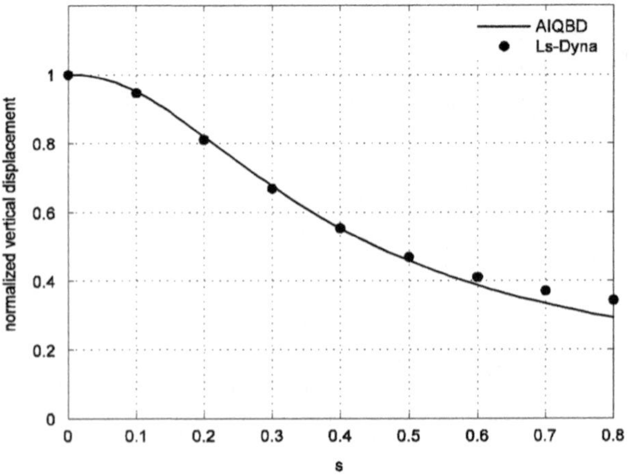

Fig. 15.8 Comparison of AIQBD element with Belytschko Bindeman element in Ls-Dyna

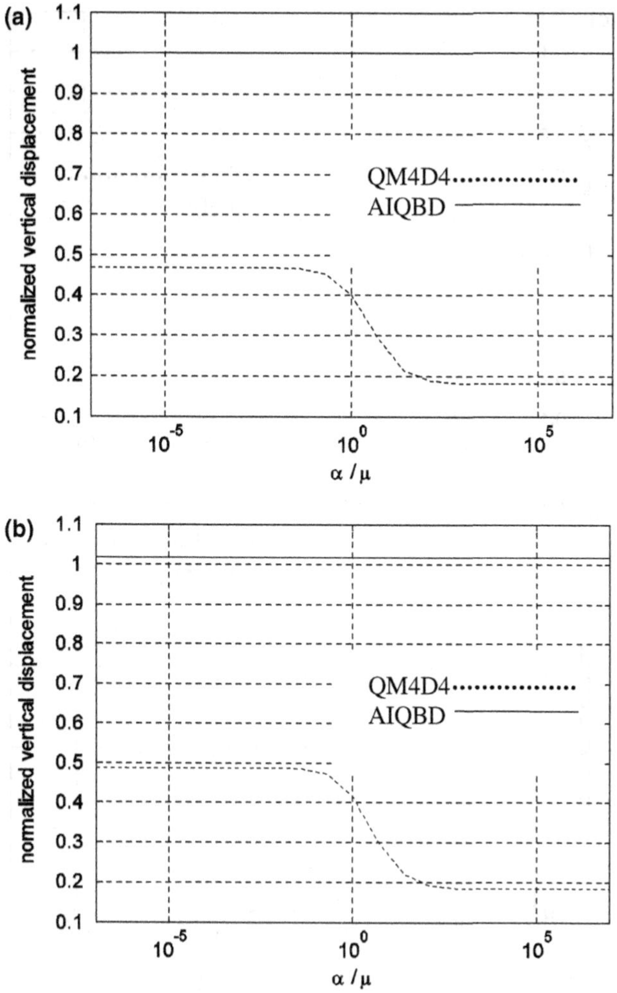

Fig. 15.9 (**a**) Normalized vertical displacements-loading by forces, (**b**) Normalized vertical displacements-loading by couples (**c**) Normalized end rotation-loading by forces (**d**) Normalized end rotation-loading by couples

No drilling degrees of freedom are restrained and the load is applied by forces on the one hand and by concentrated couples on the other hand To compare the results of our developments with the highly efficient element technology of Ls-Dyna (reduced integrated single point plane continuum element without drilling degrees of freedom and an hourglass stabilization according to Belytschko and Bindeman [27, 28]) we refer to Fig. 15.8 showing a similar behavior of both elements in terms of the element distortion parameter s.

In Figure 15.9 the sensitivity in terms of the material parameter α is depicted for a regular mesh (i.e. $s = 0$).

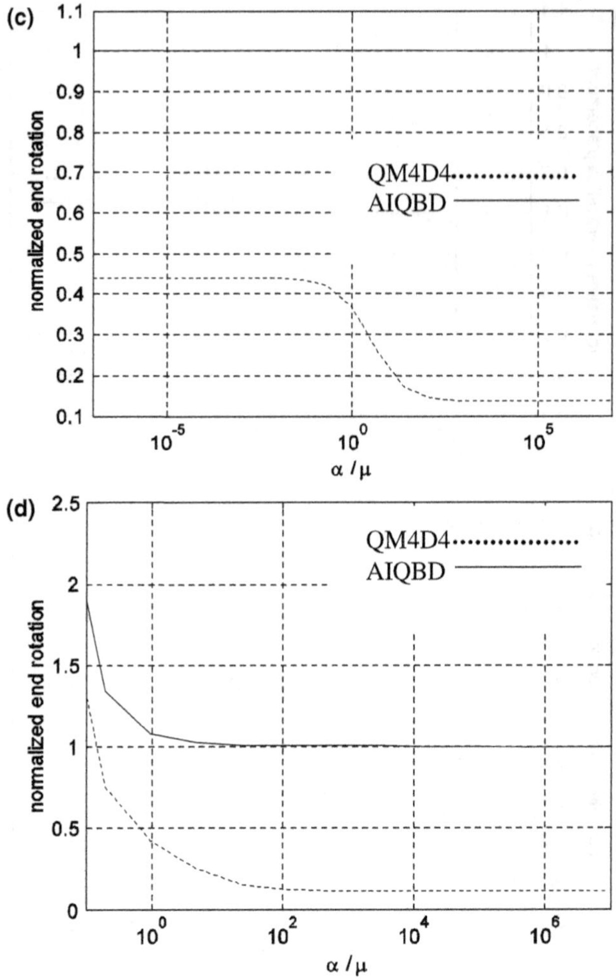

Fig. 15.9 (continued)

The reported performance of the QM4D4 element corresponds to the findings of [7]. In contrast the AIQBD element gives perfect displacements and rotations over a wide range of α. However, Fig. 15.9d indicates that higher values for α (e.g. $\alpha > 100\mu$) give better results for rotations if the loading is applied by concentrated couples. Note that this test case and the reported conclusion to use higher values for α was – at least to our knowledge – never reported in literature.

However, the theory is not very sensitive to the parameter α and an extension of the formulation to plasticity and damage seems to be straightforward since no iterations on asymmetric stress and strains have to be performed, i.e. α is unchanged during plastic motion.

References

1. MacNeal, R.H., Harder, R.L.: A refined four-noded membrane element with rotational degrees of freedom. Comput. Struct. **28**, 75–84 (1988)
2. Allman, D.J.: A compatible triangular element including vertex rotations for plane elasticity analysis. Comput. Struct. **19**, 1–8 (1984)
3. Bergan, P.G., Felippa, C.A.: A triangular membrane element with rotational degrees of freedom. Comput. Methods Appl. Mech. Eng. **50**, 25–69 (1985)
4. Reissner, E.: A note on variational principles in elasticity. Int. J. Solids Struct. **1**, 93–95 (1965)
5. Hughes, T.J.R., Brezzi, F.: On drilling degrees of freedom. Comput. Methods Appl. Mech. Eng. **72**, 105–121 (1989)
6. Zienkiewicz, O.C., Taylor, R.L.: Finite Element Method, vol. 1, The Basis (Finite Element Method). Butterworth-Heinemann, Oxford (2000)
7. Hughes, T.J.R.,Masud, A., Harari, I.: Numerical assessment of some membrane elements with drilling degrees of freedom. Comput. Struct. **55**, 297–314 (1995)
8. Ibrahimbegovic, A., Taylor, R.L., Wilson, E.L.: A robust quadrilateral membrane finite element with drilling degrees of freedom. Int. J. Numer. Methods Eng. **30**, 445–457 (1990)
9. Atluri, S.N.: On some new general and complementary energy principles for the rate problems of finite strain, classical elastoplasticity. J. Struct. Mech. **8**, 61–92 (1980)
10. Cazzani, A., Atluri, S.N.: Four-noded mixed finite elements, using unsymmetric stresses for linear and nonlinear analysis of membranes. Comput. Mech. **11**, 229–251 (1993)
11. Iura, M., Atluri, S.N.: Formulation of a membrane finite element with drilling degrees of freedom. Comput. Mech. **9**, 417–428 (1992)
12. Ibrahimbegovic, A., Frey, F.: Membrane quadrilateral finite elements with rotational degrees of freedom. Eng. Fract. Mec. **43**, 12–24 (1992)
13. Ibrahimbegovic, A., Frey, F.: Geometrically non-linear method of incompatible modes in application to finite elasticity with independent rotations. Int. J. Numer. Methods Eng. **36**, 4185–4200 (1993)
14. Ibrahimbegovic, A.: Mixed finite element with drilling rotations for plane problems in finite elasticity. Comput. Methods Appl. Mech. Eng. **107**, 225–238 (1993)
15. Ibrahimbegovic, A., Frey, F.: Stress resultant geometrically non-linear shell theory with drilling rotations. part iii: linearized kinematics. Int. J. Numer. Methods Eng. **37**, 3659–3683 (1994)
16. Ibrahimbegovic, A., Frey, F.: Variational principles and membrane finite elements with drilling rotations for geometrically non-linear elasticity. Int. J. Numer. Methods Eng. **38**, 1885–1900 (1995)
17. Pian, T.H.H., Sumihara, K.: Rational approach for assumed stress finite elements. Int. J. Numer. Methods Eng. **20**, 1685–1695 (1984)
18. Simo, J.C., Hughes, T.J.R.: On the variational foundations of assumed strain methods. Comput. Methods Appl. Mech. Eng. **53**, 1685–1695 (1986)
19. Belytschko, T., Bachrach, W.E.: Efficient implementation of quadrilaterals with high coarse-mesh accuracy. Comput. Methods Appl. Mech. Eng. **54**, 279–301 (1986)
20. Belytschko, T., Leviathan, I.: Physical stabilization of the 4-node shell element with one point quadrature. Comput. Methods Appl. Mech. Eng. **113**, 321–350 (1994)
21. Kugler, St., Fotiu, P.A., Murin, J.: A new quadrilateral membrane finite element with drilling degrees of freedom. In: CMAS 2009. Computational Modelling and Advanced Simulations: International Conference, Bratislava, 30 June to 3 July 2009, STU v Bratislava FEI, 2009. ISBN 978-80-227-3067-9 – CD-Rom
22. Cosserat, E., Cosserat, F.: Théorie des corps déformables. A. Hermann, Paris (1909)
23. Nowacki, W.: Theory of Micropolar Elasticity. Courses and Lectures-No. 25. Springer, Berlin (1970)
24. Eringen, A.C.: Nonlocal Continuum Field Theories. Springer, New York (2002)
25. Hughes, T.J.R.: The Finite Element Method, Linear Static and Dynamic Finite Element Analysis. Dover Publications, Mineola (2000)

26. Belytschko, T., Liu, W.K., Moran, B.: Nonlinear Finite Elements for Continua and Structures. Wiley, New York, NY (2000)
27. Belytschko, T., Bindeman, L.P.: Assumed strain stabilization of the eight node hexahedral. Comput. Methods Appl. Mech. Eng. **105**, 225–260 (1993)
28. Belytschko, T., Wong, B., Chiang, H.: Advances in one-point quadrature shell elements. Comput. Methods Appl. Mech. Eng. **96**, 93–107 (1992)

Chapter 16
Hybrid System for Optimal Design of Mechanical Properties of Composites

J. Wiśniewski and K. Dems

Abstract The results of investigation in the area of designing of two-dimensional composite materials subjected to service loading are presented. To solve the problem of optimal design of these structures, the series hybrid optimization algorithm composed of a sequence of evolution and gradient-oriented procedures is proposed. The detailed description of this system is presented in the paper. To illustrate the applicability of the hybrid optimization technique for the design process of the composite materials, some simple numerical examples are discussed. The results can be treated as a starting point for computer-oriented design procedure of the real composite disks subjected to actual loading conditions. Such procedure can allow to avoid of the expensive experimental testing, which can be reduced to the final phase of structural design.

Keywords Hybrid system · Design process · Composite material · Fibre layout

16.1 Introduction

During the last decades there has been a growing interest in using computational optimization methods for the design process of structural components made of composite materials. These techniques allow to avoid of expensive experimental tests of real composite structures which can be reduced to the final phase of structural design.

The choice of a relevant optimization strategy is crucial for the successful solution of the design problem. There are many important parameters such as the number and the type of design variables, shape of feasible design space, the type of objective function, constrained or unconstrained problem, availability of the first- and second-order derivatives of objective function and constraints, the cost of each simulation, etc.

J. Wiśniewski (✉)
Department of Technical Mechanics and Informatics, Faculty of Material Technologies and Textile Design, Technical University of Lodz, Zeromskiego 116, 90-924 Lodz, Poland
e-mail: jacek.wisniewski@p.lodz.pl

J. Murín et al. (eds.), *Computational Modelling and Advanced Simulations*, Computational Methods in Applied Sciences 24, DOI 10.1007/978-94-007-0317-9_16,

The use of different, optimization techniques in the design process of composite structures is the subject of many scientific papers. All these methods can be divided into two main groups.

The first group is the group of deterministic methods. These methods use the objective function and/or the gradient information from objective function to construct mathematical approximation of the functions, and then they find an optimum point employing hill-climbing methods. These methods work normally with continuous design variables and need a small number of function evaluations. The deterministic methods are the most effective optimization techniques, but they often fall into a local optimum. The use of the deterministic methods in the optimal design of composite materials and structures were discussed, for instance, in [1–3].

The second group of the optimization methods, which is very often applied in the field of designing of structures made of composite, is the group of stochastic methods. The most common techniques in this class are random search (e.g., [4, 5]), genetic algorithms (e.g., [6–8]) or evolutionary algorithms and strategies (e.g., [9–11]). These methods are a simple, powerful and effective tool used for finding the best solution in a complicated space of design parameters. They can deal with discrete and/or continuous design variables and they are not limited by restrictive assumption about the research space. The stochastic methods, in comparison with the gradient-oriented methods, only need the information based on the objective function and they always find the global optimum. However, the number of function evaluations is high. Besides these methods are generally suited for unconstrained optimization problems.

In this research work, the hybrid, evolution-gradient oriented, system for the design process of composite structures is proposed. The main advantage of this system is its flexibility in using different approaches for solving optimization problem. Besides, the two-stage algorithm can be an alternative technique for above mentioned methods or can supplement these methods. The present paper is continuation of the previous Authors' investigation [12–14] in the area of designing of structures made of composite materials.

16.2 Object of Analysis

A thin, two-dimensional and linearly elastic disk (see Fig. 16.1) is considered. The disk is supported on the boundary portion S_U with prescribed displacements $\mathbf{u}^0$ and loaded by body forces $\mathbf{f}^0$ within domain A and by external traction $\mathbf{T}^0$ acting along the boundary portion S_T. The material of the disk is a composite made of a matrix reinforced with a ply of unidirectional and long fibers of arbitrary shape and assumed cross-section. The bonds between fibers and matrix are perfect. The matrix and the reinforcing fibers are homogeneous, isotropic and linearly elastic, and their mechanical properties are E_m, ν_m and E_w, ν_w, respectively.

The fibers are regularly spaced and perfectly aligned in the matrix with density ρ_w and their layout at any point of the composite material is defined by fiber orientation angle θ with respect to the global coordinate system x-y. This angle

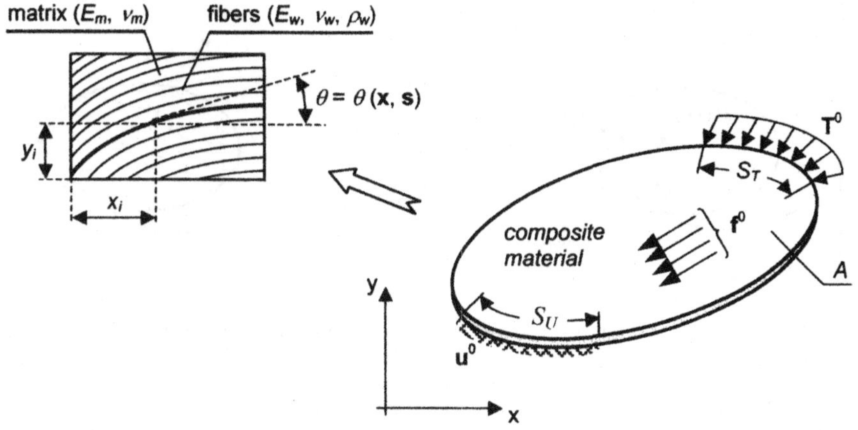

Fig. 16.1 Two dimensional composite disk subjected to service loading

can be constant in the composite and then the fibers are rectilinearly spaced in the matrix or can vary through the composite domain. In this last case, the fibers are placed curvilinearly in the matrix. In general, the layout of reinforcing fibers can be described by the shape of the so-called directional fiber using, for instance, the polynomial, spline or Bezier representation. Thus, the fiber orientation at any point of the composite material depends on a set of fiber shape parameters **s** defining these particular representation, i.e. $\theta = \theta(\mathbf{x}, \mathbf{s})$, as it is shown in Fig. 16.1

The first case of fibers layout deals with one family of straight fibers (see Fig. 16.2). In this case, the shape of the directional fiber is described by:

$$y(\mathbf{s}) = s_1 x + s_2 \tag{16.1}$$

where s_1 and s_2 denote the parameters of straight fiber. The fiber orientation angle θ for straight fiber at any point of the composite is related to the parameter s_1 as follows:

$$tg\theta = s_1 \tag{16.2}$$

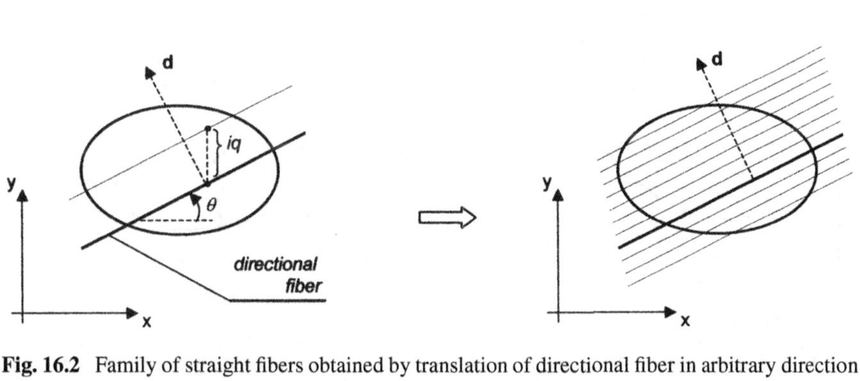

Fig. 16.2 Family of straight fibers obtained by translation of directional fiber in arbitrary direction

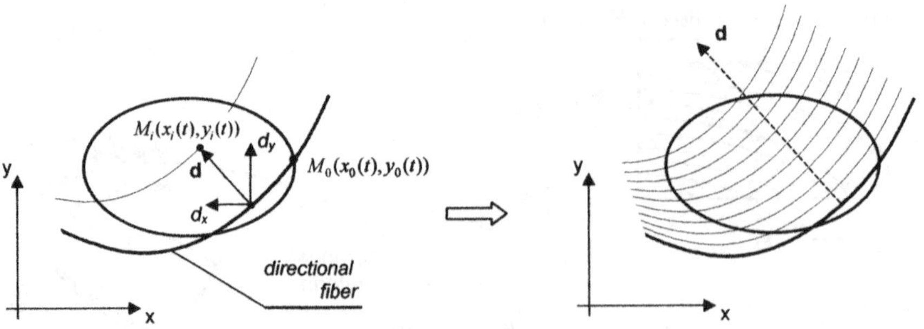

Fig. 16.3 Family of curvilinear fibers obtained by translation of directional fiber in arbitrary direction

or it is directly defined by the angle between the fiber line and the x-axis of the global coordinate system. All other fibers in the family are obtained by translation of the directional fiber in the **d**-direction, according to the rule:

$$y_i = y(\mathbf{s}) + iq \tag{16.3}$$

where i is a number of a current fiber in the family and q is the distance between two adjacent fibers. As a result, the one family of reinforcing fibers of the same shape and constant fiber density is created.

Besides the family of straight fibers, also the family of curvilinear fibers can be created (see Figs. 16.3 and 16.4). In this case the shape of the directional fiber can be defined using the arbitrary parametric description, namely:

$$\begin{cases} x = x(\mathbf{s}, t) \\ y = y(\mathbf{s}, t) \end{cases} \tag{16.4}$$

where $\mathbf{s}$ is a set of the fiber shape parameters defining the curvilinear, directional fiber and t is a real parameter varying in the range $\langle \alpha, \beta \rangle$. The fiber orientation

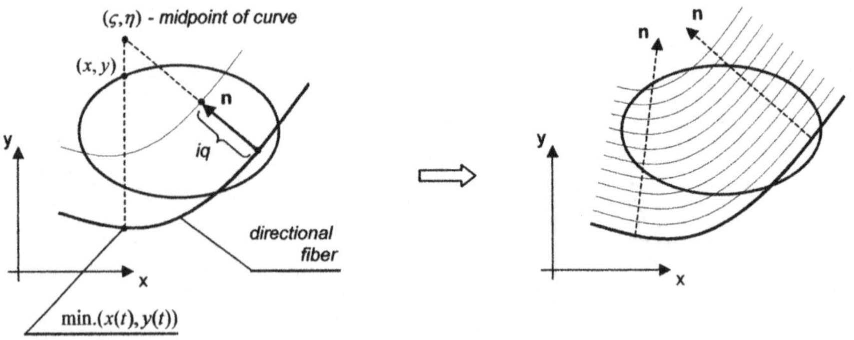

Fig. 16.4 Family of curvilinear fibers obtained by shifting of directional fiber in normal direction

angle θ at any point of the composite material is defined now by:

$$tg\theta = \frac{y(\mathbf{s},t)_{,t}}{x(\mathbf{s},t)_{,t}} = \left(\frac{y_{,t}}{x_{,t}}\right) \text{ and } x_{,t} \neq 0 \tag{16.5}$$

where $y_{,t}$ and $x_{,t}$ denote the derivatives of the function $y(\mathbf{s},t)$ and $x(\mathbf{s},t)$ with respect to parameter t.

For the case of curvilinear fibers, the layout of all other fibers in the family can be obtained by the translation of the directional fiber in arbitrary direction (see Fig. 16.3), or by its shifting in direction normal to the middle line of the directional fiber (see Fig. 16.4).

In order to create the family of curvilinear fibers by translation of the directional fiber (see Fig. 16.3), the fiber is translated in the **d**-direction according to the rule:

$$\begin{cases} x_i = x(\mathbf{s},t) + id_x \\ y_i = y(\mathbf{s},t) + id_y \end{cases} \tag{16.6}$$

where i is a number of a current fiber in the family, while d_x and d_y denote the components of distance vector between two adjacent fibers in the direction of translation. All fibers constitute one parameter family of fibers of the same shape, but the local fiber density is varying in the composite domain. The fiber density at any point $M_i(x_i(t), y_i(t))$ is specified as follows:

$$\rho_{w(i)} = \rho_{w(0)} \frac{\mu\, x_{0,t} - y_{0,t}}{\mu\, x_{,t} - y_{,t}} \sqrt{\frac{(x_{,t})^2 + (y_{,t})^2}{(x_{0,t})^2 + (y_{0,t})^2}} \quad and \quad \mu = dx/dy \tag{16.7}$$

where $\rho_{w(0)}$ denotes the fiber density in the assumed point $M_0(x_0(t), y_0(t))$.

The family of curvilinear reinforcing fibers can be also obtained by shifting of the directional fiber in the direction normal to its middle line, as it is shown in Fig. 16.4. In this case, the local fiber density is constant in the composite domain, but each particular fiber in this family has slightly different shape described by the following expressions:

$$\begin{cases} x_i = x(\mathbf{s},t) + iq\left(-\dfrac{y_{,t}}{\sqrt{(x_{,t})^2 + (y_{,t})^2}}\right) \\ y_i = y(\mathbf{s},t) + iq\left(\dfrac{x_{,t}}{\sqrt{(x_{,t})^2 + (y_{,t})^2}}\right) \end{cases} \tag{16.8}$$

where i is a number of a current fiber in the family and q is the distance between two adjacent fibers. Moreover, the creation procedure of family of curvilinear fibers by shifting of the directional fiber has to be subjected to the following condition:

$$\sqrt{[\varsigma - x(t)_{\min}]^2 + [\eta - y(t)_{\min}]^2} \geq \sqrt{[x - x(t)_{\min}]^2 + [y - y(t)_{\min}]^2} \tag{16.9}$$

preserving the fibers from intersection their middle lines.

16.3 Optimization Problem

The general idea of the optimal design of composite materials regards the modification of structure parameters of the composite, such as mechanical properties of reinforcing fibers and matrix, percentage participation of fibers in the material, as well as fiber shape and orientation. Each of these parameters influences the mechanical properties of the composite material and can be treated as the design variable during the optimal design of structural components made of composite. In analysis presented here, the mechanical properties of components of composite material are given in advance, whereas the parameters defining shape of middle line of directional fiber, as well as the fiber density are selected as a vector of design variables, i.e. $\mathbf{b} = \{\mathbf{s}, \rho_w\}$.

The problem of optimal design of fibers layout in the composite is considered in the form, for which the disk made of this material should satisfy assumed requirements in the range of its mechanical properties. Thus, the typical optimization problem can be formulated as minimization or maximization of the proper quality index of structural behavior, namely:

$$\min\,(or\ \max)\, F_c = \int_A \Gamma(\boldsymbol{\sigma}, \mathbf{e}, \mathbf{u}, \mathbf{b})\, dA + \int_{S_T} \Psi(\mathbf{T}^0, \mathbf{u})\, dS_T \tag{16.10}$$

subjected to the global or the local behavioral constraints:

$$G_i = \left(\int_A \Gamma(\boldsymbol{\sigma}, \mathbf{e}, \mathbf{u}, \mathbf{b})\, dA + \int_{S_T} \Psi(\mathbf{T}^0, \mathbf{u})\, dS_T \right) - G_0 \le 0 \tag{16.11}$$

and/or the constraint imposed on the total cost of composite structure:

$$[c_w \rho_w + c_m(1 - \rho_w)] \cdot V - C_0 \le 0 \tag{16.12}$$

where Γ and Ψ are the continuous functions depending on the state fields induced in deformed structure and on the design vector, while c_w and c_m are the cost per unit volume of fiber and matrix materials, respectively. It should be noted that the objective functional can express global measure of stress, strain or displacement intensity within the disk domain or along its boundary, or even the local quantities such as stress, strain or displacement components at a given point of the disk. When this functional coincides with the total potential or complementary energies then it is a measure of the disk mean stiffness or compliance. Some particular forms of the optimization problem (16.10), (16.11) and (16.12) is introduced in the successive examples in Section 16.5.

16.4 Series Hybrid Optimization System

To perform the optimization task, defined by (16.10), (16.11) and (16.12), a hybrid, evolution-gradient oriented optimization system is proposed. This system consists of series connection of two main modules, as it is shown in Fig. 16.5.

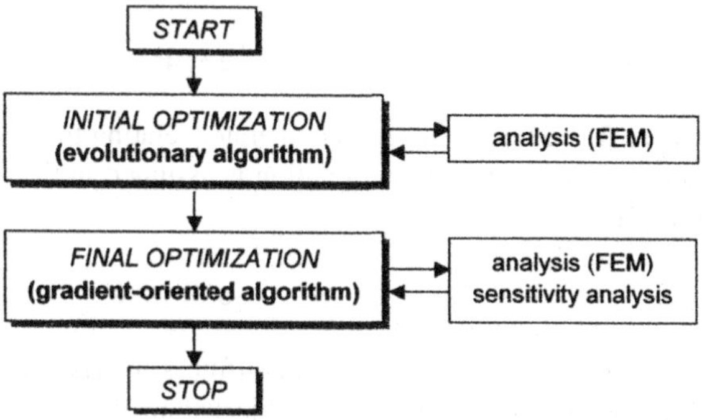

Fig. 16.5 Flow chart of series hybrid optimization system

The first module performs the initial optimization using the evolutionary algorithm starting from randomly selecting initial solution. The evolution simulation is continued when a specified number of generations is attained. Next, in order to increase efficiency of optimization process, the gradient-oriented algorithm starting from the last, best solution generated by evolutionary algorithm is used in the second, final module of this optimization system. The finite element method is applied in both modules in order to perform the analysis step of structural behavior. In addition, this method is also used in the final optimization module for performing the sensitivity analysis of the state fields.

It should be noted, that for the case of the optimization problem with constraints, the penalty function approach [15] is applied in the proposed system. Using this approach, the constrained problem is transformed to unconstrained one as follows:

$$\min Z(\mathbf{b},\boldsymbol{\alpha}) = \min\left[F_c(\mathbf{b}) + \frac{1}{2}\sum_{i=1}^{n_g}\alpha_i[\max.(0; G_i(\mathbf{b}))]^2\right] \quad (16.13)$$

where $\boldsymbol{\alpha}$ is a vector of positive coefficients of penalty functions, and n_g is a number of inequality constraints in the constrained problem.

16.4.1 Analysis of Structural Behavior

Under the applied load, the structure shown in Fig. 16.1 undergoes some deformations described by the displacement field **u** as well as by the strain field **e** and the stress field σ. Thus, the behavior of the structure can be described by the equilibrium equation given in the form [16]:

$$div\,\sigma + \mathbf{f}^0 = 0 \tag{16.14}$$

as well as the kinematics relation between the strain and the displacement fields, expressed as [16]:

$$\mathbf{e} = \mathbf{B} \cdot \mathbf{u} \tag{16.15}$$

where **B** is a linear differential operator relating the displacement field to the strain field. A linear stress-strain relation is assumed in the form of generalized Hooke's law [16]:

$$\sigma = \mathbf{D} \cdot \mathbf{e} \tag{16.16}$$

where **D** denotes the extensional stiffness matrix for a model of composite material. Besides, the structure is subjected to the boundary conditions specified as follows [16]:

$$\begin{cases} \sigma \cdot \mathbf{n} = T^0 & \text{on } S_T \\ \mathbf{u} = \mathbf{u}^0 & \text{on } S_U \end{cases} \tag{16.17}$$

where **n** is the unit normal vector on the external boundary S of the disk.

To solve the set of (16.14), (16.15), (16.16) and (16.17), the finite element method (FEM) is proposed. In this numerical method of analysis of structural behavior, a region defining a continuum is discretized into simple geometric shape called finite elements, which are connected at nodes each other. In the proposed algorithm, the two-dimensional four-node quadrilateral elements are used (see Fig. 16.6).

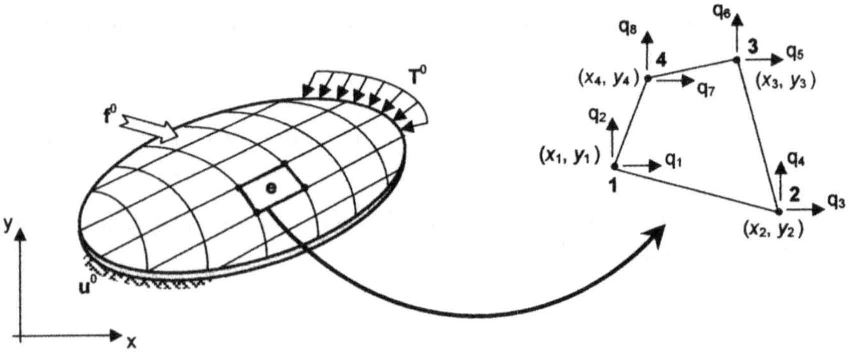

Fig. 16.6 Discretization of disk domain using finite elements

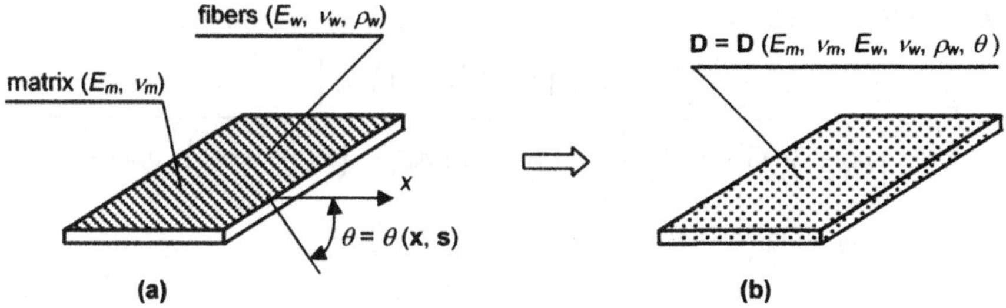

Fig. 16.7 Fiber reinforced composite material: (**a**) real composite, (**b**) model of composite

The detailed description of the finite element method is presented, for instance, in [16].

In analysis step of structural behavior, the microscopically non-homogeneous composite material is modeled by a plane, homogeneous, orthotropic and linearly elastic material (see Fig. 16.7). The purpose of the modeling process is to determine of the extensional stiffness matrix **D**, appearing in (16.16), and to express its components in terms of mechanical and geometrical properties of the matrix and the reinforcing fibers.

The extensional stiffness matrix **D** for the assumed model of the composite in the global coordinate system *x*-*y* is expressed by [17]:

$$\mathbf{D} = \mathbf{T}^{-1} \cdot \mathbf{C} \cdot \mathbf{T}^{-\mathrm{T}} \tag{16.18}$$

The matrix **C** denotes here the stiffness matrix for the composite with respect to the material axes 1–2, coinciding with the fiber direction and the direction perpendicular to the fiber. This matrix in terms of the so-called engineering constants for the plane and orthotropic material has the following form [17]:

$$\mathbf{C} = \begin{bmatrix} \dfrac{E_1}{1-\nu_{12}\nu_{21}} & \dfrac{E_1\nu_{21}}{1-\nu_{12}\nu_{21}} & 0 \\ \dfrac{E_2\nu_{12}}{1-\nu_{12}\nu_{21}} & \dfrac{E_2}{1-\nu_{12}\nu_{21}} & 0 \\ 0 & 0 & G_{12} \end{bmatrix} \tag{16.19}$$

where E_1 and E_2 are the apparent Young's moduli in the fiber direction and in the direction transverse to the fiber, respectively, while ν_{12} is the major and ν_{21} minor Poisson's ratios, and G_{12} denotes the in-plane shear modulus for the composite. Using the model presented in [18, 19] these engineering constants can be expressed as follows:

$$
\begin{aligned}
E_1 &= E_w\rho_w + E_m(1-\rho_w) \\
E_2 &= \frac{E_w\left[1+\left(\frac{E_w}{E_m}-1\right)\rho_w\right]}{\left[\rho_w+\frac{E_w}{E_m}(1-\rho_w)\right]\left[1+\left(\frac{E_w}{E_m}-1\right)\rho_w\right]-\left(\nu_m\frac{E_w}{E_m}-\nu_w\right)^2\rho_w(1-\rho_w)} \\
\nu_{12} &= \nu_w\rho_w + \nu_m(1-\rho_w) \qquad \text{and} \qquad \nu_{21}=\nu_{12}\frac{E_2}{E_1} \\
G_{12} &= \frac{E_m\left[\frac{E_w}{E_m}(1+\nu_m)(1+\rho_w)+(1+\nu_w)(1-\rho_w)\right]}{2(1+\nu_m)\left[\frac{E_w}{E_m}(1+\nu_m)(1-\rho_w)+(1+\nu_w)(1+\rho_w)\right]}
\end{aligned}
\tag{16.20}
$$

The matrix **T** in (16.18) denotes the transformation matrix from the global coordinate system *x*-*y* to the material axes 1–2 and has the form [17]:

$$
\mathbf{T}=\begin{bmatrix} \cos^2\theta & \sin^2\theta & 2\sin\theta\cos\theta \\ \sin^2\theta & \cos^2\theta & -2\sin\theta\cos\theta \\ -\sin\theta\cos\theta & \sin\theta\cos\theta & \cos^2\theta-\sin^2\theta \end{bmatrix} \tag{16.21}
$$

This matrix is considered as the matrix function of fiber orientation angle θ. When the fiber line is described by (16.1), its components are simply defined by:

$$
\begin{aligned}
\sin\theta &= \frac{tg\theta}{\sqrt{1+tg^2\theta}} = \frac{s_1}{\sqrt{1+s_1^2}} \\
\cos\theta &= \frac{1}{\sqrt{1+tg^2\theta}} = \frac{1}{\sqrt{1+s_1^2}}
\end{aligned}
\tag{16.22}
$$

or they can be explicitly defined by the angle between the fiber line and the *x*-axis of the global coordinate system (see Fig. 16.2). For the case of curvilinear fibers sinθ and cosθ, appearing in matrix **T**, follow from (16.5) and equal:

$$
\begin{aligned}
\sin\theta &= \frac{tg\theta}{\sqrt{1+tg^2\theta}} = \frac{y_{,t}}{\sqrt{(x_{,t})^2+(y_{,t})^2}} \\
\cos\theta &= \frac{1}{\sqrt{1+tg^2\theta}} = \frac{x_{,t}}{\sqrt{(x_{,t})^2+(y_{,t})^2}}
\end{aligned}
\tag{16.23}
$$

16.4.2 Initial Optimization

The evolutionary algorithm [20], based on imitation of the evolution processes occurring in nature, is used in the first step of the proposed optimization system. Its flow chart is shown in Fig. 16.8.

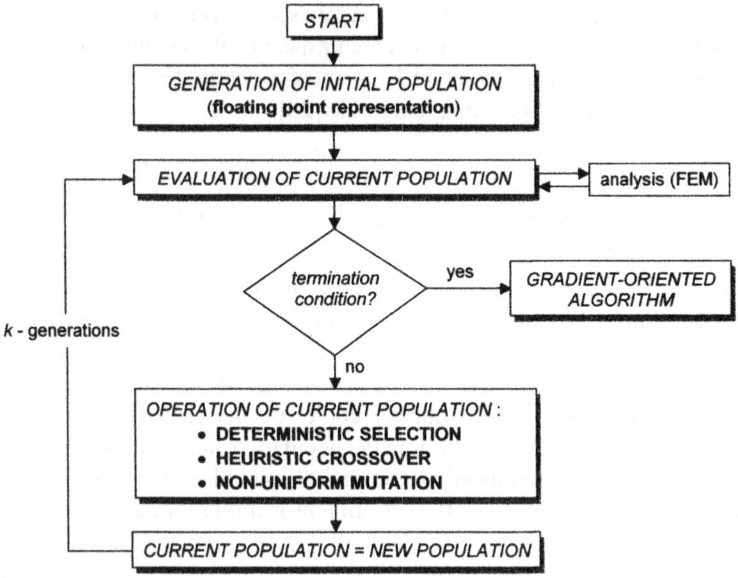

Fig. 16.8 Flow chart of evolutionary algorithm

An initial population of N chromosomes is created in order to start the evolution simulation. The vector of design parameters $\mathbf{b}$ defining the layout of reinforcing fibers in the composite is coded as a chromosome. Each chromosome in this population represents a point in design space and describes a possible solution to the given problem. All chromosomes are created randomly, and then this thus guarantees a very great population diversity. The floating-point representation is applied in this step of the evolutionary algorithm [20]:

$$ch_j \rightarrow \mathbf{b}_j = [b_1, b_2, \ldots, b_p] \text{ where } \underset{i=1\ldots p}{\Lambda} \; b_i = b_{i(\min)} + r(b_{i(\max)} - b_{i(\min)}) \quad (16.24)$$

The notations $b_{i(\min)}$ and $b_{i(\max)}$ are the variable bounds for the i-th design parameter of the vector $\mathbf{b}$, while r is a random number in range $\langle 0, 1\rangle$.

Thereafter, all chromosomes in the current population are evaluated using the objective functional:

$$\underset{j=1\ldots N}{\Lambda} \; \nu(ch_j) \equiv F_c(\mathbf{b}_j) \quad (16.25)$$

where $\nu(ch_j)$ denotes a fitness value of the j-th chromosome. This value is related to the value of the objective functional F_c for the j-th vector of design parameters $\mathbf{b}_j$. The values of state fields appearing in functional F_c are calculated using the finite element method (see Section 16.4.1).

The current population is processed by three main operators of the evolutionary algorithm. They are deterministic selection, heuristic crossover and non-uniform mutation. Before these processes, the power law scaling is applied in the algorithm. It is mainly responsible for the better search aspect of the evolutionary algorithm. At this point, the objective functional F_c is transformed to the new form, so-called the fitness functional F_p, according to the rule [20]:

$$F_p(\mathbf{b}) = e^{\left(-a\frac{F_c(\mathbf{b}) - F_{c(\min)}(\mathbf{b})}{F_{c(\max)}(\mathbf{b}) - F_{c(\min)}(\mathbf{b})}\right)} \tag{16.26}$$

where $F_{c(\min)}(\mathbf{b})$ and $F_{c(\max)}(\mathbf{b})$ denote minimal and maximal value of the objective functional F_c in the current population, respectively, while a is a positive scaling parameter.

The main idea of the selection operator is that "good" chromosomes are picked from the current population and multiple copies of them are created. As a result, "bad" chromosomes are eliminated from the population and do not undergo any further modification. The deterministic selection [20] is used in the proposed version of the evolutionary algorithm. First, the selection probability for each chromosome p_j is calculated as follows:

$$\underset{j=1\ldots N}{\Lambda} \; p_j = \frac{F_p(\mathbf{b}_j)}{\sum\limits_{j=1}^{N} F_p(\mathbf{b}_j)} \tag{16.27}$$

and next the number of its expected copies $olk_{(ch)}$ is obtained:

$$\underset{j=1\ldots N}{\Lambda} \qquad olk_{(ch_j)} = p_j * N \tag{16.28}$$

The number of copies of each chromosome is equal to integer part of $olk_{(ch)}$. Finally, the chromosomes are placed in the population according to the fractional part of $olk_{(ch)}$ and the empty places in this population are filled with the copies of chromosomes from the beginning of the population.

The heuristic crossover [20] operator recombines randomly chosen two parents chromosomes ch_1 and ch_2 to better child chromosome ch', according to the following scheme:

$$ch' = r\ (ch_2 - ch_1) + ch_2 \qquad \text{where :} \quad F_p(ch_2) \geq F_p(ch_1) \tag{16.29}$$

where r is a random number in range $\langle 0,1 \rangle$. This operation is carried out with a crossover probability p_c. It must be noted that the child chromosome can be an infeasible solution. In this case, the crossover operator is repeated.

As the last operator, the non-uniform mutation [20] is applied. This operator alters a chromosome locally:

$$ch_j = [\,b_1,\ \ldots\ ,b_i,\ \ldots\ ,b_n] \quad \rightarrow \quad ch_j' = [\,b_1,\ \ldots\ ,b_i',\ \ldots\ ,b_n] \tag{16.30}$$

The component of the chromosome b_i' is chosen with a mutation probability p_m. Its new value is calculated from the following relationship:

$$b_i' = \begin{cases} b_i + \left(1 - r^{(1-k/LP)s}\right)\left(b_{i(\max)} - b_i\right) \text{ if } l = 0 \\ b_i - \left(1 - r^{(1-k/LP)s}\right)\left(b_i - b_{i(\min)}\right) \text{ if } l = 1 \end{cases} \tag{16.31}$$

where $b_{i(\min)}$ and $b_{i(\max)}$ are the variable bounds for the i-th design parameter b_i, r denotes a random number in range $\langle 0,1 \rangle$, k is a iteration number and LP is a maximal number of generations, while s defines a degree of non-uniform.

Finally, the new population of solutions is created and the single cycle of the evolutionary algorithm, which is known as a generation, comes to the end. This new population is again modified by means of the above three operators. Each successive generation contains increasingly better "partial solutions" than previous generations, and converges towards the global optimum. This evolution procedure is continued until there is no substantial improvement of the best or average population statistics for a few consecutive generations or when a specified number of generations is attained.

16.4.3 Final Optimization

In order to increase efficiency of optimization process, the gradient-oriented algorithm starting from the last, best solution generated by evolutionary algorithm is used in the second, final module of the hybrid optimization system.

On this stage of optimization procedure the variable metric method [15] is proposed (see Fig. 16.9). In this method, the estimation of Hesian matrix inverse $\mathbf{V}_k$ in the k-th iteration is obtained according to the rule:

$$\mathbf{V}_k = \mathbf{V}_{k-1} + \Delta\mathbf{V}_k = \mathbf{V}_{k-1} + \left(\frac{\mathbf{s}^{(k)}(\mathbf{s}^{(k)})^{\mathrm{T}}}{(\mathbf{s}^{(k)})^{\mathrm{T}}\mathbf{r}^{(k)}} - \frac{\mathbf{V}_{k-1}\mathbf{r}^{(k)}(\mathbf{r}^{(k)})^{\mathrm{T}}\mathbf{V}_{k-1}}{(\mathbf{r}^{(k)})^{\mathrm{T}}\mathbf{V}_{k-1}\mathbf{r}^{(k)}} \right) \tag{16.32}$$

where:

$$\mathbf{s}^{(k)} = \mathbf{b}^{(k)} - \mathbf{b}^{(k-1)} \text{ and } \mathbf{r}^{(k)} = \nabla F_c(\mathbf{b}^{(k)}) - \nabla F_c(\mathbf{b}^{(k-1)}) \tag{16.33}$$

It must be noted, that the unit matrix $\mathbf{I}$ is used as the first estimation. The feasible direction $\mathbf{d}^{(k)}$ in consecutive points $\mathbf{b}^{(k)}$ generated by the algorithm, is derived as follows:

$$\mathbf{d}^{(k)} = -\mathbf{V}_k \nabla F_c(\mathbf{b}^{(k)}) \tag{16.34}$$

and next point $\mathbf{b}^{(k+1)}$ is obtained as the result of directional minimization, namely:

$$\mathbf{b}^{(k+1)} = \mathbf{b}^{(k)} + \hat{\tau}\mathbf{d}^{(k)} \tag{16.35}$$

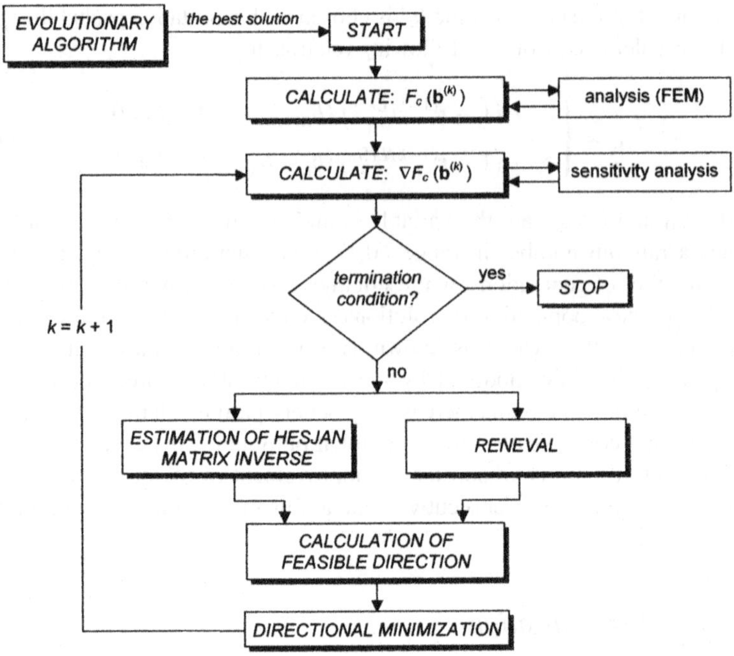

Fig. 16.9 Flow chart of variable metric method

where:

$$F_c(\hat{\tau}) = \min_{\tau \geq 0}(\mathbf{b}^{(k)} + \tau \mathbf{d}^{(k)}) \tag{16.36}$$

As the stop criterion of the algorithm, the following condition is used:

$$\left\| \nabla F_c(\mathbf{b}^{(k)}) \right\| \leq \varepsilon \tag{16.37}$$

where ε denotes the accuracy of calculations.

16.4.4 Sensitivity Analysis of Structural Behavior

The necessity of deriving the gradient information of objective functional and any constraint constitutes the integral part of any gradient-oriented algorithm. In the proposed algorithm the direct approach to sensitivity analysis [21] is applied in order to gather the necessary gradient information.

The first-order sensitivity for an arbitrary functional F_c, given in (16.10), with respect to i-th design parameter of the vector **b** has the following form:

$$\underset{i=1\ldots p}{\Lambda} \ F_{c},_{b_i} = \int_A (\Gamma,_{b_i} + \Gamma,_{\boldsymbol{\sigma}} \cdot \boldsymbol{\sigma},_{b_i} + \Gamma,_{\mathbf{e}} \cdot \mathbf{e},_{b_i} + \Gamma,_{\mathbf{u}} \cdot \mathbf{u},_{b_i})\, dA + \int_{S_T} (\Psi,_{\mathbf{T}^0} \cdot \mathbf{T},^0_{b_i} + \Psi,_{\mathbf{u}} \cdot \mathbf{u},_{b_i})\, dS_T \tag{16.38}$$

As can be seen from (16.38), to calculate $F_{c,\,\mathbf{b}}$ the knowledge of sensitivities of state fields is necessary. Using the direct approach in order to obtain the desired state fields sensitivities, the set of (16.14), (16.15), (16.16) and (16.17) constituting the boundary problem is differentiated with respect to the design parameter b_i, and next is solved with respect to the sensitivities required. Thus, by differentiating these equations we obtain respectively:

$$div\, \boldsymbol{\sigma},_{bi} = 0 \tag{16.39}$$

$$\mathbf{e},_{b_i} = \mathbf{B} \cdot \mathbf{u},_{b_i} \tag{16.40}$$

$$\boldsymbol{\sigma},_{b_i} = \mathbf{D} \cdot \mathbf{e},_{b_i} + \mathbf{D},_{b_i} \cdot \mathbf{e} = \mathbf{D} \cdot \mathbf{e},_{b_i} + \boldsymbol{\sigma}_i^0 \tag{16.41}$$

$$\begin{cases} \boldsymbol{\sigma},_{b_i} \cdot \mathbf{n} = 0 \text{ on } S_T \\ \mathbf{u},_{b_i} = 0 \quad \text{ on } S_U \end{cases} \tag{16.42}$$

The set of (16.39), (16.40), (16.41) and (16.42) has a similar form as (16.14), (16.15), (16.16) and (16.17) and it constitutes the boundary problem for the so-called i-th additional disk (see Fig. 16.10).

The additional disk is subjected to the field of generalized initial stress $\boldsymbol{\sigma}_i^0$ within its domain. This field is due to a change of the extensional stiffness matrix **D** with

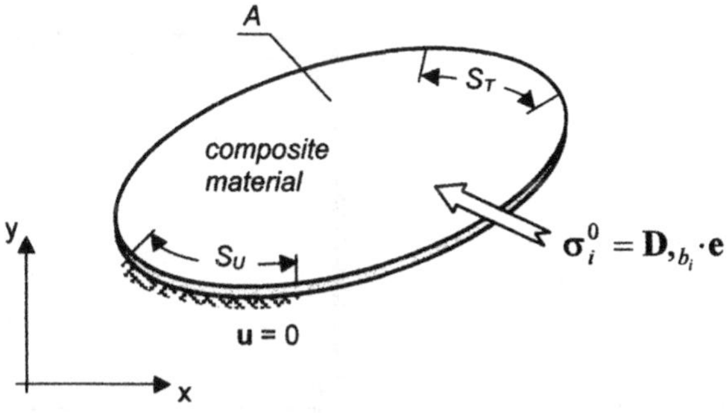

Fig. 16.10 *i*-th additional composite disk

respect to i-th design parameter b_i. The sensitivity of **D** follows from (16.18) and is expressed as:

$$\mathbf{D}_{,b_i=s_i} = (\mathbf{T}^{-1})_{,s_i} \cdot \mathbf{C} \cdot \mathbf{T}^{-\mathrm{T}} + \mathbf{T}^{-1} \cdot \mathbf{C} \cdot (\mathbf{T}^{-\mathrm{T}})_{,s_i} \tag{16.43}$$

and

$$\mathbf{D}_{,b_i=\rho_w} = \mathbf{T}^{-1} \cdot (\mathbf{C})_{,\rho_w} \cdot \mathbf{T}^{-\mathrm{T}} \tag{16.44}$$

Thus, the first-order sensitivities of the functional F_c can be now calculated using (16.38), where state fields of primary composite disk and their sensitivities are obtained as solutions for primary composite disk (see Fig. 16.1) and p additional composite disks (see Fig. 16.10).

It should be noted, that to perform the sensitivity analysis of state fields, the finite element method (FEM) with the two-dimensional four-node quadrilateral elements is applied [16].

16.5 Numerical Examples

To illustrate the applicability of the proposed series hybrid system for the design process of mechanical properties of composite disks, some simple numerical examples are presented in this Section.

16.5.1 Example 1

As a first example, let us consider a thin, rectangular disk supported along the edge AD and uniformly loaded by traction along the edge CD (see Fig. 16.11). The material of the disk is a composite made of a epoxy matrix reinforced with

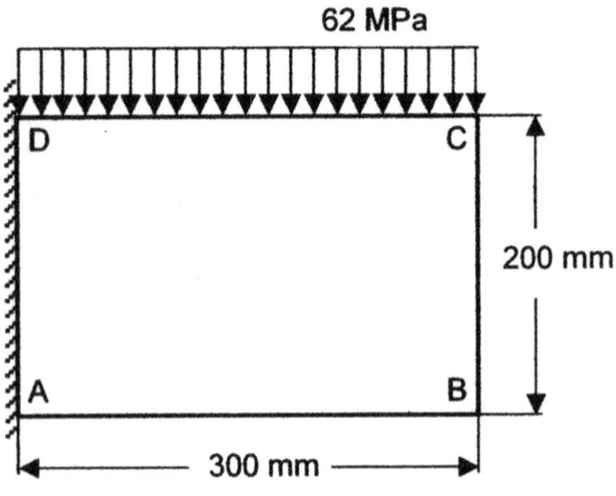

Fig. 16.11 Composite disk subjected to load and boundary conditions

Table 16.1 Material data of components of composite

	E [GPa]	ν	ρ [%]
Fibers (glass E)	75.0	0.22	45
Matrix (epoxy)	3.5	0.38	55

Table 16.2 Allowable stress level for composite

	Strength characteristic
$R_{r1} = 1,062$ [MPa]	Longitudinal tensile strength
$R_{c1} = 610$ [MPa]	Longitudinal compressive strength
$R_{r2} = 31$ [MPa]	Transverse tensile strength
$R_{c2} = 118$ [MPa]	Transverse compressive strength
$R_s = 72$ [MPa]	Shear strength

glass fibers. The mechanical properties for the components of composite material are given in Tables 16.1 and 16.2.

The problem deals with the optimal layout of reinforcing fibers in the composite for the case of mean stiffness design of the disk. Thus, the minimization of work done by external forces is selected as the objective functional and the optimization problem can be formulated in the following form:

$$\min .F_c(\mathbf{b}) = \int\limits_{S_T} \mathbf{u}^{\mathrm{T}} \cdot \mathbf{T}^0 dS_T \tag{16.45}$$

where **b** is a vector of design parameters defining the shape of middle line of directional fiber.

The above formulated design problem was considered for the case of one family of straight fibers. Thus, an angle of fiber orientation θ was selected as the design parameter (see Fig. 16.12a), i.e. $\mathbf{b} = \{\theta\}$, and it was determined during

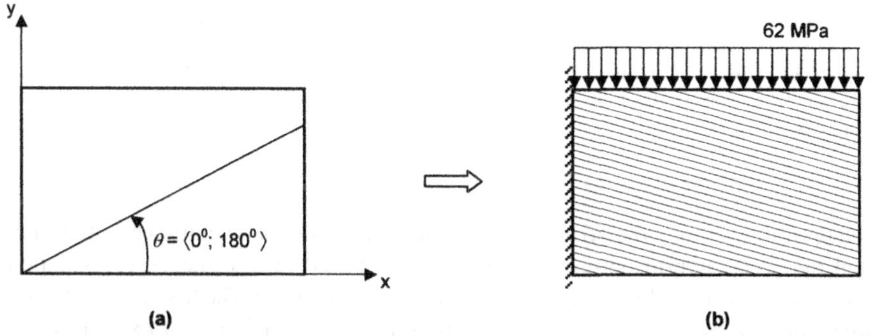

Fig. 16.12 Optimization process of fiber layout in composite disk: (**a**) design parameter for directional fiber, (**b**) results of optimization process

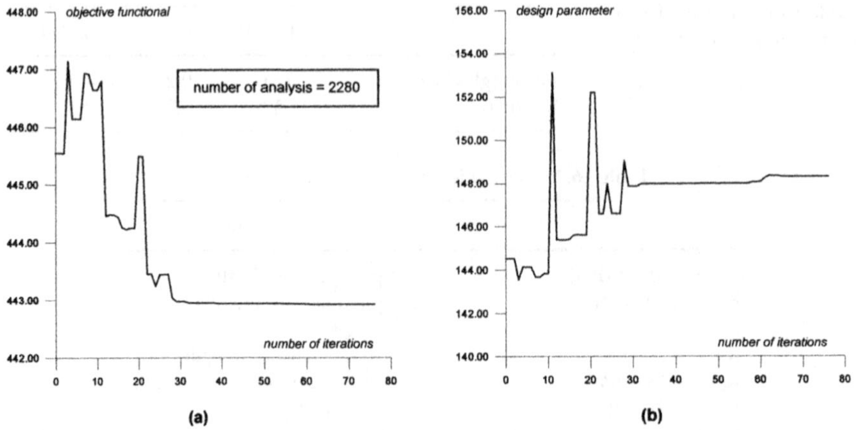

Fig. 16.13 History of optimization process using evolutionary algorithm: (**a**) dependence of objective functional versus iteration numbers, (**b**) dependence of design parameter versus iteration numbers

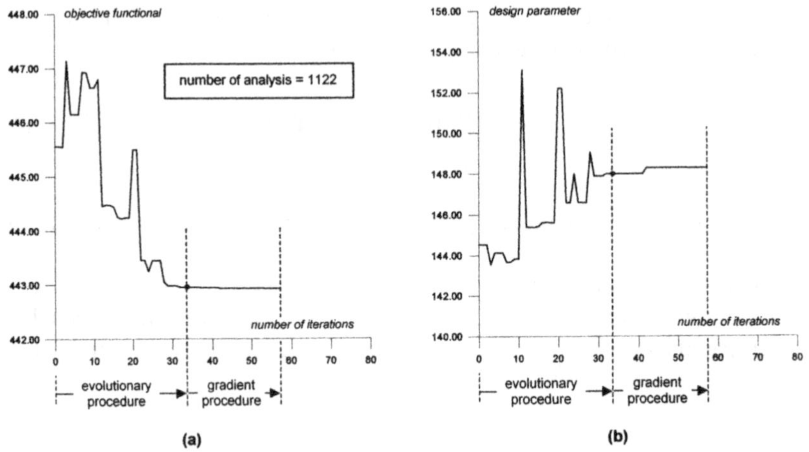

Fig. 16.14 History of optimization process using series hybrid system: (**a**) dependence of objective functional versus iteration numbers, (**b**) dependence of design parameter versus iteration numbers

the optimization process. To solve this design problem, the evolutionary algorithm and the series hybrid optimization system composed of a sequence of evolution and gradient-oriented procedures were used simultaneously. The history of the optimization processes using these methods is presented in Figs. 16.13 and 16.14, respectively and the solutions are presented in Table 16.3. The optimal composite disk with one family of straight fibers is shown in Fig. 16.12b.

Table 16.3 Results of optimization process

	Design parameter	Objective functional
Evolutionary algorithm	$\theta = 148.29°$	$F_c = 442.9291$ [J]
Series hybrid system	$\theta = 148.25°$	$F_c = 442.9287$ [J]

Additionally, the efficiency of the evolutionary algorithm and the proposed series hybrid system was verified in 10 runs of optimization process. The comparison of obtained results using these approaches is presented in Figs. 16.15 and 16.16

We can observe from Figs. 16.13, 16.14, 16.15 and 16.16 that the hybrid algorithm and the evolutionary algorithm find practically this same solution for the discussed problem. However, the hybrid algorithm used less number of iterations and the total time of calculations was shorter. It was due to less number of analysis need in the second part of optimization procedure, where the gradient-oriented method was used. It must be also noted, that the gradient-oriented algorithm starting from the last, best solution generated by evolutionary algorithm in the initial procedure of the hybrid system does not fall into a local optimum.

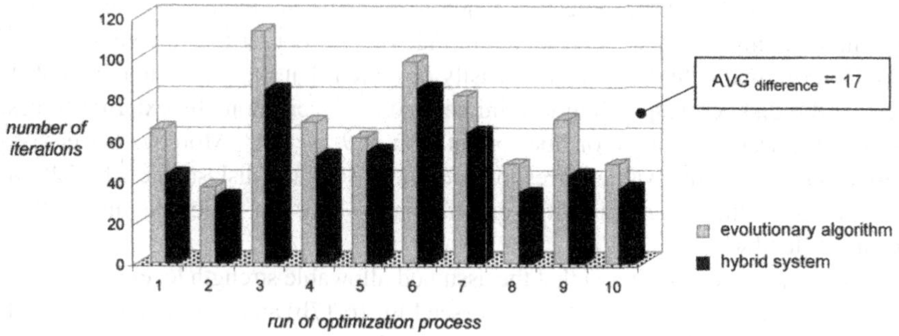

Fig. 16.15 Difference between number of iterations in successive optimization processes

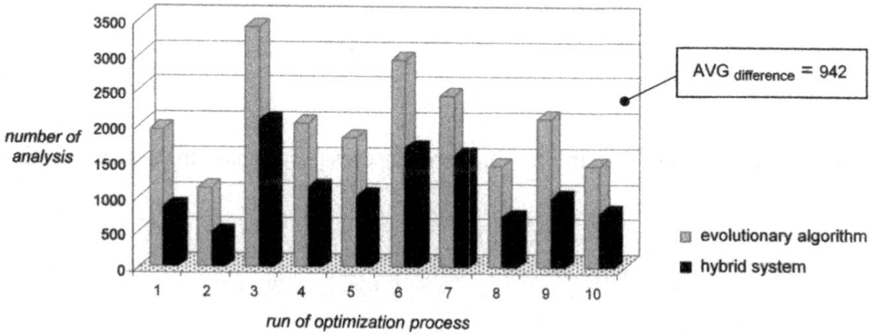

Fig. 16.16 Difference between number of analysis in successive optimization processes

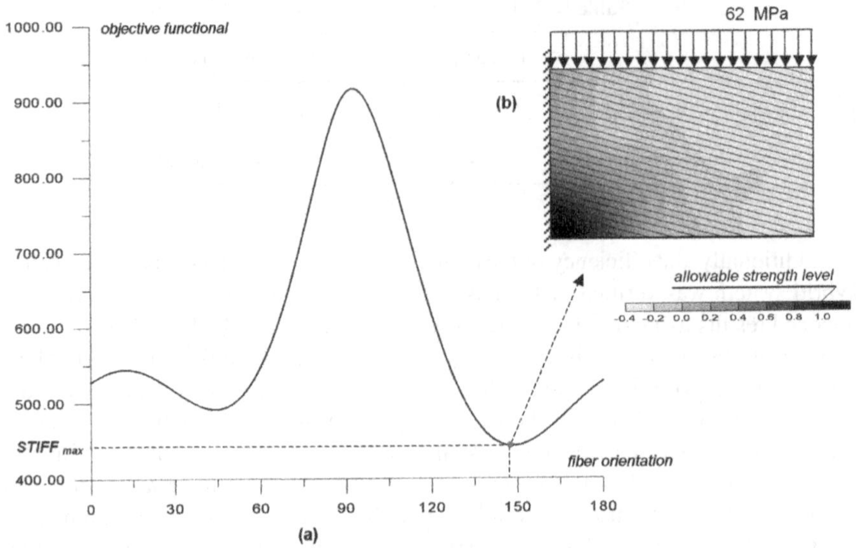

Fig. 16.17 (**a**) plot of objective functional versus fiber orientation in composite disk, (**b**) plot of strength function in the most stiffness composite disk

Finally, the plot of the objective functional versus the angles of fiber orientation in the composite disk is presented in Fig. 16.17ain order to evaluate quality of obtained optimal solution. It can be easily observed that the global maximal stiffness of the disk, corresponding to minimal value of work done by external forces, is obtained when the angle of fiber orientation is $\theta = 148°$. Moreover, is easy to verify that this optimal value corresponds to increase of the disk stiffness by 52% in comparison to the design with fiber orientation $\theta = 93°$ for which the disk will be the most flexible one.

However, one can be noted that the assumed allowable strength level is exceeded in the lower, left sub-domain of the disk (see Fig. 16.17b) and consequently the disk will be destroyed. Thus, the optimization problem should be reformulated to the following form:

$$\min .F_c(\mathbf{b}) = \int_{S_T} \mathbf{u}^{\mathrm{T}} \cdot \mathbf{T}^0 dS_T \tag{16.46}$$

subjected to the local constraints imposed on cross-sectional stresses in the disk domain:

$$\Phi_{ij}(\sigma, \mathbf{b}) - 1 \leq 0 \text{ for } \begin{cases} i = 1, \ldots, el \\ j = 1, \ldots k_g \end{cases} \tag{16.47}$$

where el denotes the number of elements appearing in the finite element model of the disk and k_g is the number of Gauss integration points in each element. The functions ϕ, appearing in constraints (16.47), is the strength function for orthotropic materials. According to the bi-axial Tsai-Wu theory [22], it should satisfy the following relationship at any point of structure:

$$\Phi(\sigma) = F_1\sigma_1 + F_2\sigma_2 + F_{11}\sigma_1^2 + 2F_{12}\sigma_1\sigma_2 + F_{22}\sigma_2^2 + F_{33}\tau_{12}^2 \leq 1 \quad (16.48)$$

where F_i and F_{ij} denote the strength coefficients and they depend on the strength characteristic of composite [22]:

$$F_1 = \frac{1}{R_{r1}} - \frac{1}{R_{c1}}; \qquad F_2 = \frac{1}{R_{r2}} - \frac{1}{R_{c2}}$$

$$F_{11} = \frac{1}{R_{r1}R_{c1}}; \quad F_{22} = \frac{1}{R_{r2}R_{c2}}; \qquad F_{33} = \frac{1}{R_s^2}; \qquad F_{12} = -\frac{\sqrt{F_{11}F_{22}}}{2} \quad (16.49)$$

while σ_1, σ_2 and τ_{12} are components of the stress field with respect to the material axes of the composite, coinciding with the fiber direction and the direction perpendicular to the fiber. They are obtained using the following stress-transformation rule [22]:

$$\begin{Bmatrix} \sigma_1 \\ \sigma_2 \\ \tau_{12} \end{Bmatrix} = \begin{bmatrix} \cos^2\theta & \sin^2\theta & 2\sin\theta\cos\theta \\ \sin^2\theta & \cos^2\theta & -2\sin\theta\cos\theta \\ -\sin\theta\cos\theta & \sin\theta\cos\theta & \cos^2\theta - \sin^2\theta \end{bmatrix} \cdot \begin{Bmatrix} \sigma_x \\ \sigma_y \\ \tau_{xy} \end{Bmatrix} \quad (16.50)$$

The design problem of fiber layout in the composite disk, defined by (16.46) and (16.47) was discussed here for the case of family of curvilinear fibers for which the shape of middle line of directional fiber was described by Bezier curve [23] in the form:

$$\begin{Bmatrix} x(t) \\ y(t) \end{Bmatrix} = \sum_{i=0}^{m} \begin{Bmatrix} x_i \\ y_i \end{Bmatrix} \binom{m}{i} t^i(1-t)^{m-i} \quad \text{where} \quad 0 \leq t \leq 1 \quad (16.51)$$

In this case, the coordinates of three nodes of the Bezier polygon were chosen as the design parameters, i.e. $\mathbf{b} = \{x_i, y_i\}$ and $i = 0, 1, 2$ (see Fig. 16.18a). The layout of all other fibers in the family was obtained by shifting of the optimal directional fiber in direction normal to its middle line.

To solve this design problem, as previously, the series hybrid optimization system was used. The results of optimization process are given in Table 16.4, and the optimal layout of reinforcing fibers in the disk domain is shown in Fig. 16.18b. We can observe now, that the disk is more stiffness and the distribution of the strength function F is considerable reduced when compared to the case presented in Fig. 16.17b. It must be also noted, that the values of the local strength function violating allowable strength level are eliminated after optimization.

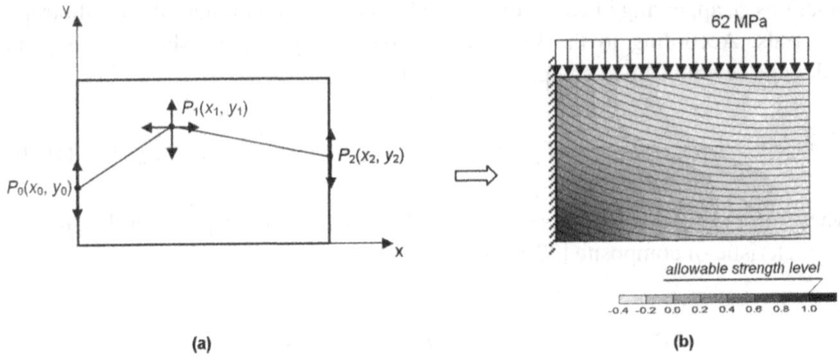

Fig. 16.18 Optimization process of fiber layout in composite disk: (**a**) design parameter for directional fiber, (**b**) results of optimization process

Table 16.4 Results of optimization process

	Design parameters	Objective functional	Value of local strength function
The most stiffness disk satisfying constraints imposed on stresses	P_0 (0; 150)	$F_c = 413.77$ [J]	$\Phi_{max.} = 0.99$
	P_1 (124; 102)		$\Phi_{min.} = -0.53$
	P_2 (300; 108)		
The most stiffness disk with no constraints imposed on stresses	$\theta = 148.25°$	$F_c = 442.92$ [J]	$\Phi_{max.} = 1.27$
			$\Phi_{min.} = -0.48$

16.5.2 Example 2

Let us now consider a disk subjected to load and boundary conditions as it is shown in Fig. 16.19. The disk has a unitary thickness and is made of composite material with the mechanical properties given in Tables 16.5 and 16.6.

The problem discussed in the example regards the optimal design of fiber shape and orientation in the composite so that the disk made of this material should be as tough as possible. The main idea of strength design of composite structures results in reduction of the values of strength function Φ violating criterion (16.48) and redistribution stress fields in the composite disk domain. Thus, the global measure of local strength criterion is introduced as the behavioral functional in this design problem and the optimization problem can be formulated in the form:

$$\min .F_c(\mathbf{b}) = \left[\int_A [\Phi(\sigma)]^n dA \right]^{\frac{1}{n}} \tag{16.52}$$

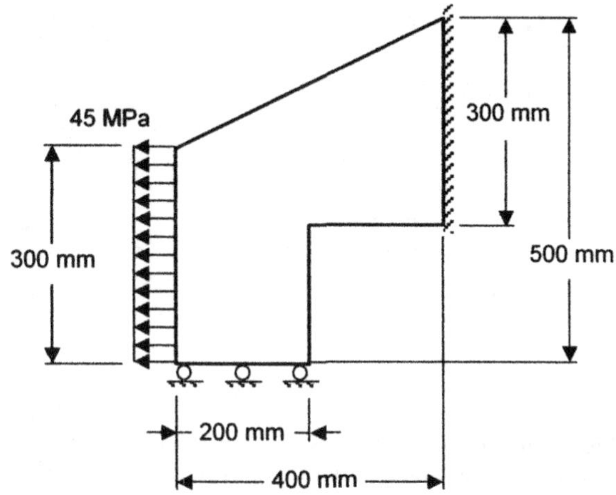

Fig. 16.19 Composite disk subjected to load and boundary conditions

Table 16.5 Material data of components of composite

	E [GPa]	ν	ρ [%]
Fibers (graphite HS)	230.0	0.25	60
Matrix (epoxy)	3.5	0.38	40

Table 16.6 Allowable stress level for composite

	Strength characteristic
$R_{r1} = 1,531$ [MPa]	Longitudinal tensile strength
$R_{c1} = 1,390$ [MPa]	Longitudinal compressive strength
$R_{r2} = 41$ [MPa]	Transverse tensile strength
$R_{c2} = 145$ [MPa]	Transverse compressive strength
$R_s = 98$ [MPa]	Shear strength

where A denotes the area of structure domain and **b** is a vector of design parameters defining the shape or orientation of reinforcing fibers in the composite material. The notations n is here the natural even number and when $n \to \infty$, the functional is a strict measure of maximum local values of strength criterion. In practice the values of n does not exceed the upper bound limit following from numerical constraints. In this example the value of n is set to 20.

The optimization problem, defined by (16.52), was discussed for two classes of fiber shape. In the first case, the composite material was reinforced with one family of straight fibers (see Fig. 16.20a) for which an angle of orientation θ for the directional fiber was selected as the design parameter, i.e. $\mathbf{b} = \{\theta\}$. Next, the case of one family of parabolic fibers was considered (see Fig. 16.21a). In this case, the shape of middle line of the directional fiber was assumed in the form $x = ay^2$, where the coefficient a was treated as the design parameter, i.e. $\mathbf{b} = \{a\}$. All other

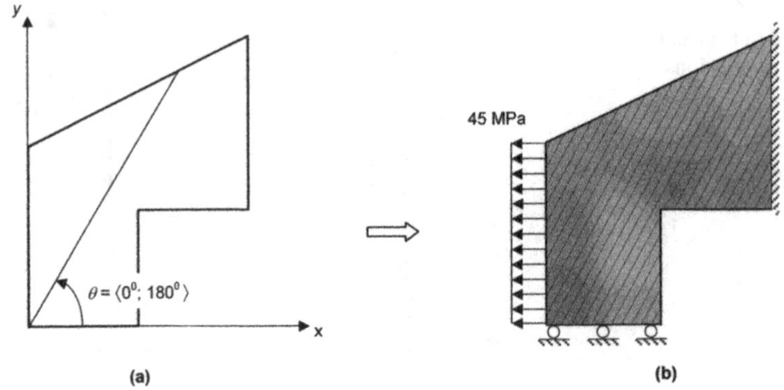

Fig. 16.20 Optimization process for case of family of straight fibers in composite disk: (**a**) design parameter for directional fiber, (**b**) results of optimization process

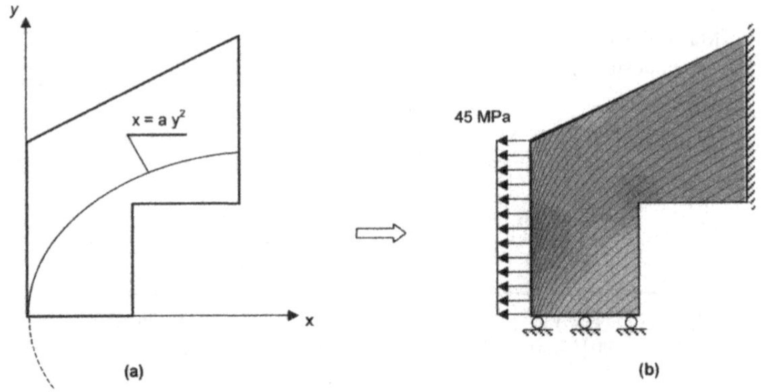

Fig. 16.21 Optimization process for case of family of parabolic fibers in composite disk: (**a**) design parameter for directional fiber, (**b**) results of optimization process

Table 16.7 Results of optimization process

	Design parameters	Value of local strength function
Optimal disk with family of straight fibers	$\theta_{opt.} = 62.17°$	$\Phi_{max.} = 0.78$ $\Phi_{min.} = -0.14$
Optimal disk with family of parabolic fibers	$a_{opt.} = 14.93$	$\Phi_{max.} = 0.26$ $\Phi_{min.} = -0.22$
Reference disk	$\theta. = 0°$	$\Phi_{max.} = 1.09$ $\Phi_{min.} = -0.37$

fibers in the both families were obtained by translation of the directional fiber in the x-direction. To solve the design problem, the series hybrid optimization system was used. The results of optimization process are given in Table 16.7.

The optimal layouts of reinforcing fibers in the disk for the case of straight and parabolic fibers are shown in Figs. 16.20b and 16.21b, respectively. These optimal

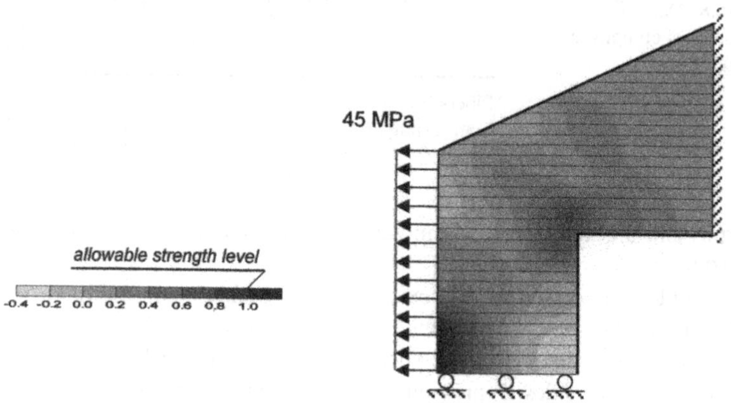

Fig. 16.22 Reference composite disk with family of straight fibers

solutions were also compared with the reference solution for the disk with one family of straight fibers parallel to the x-axis (see Fig. 16.22) in order to qualify results of optimization process. We can observe from plots shown in Figs. 16.20, 16.21 and 16.22 and maximal values of strength function presented in Table 16.7, that the distribution of strength function Φ in reference disk is considerable reduced after optimization process and the values of the local strength function violating the allowable strength level are eliminated for both optimal solutions. It must be also noted that the distribution of the stresses in the optimal disks is practically uniform what allows for better usage of composite material or allows to increase the upper level of service loading.

16.5.3 Example 3

Finally, let us consider a thin disk shown in Fig. 16.23. The disk is made of a polyester matrix reinforced with one family of straight, glass fibers. The mechanical

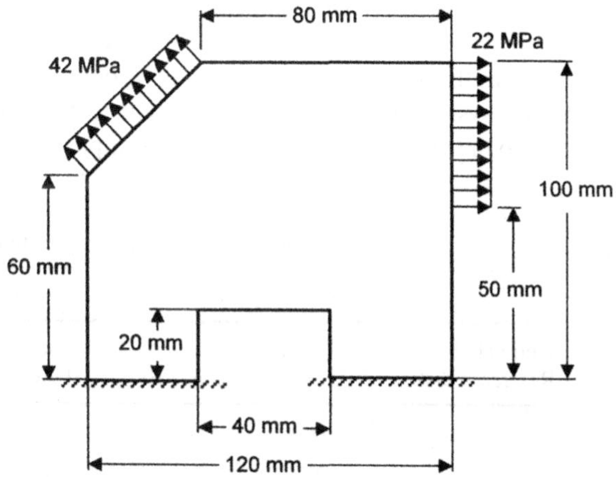

Fig. 16.23 Composite disk subjected to load and boundary conditions

Table 16.8 Data of components of composite material

	E [GPa]	ν	Cost [PLN]
Fibers (glass E)	75.0	0.22	30
Matrix (polyester)	3.2	0.41	12

properties and the unit cost of components of the composite material are given in Table 16.8.

The problem discussed in this example corresponds to the optimal design of fibers layout in the composite for the case of minimization of the total cost of the disk. It was also assumed that the stiffness of reinforced disk $STIFF_0$ should be equals to 170 [J]. This design problem can be written as follows:

$$\min .C(\mathbf{b}) = [c_w\rho_w + c_m(1 - \rho_w)] \cdot V \tag{16.53}$$

with the imposed global behavioral constraint:

$$\left(\int_{S_T} \mathbf{u}^{\mathbf{T}} \cdot \mathbf{T}^0 dS_T \right) - STIFF_0 \leq 0 \tag{16.54}$$

where the angle of fiber orientation θ and fiber density P_w were selected as the design parameters, i.e. $\mathbf{b} = \{\theta, \rho_w\}$, and they were determined during the optimization process. The above formulated problem, once again, was solved using the proposed the hybrid optimization system. The results, obtained after the optimization process, are given in Table 16.9.

Finally, the optimal solution, shown in Fig. 16.24a, was compared with solution for the same disk reinforced with one family of straight fibers parallel to the y-axis (see Fig. 16.24b). The results allow us to state, that the optimal layout of fibers decreases the total cost of the disk by 17%, when compared to the reference disk. Additionally, in order to verify results of optimization process, the plot of the work done by external forces versus the angles of fiber orientation in the composite disks with fiber density $\rho_w = 0.52$ and $\rho_w = 0.70$ is shown in Fig. 16.24c. The results presented in this plot confirm the results of optimization process, and once again, the hybrid system was a very effective method for quickly finding the reasonable solution to the discussed problem in this example.

Table 16.9 Results of optimization

	Design parameters	Cost [PLN]	Stiffness [J]
Optimal disk	$\theta = 134°$; $\rho_w = 0.52$	2.48	170.0
Reference disk	$\theta = 90°$; $\rho_w = 0.70$	2.96	170.0

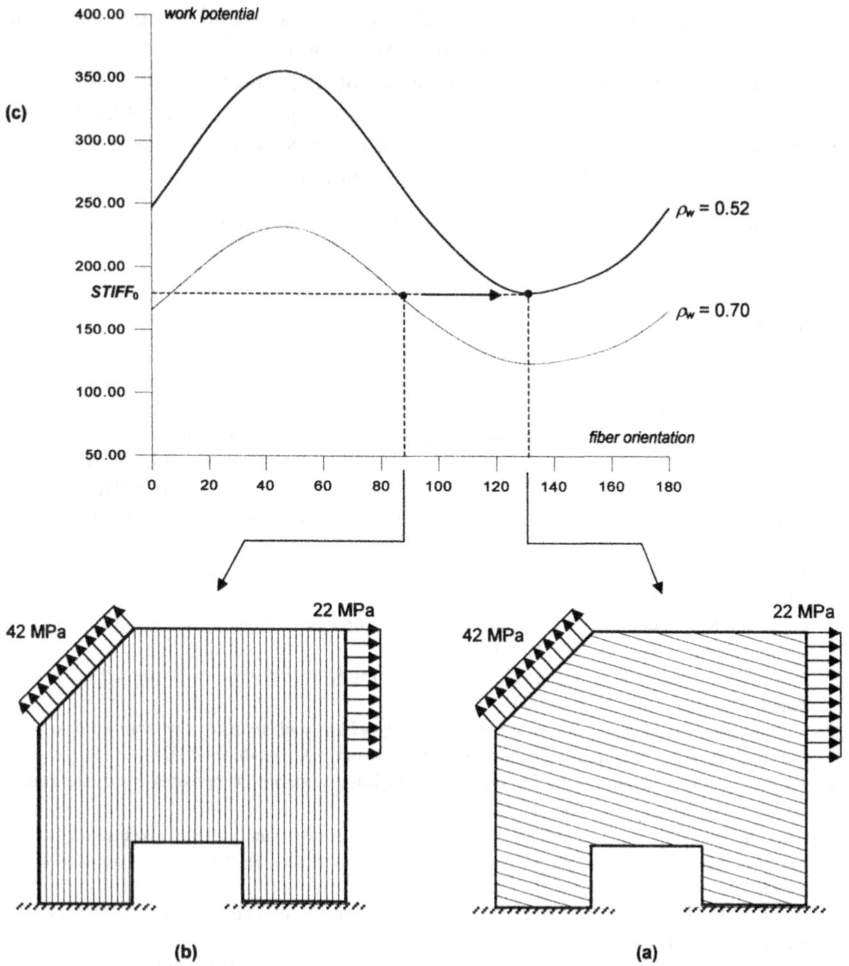

Fig. 16.24 Disk reinforced with one family of straight fibers: (**a**) optimal solution, (**b**) reference solution (**c**) plot of work done by external forces versus fiber orientation for these disks

16.6 Conclusions

The series hybrid system, composed of a sequence of evolution and gradient-oriented procedures, for optimal design of mechanical properties of composite structures was presented in the paper. The applicability of this system was verified in some simple numerical examples. The results of the research allow us to state, that the proposed algorithm is a very effective tool for finding a reasonable solution to the problem of optimal design of composite materials. The main advantage of the hybrid system is its flexibility in using different approaches for solving optimization problem. Thus, this method can constitute an alternative technique for the classical methods applied in optimization of structural components made of composite material, or can supplement these methods.

The presented analysis can be treated as a starting point for computer-oriented design procedures of the real composite structure subjected to actual loading conditions. Such procedure can allow to avoid of the expensive experimental testing, which can be reduced to the final phase of structural design.

Acknowledgements This research work is supported by Grant No. 501 060 32/3955 of the Polish Ministry of Education and Science.

References

1. Cheng, K.: Sensitivity analysis and a mixed approach to the optimization of symmetric layered composite panels. Eng. Optim. **9**, 233–248 (1986)
2. Thomsen, J.: Optimization of composite discs. Struct. Optim. **3**, 89–98 (1991)
3. Matheus, H.C., Mota Soares, C.M., et al.: Sensitivity analysis and optimal design of thin laminated composite structures. Proceedings of Course on CAOD of Structures, Trento (1991)
4. Graesser, D.L., Zabinsky, Z.B., et al.: Designing laminated composites using random search techniques. Compos. Struct. **18**, 311–325 (1991)
5. Fang, C., Springer, G.S.: Design of composite laminates by a Monte Carlo method. J. Compos. Mater. **27**, 721–753 (1993)
6. Callahan, K.J., Weeks, G.E.: Optimum design of composite laminates using genetic algorithm. Compos. Eng. **2**(3), 149–160 (1992)
7. Gurdal, Z., Haftka, R.T., et al.: Genetic algorithms for the design of laminated composite panel. SAMPE J. **30**, 29–35 (1994)
8. Gantownik, V.B., Gurdal, Z., et al.: A genetic algorithm with memory for optimal design of laminated sandwich composite panels. Compos. Struct. **53**, 513–520 (2002)
9. Richie Le, R.G., Gaudin. J.: Design of dimensionally stable composites by evolutionary optimization. Compos. Struct. **41**, 97–111 (1998)
10. Kaletta, P., Wolf, K.: Optimization of composite aircraft panels using evolutionary computation methods. Proceedings of 22nd ICAS Conference, Harrogate, pp. 411.1–411.10 (2000)
11. Spallino, R., Thierauf, G.: Thermal buckling optimization of composite laminates by evolution strategies. Comput. Struct. **78**, 691–697 (2000)
12. Dems, K.: Sensitivity analysis and optimal design for fiber reinforced composite disks. Struct Optim. **11**, 178–186 (1996)
13. Wiśniewski, J.: Optimal design of reinforcing fibers in multilayer composites using genetic algorithms. Fibers Textiles East. Eur. **12**, 58–63 (2004)
14. Dems, K., Wiśniewski, J.: Optimal design of fiber-reinforced composite disks. J. Theor. Appl. Mech. **47**, 515–535 (2009)
15. Findeisen, W., Szymanowski, J.: Teoria i metody obliczeniowe optymalizacji. Państwowe Wydawnictwo Naukowe, Warszawa (1980)
16. Zienkiewicz, O.C.: The Finite Element Method in Engineering Science. McGraw-Hill, London (1971)
17. German, J.: Podstawy mechaniki kompozytów włóknistych. Wydawnictwa Politechniki Krakowskiej, Kraków (1996)
18. Malmajster, A.K., Tamuz, V., et al.: Soprotivlenie źestkich polimernych materialow. Zinatne, Ryga (1967)
19. Hashin, Z., Rosen, B.: The elastic moduli of fiber-reinforced materials. J. Appl. Mech. **31,** 223–232 (1964)
20. Michalewicz, Z.: Genetic Algorithms + Data Structures = Evolution Programs. Springer, Berlin (1996)

21. Dems, K., Mróz, Z.: Variational approach to first- and second order sensitivity analysis of elastic structures. Int. J. Numer. Methods Eng. **21**, 637–661 (1985)
22. Tsai, S.W., Wu, E.M.: A general theory of strength for anisotropic materials. J. Compos. Mater. **5**, 58–80 (1971)
23. Kiciak, P.: Podstawy modelowania krzywych i powierzchni – zastosowanie w grafice komputerowej. Wydawnictwa Naukowo-Techniczne, Warszawa (2000)

Chapter 17
Analysis of Representative Volume Elements with Random Microcracks

P. Fedelinski

Abstract The paper presents the computation of effective elastic properties and an analysis of stress intensity factors for representative volume elements (RVE) with randomly distributed microcracks. The RVEs are subjected to static and dynamic loadings. The microcracks having the same length, randomly distributed, parallel or randomly oriented, are considered. The structures with microcracks are modelled by using the boundary element method (BEM). The time dependent problems are solved using the Laplace transform method. In the BEM the boundaries are discretized and it is very easy to modify positions and directions of microcracks. The influence of density of microcracks on the effective Young modulus, the effective Poisson ratio, stress intensity factors and speed of the wave travelling through the cracked structure is investigated. The numerically computed Young moduli are compared with the solutions obtained by analytical methods: the non-interacting method, the self-consistent method and the differential method.

Keywords Representative volume element · Effective properties · Microcrack · Micromechanics · Boundary element method

17.1 Introduction

Representative volume elements (RVE) [1, 2] are used to compute effective material properties of complex structures containing voids, inclusions and cracks. The size of the RVE depends on the type of material and should represent a typical structure of material. The RVE is properly chosen if the overall properties are the same for the RVE subjected to displacement and traction boundary conditions. The RVE should contain a large number of heterogeneities. Therefore the analysis of the RVE, which is a part of the homogenization procedure, is very complex but simpler than the analysis of the whole structure. In order to obtain statistically representative results it is necessary to analyze many RVE.

P. Fedelinski (✉)
Department of Strength of Materials and Computational Mechanics, Silesian University of Technology, Konarskiego 18A, 44-100 Gliwice, Poland
e-mail: Piotr.Fedelinski@polsl.pl

J. Murín et al. (eds.), *Computational Modelling and Advanced Simulations*,
Computational Methods in Applied Sciences 24, DOI 10.1007/978-94-007-0317-9_17,

Defects in materials have the form of cracks, voids and inclusions. They can reduce significantly stiffness and strength of materials. Therefore it is very important to know the influence of these defects on the overall material properties. Additionally, knowledge about the effective properties can be used to estimate density of defects.

One of the aims of micromechanical modeling is the determination of effective material properties of bodies containing microcracks [3, 4]. There are several theoretical approaches to analyze these problems. In the *non-interacting method*, it is assumed that the microcracks are isolated in the initial undamaged material. In the *self-consistent method*, the microcracks are embedded in the effective medium. The influence of interactions is simulated by reducing stiffness of the surrounding material. In the *differential method*, similar to the self-consistent method, one isolated crack is considered in the effective material. The interactions are taken into account through an incremental increase of crack densities and the effective modulus of the matrix is recalculated at each iteration. The interactions cause softening and effective moduli are lower than predicted by the non-interacting method. These theoretical methods usually give accurate results for low crack densities.

More general cases can be considered using numerical methods. Renaud et al. [5] applied the indirect boundary element method (BEM) – the displacement discontinuity method to compute effective moduli of brittle materials weakened by microcracks. Structures with microcracks of different size, location and orientation were investigated. The structures were subjected to tensile and compressive loadings. In the last case an iterative algorithm was used to analyze cracks in contact with friction. The numerical results were compared with theoretical approximations. For microcracks in finite bodies good agreement with the differential method was obtained. Contact with friction decreases the effect of randomly distributed cracks on effective compliance. Huang et al. [6] used the boundary element method and the unit cell method to calculate effective properties of solids with randomly distributed and parallel microcracks. In the BEM the modified fundamental solutions were used. The method does not require integration along the crack surfaces. Traction free cracks were considered. The results agree well with the differential method for low crack densities and with the generalized self-consistent method for high crack densities. Zhan et al. [7] applied series expansions of complex potentials and the superposition technique. The governing equations were solved numerically using the boundary collocation procedure. The effective Young moduli were calculated for randomly distributed and parallel cracks. The results were compared with various micromechanical models and experimental results. For randomly oriented cracks the results agree with the differential method and for parallel cracks the moduli are below the non-interacting solution and above the differential results.

Su et al. [8] used the finite element method (FEM) to analyze the influence of the applied tensile and compressive load, the angular distribution of randomly located cracks and the wavelength on the propagating wave speed and the output signal strength.

The effective moduli were also determined experimentally by Carvalho and Labuz [9]. Artificially cracked aluminum plates with randomly distributed slots were subjected to tension. The experimental results agree well with the non-interacting approximation even for high density of cracks, where interactions are expected to occur. The authors found that the number of slots should be about 20 to guarantee the randomness of distribution.

Structures with high density of cracks have small Young moduli, however there is no direct quantitative correlation between the stiffness and damage [1]. Local positions of microcracks have strong influence on stress intensity factors while the effective modulus, which is a volume average quantity, is insensitive to such distributions.

The aim of this work is to apply another BEM approach, called the dual boundary element method (DBEM). This method is used not only to calculate effective Young's modulus but also to compute overall Poisson's ratio, stress intensity factors (SIF) of cracked materials and the influence of microcracks on waves traveling through the body.

17.2 Effective Properties of Solids with Microcracks

The effective properties of solids with microcracks are computed by using square representative volume elements subjected to uniformly distributed tractions. The average strains in two directions are computed and used to calculate the effective Young modulus and Poisson ratio. The influence of density of microcracks on overall properties is investigated. The density of microcracks is defined as

$$\rho = \frac{1}{A}\sum_{i=1}^{n} a_i^2 \tag{17.1}$$

where A is the area of the RVE, n is the number of microcracks and a_i is the half-length of the microcrack [3]. The effective Young moduli are compared with the results given by analytical methods [3, 5].

17.3 The Boundary Element Method for Crack Problems

In the present work two-dimensional, linear-elastic, isotropic and homogenous solids with microcracks are analyzed using the dual boundary element method (DBEM) [10, 11]. In this approach only boundaries of the body and crack surfaces are divided into boundary elements. The variations of boundary coordinates, displacements and tractions are interpolated using shape functions and nodal values. In the DBEM the relations between boundary displacements and tractions are expressed by the displacement and traction boundary integral equations. The displacement equation is applied for nodes on the external boundary and both equations for nodes on crack surfaces. The time dependent problems are solved using the Laplace transform method.

In the DBEM the boundary displacements and tractions are computed directly. Stress intensity factors (SIF) are calculated very accurately using path independent integrals. The DBEM is very efficient method for analysis of effective properties of solids with multiple microcracks. It is very easy to generate solids with randomly distributed microcracks because in this approach only crack surfaces are discretized.

17.4 Numerical Results

17.4.1 Representative Volume Elements Subjected to Static Loading

A square representative volume element (RVE), which contains 20 microcracks, having the same lengths and random positions, is investigated (Fig. 17.1a). The length of the edges of the RVE is l, the length of each crack is $2a$ and the minimal distance between the microcracks and also between the external boundary is b (Fig. 17.1b). This distance is such that $b/l = 0.05$. The RVE is made of linear elastic, isotropic and homogenous material in plane stress with the Poisson ratio $\nu = 0.3$. The RVE is simply supported and subjected to uniformly distributed static tractions t acting in the horizontal or vertical direction. The boundaries of the RVE and the cracks are divided into 400 quadratic boundary elements.

The effective properties are calculated for 5 different densities, 5 different random distributions of microcracks and 2 directions of loading. The normalized lengths of microcracks are $a/l = 0.025, 0.050, 0.075, 0.100$ and 0.125 and the corresponding densities are $\rho = 0.0125, 0.0500, 0.1125, 0.2000$ and 0.3125. For each RVE the relative Young modulus E/E_o and the relative Poisson ratio ν/ν_o, where E_o and ν_o are properties of the undamaged material, are computed and shown as a function of density of microcracks. The numerical results are approximated by the third order polynomial.

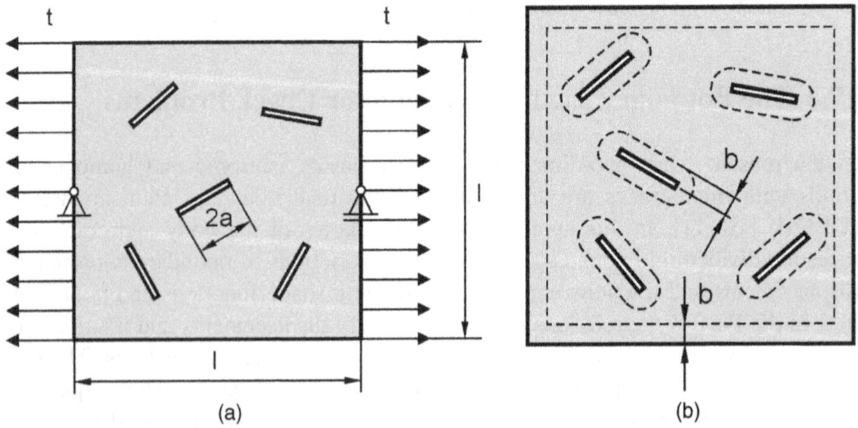

Fig. 17.1 Representative volume element with microcracks subjected to static tension (**a**) dimensions and boundary conditions, (**b**) minimal distance between microcracks and the external boundary

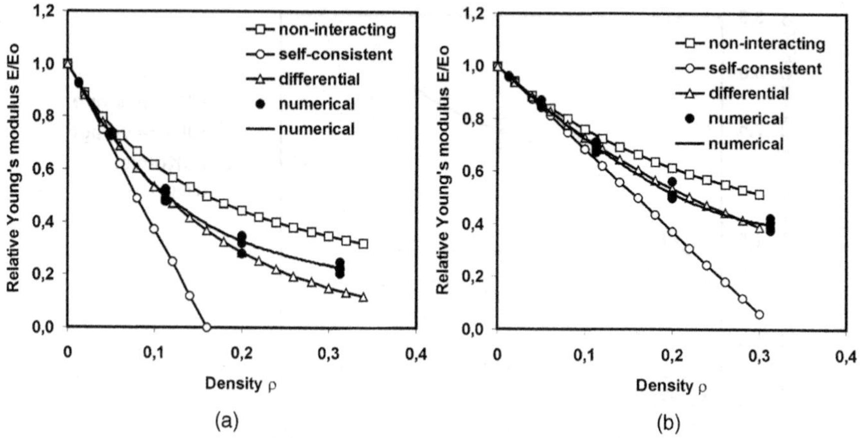

Fig. 17.2 Relative Young's modulus versus density of microcracks – comparison of numerical with analytical results (**a**) parallel microcracks, (**b**) random microcracks

The relative Young moduli computed numerically are compared with the results given by the non-interacting, the self-consistent and the differential method in Fig. 17.2.

The numerical results agree well with the solutions obtained by using the differential method. The relative Young moduli for parallel microcracks are smaller than for random directions of microcracks. The locations and directions of microcracks, and the direction of loading have strong influence on E/E_o for high density of microcracks. Therefore the results for the same density are different.

The relative Poisson ratio is shown in Fig. 17.3. The locations and directions of microcracks and the direction of loading have strong influence on ν/ν_o for high density of microcracks. The relative Poisson ratios are not shown for parallel

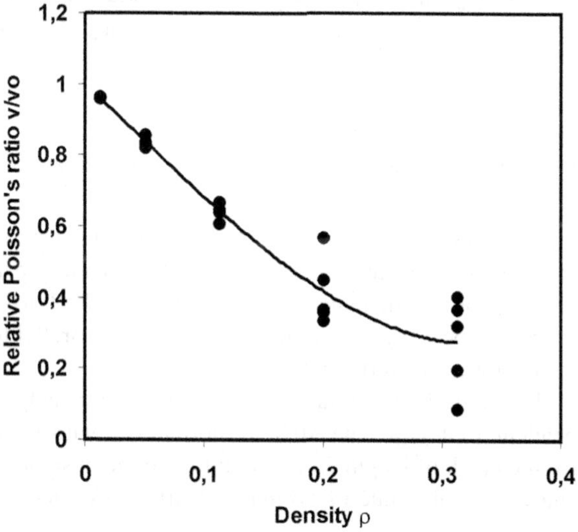

Fig. 17.3 Relative Poisson's ratio versus density of microcracks

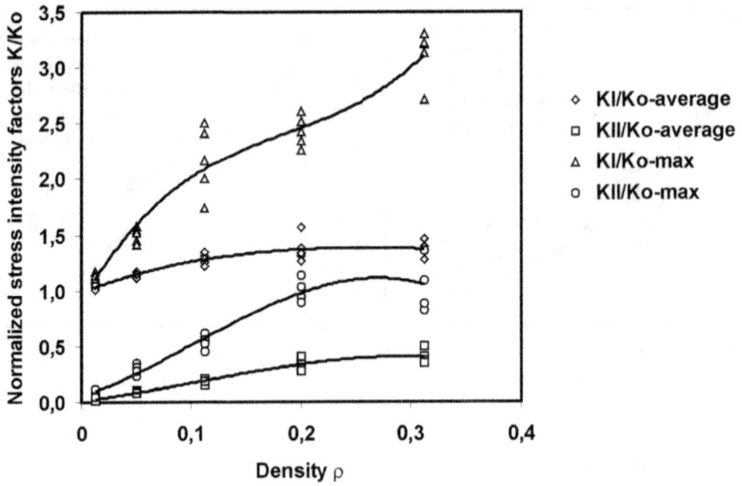

(a)

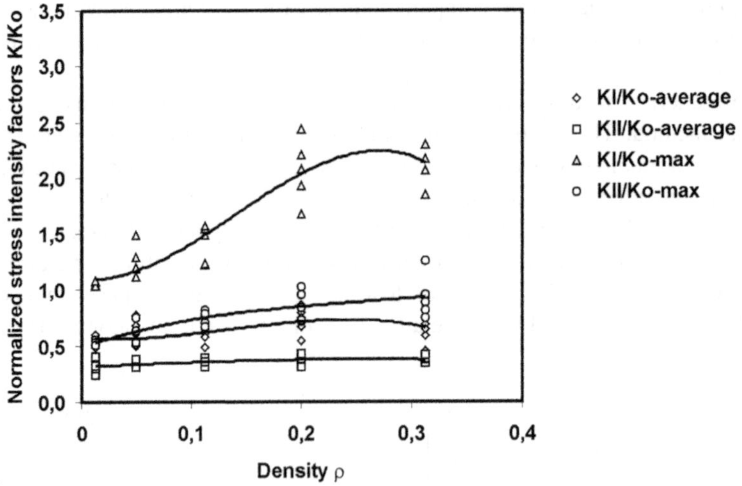

(b)

Fig. 17.4 Normalized stress intensity factors (**a**) parallel microcracks, (**b**) random microcracks

microcracks because for the loading perpendicular to the cracks they are similar to the relative Young modulus and for the loading parallel to the cracks they are the same as for the material without cracks. The Poisson ratio has these particular properties because the strains in the direction parallel to microcracks are similar to strains for the material without cracks.

The stress intensity factors are computed for each crack tip using the path independent J-integral. The SIFs are normalized with respect to K_o, which is the SIF for the crack of length $2a$ in the infinite domain subjected to tension t. The average and maximum value of normalized SIF K_I/K_o and absolute value of K_{II}/K_o are computed for each RVE. The results are shown in Fig. 17.4 for parallel and random directions of microcracks. Generally the SIFs for parallel cracks, perpendicular to the applied loading, are higher than for the random directions of microcracks. For high density of microcracks there is a small influence of density on average values of SIFs. The average SIFs K_I/K_o are higher than the average K_{II}/K_o.

17.4.2 Representative Volume Elements Subjected to Dynamic Loading

The same representative volume element, which was considered in Section 17.4.1, is subjected to dynamic loading. The left edge is constrained and the right edge is subjected to uniformly distributed tractions with the Heaviside time dependence (Fig. 17.5).

The case of randomly distributed parallel microcracks, perpendicular to the applied loading, is considered. The displacements of corner points B and D are computed and normalized with respect to the static displacement u_o. The time is normalized with respect to t_o, which is the time needed by the wave to travel from the loaded edge to the constrained edge for a one-dimensional case. The normalized factors are calculated using the following equations

$$u_o = \frac{tl}{E_o}, \quad t_o = \frac{l}{c_o}, \tag{17.2}$$

where the velocity of the wave is

$$c_o = \sqrt{\frac{E_o}{\rho}}, \tag{17.3}$$

and ρ is mass density.

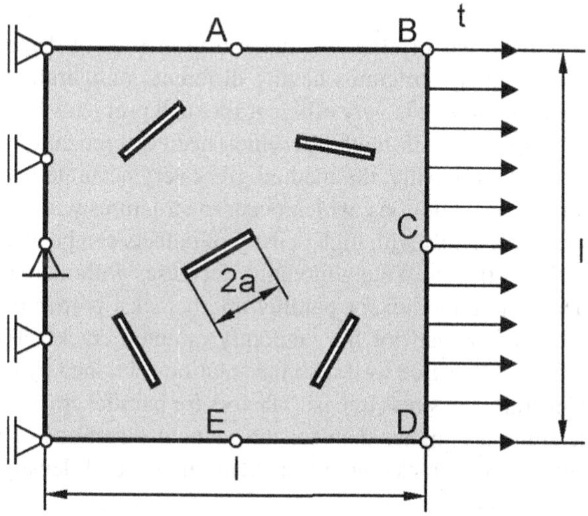

Fig. 17.5 Representative volume element with microcracks subjected to dynamic tension – dimensions and boundary conditions

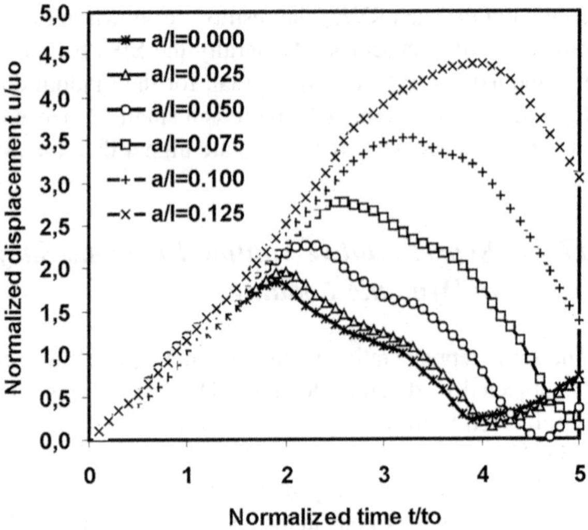

Fig. 17.6 Average normalized horizontal displacements of the corner points B and D versus normalized time

The normalized average horizontal displacements of corner points B and D versus normalized time for different lengths of microcracks are shown in Fig. 17.6. The displacements for small microcracks are similar to the displacements of the RVE without cracks. The plots for initial time (normalized time smaller than 2) are independent of density of microcracks. The maximum value of displacements and the corresponding time increase linearly with the density of microcracks and these values for the highest considered density are more than two times larger than for the RVE without cracks.

17.5 Conclusions

Computations of effective material properties of damaged structures require analysis of many structures having different configurations of defects. The boundary element method is very efficient for such problems because modification of defects is simpler than in methods, which need discretization of the whole domain of the body. Additionally, the method gives very accurate results for structures with high stress concentrations, which occur in structures with defects. Contrary to analytical methods, solids with high density of defects can be analyzed accurately.

The effective Young modulus decreases with increasing density of microcracks. The Young moduli for parallel microcracks, perpendicular to the applied tension, are smaller than for the randomly oriented cracks. The Young moduli computed numerically agree well with the solutions obtained by using the differential method. Generally the stress intensity factors for parallel cracks, perpendicular to the applied loading, are higher than for the random directions of microcracks. For high density of microcracks there is a small influence of density on average values of stress intensity factors. The average stress intensity factors for mode I are higher than the average values for mode II.

The maximum value of displacements and the corresponding time for representative volume elements subjected to impact loads increases linearly with increasing density of microcracks.

Acknowledgments The scientific research is financed by the Ministry of Science and Higher Education of Poland in years 2010–2012.

References

1. Nemat-Nasser, S., Hori, M.: Micromechanics: Overall Properties of Heterogeneous Materials. Elsevier, Amsterdam (1993)
2. Kouznetsova, V.G.: Computational Homogenization for the Multi-scale Analysis of Multi-phase Materials. Technische Universiteit Eindhoven, Eindhoven (2002)
3. Kachanov, M.: Effective elastic properties of cracked solids: critical review of some basic concepts. Appl. Mech. Rev. **45**, 304–335 (1992)
4. Kachanov, M.: Solids with cracks and non-spherical pores: proper parameters of defect density and effective elastic properties. Int. J. Fract. **97**, 1–32 (1999)
5. Renaud, V., Kondo, D., Henry, J.P.: Computations of effective moduli for microcracked materials: a boundary element approach. Comput. Mater. Sci. **5**, 227–237 (1996)
6. Huang, Y., Chandra, A., Jiang, Z.Q., Wei, X., Hu, K.X. The numerical calculation of two-dimensional effective moduli for microcracked solids. Int. J. Solids Struct. **33**, 1575–1586 (1996)
7. Zhan, S., Wang, T., Han, X.: Analysis of two-dimensional finite solids with microcracks. Int. J. Solids Struct. **36**, 3735–3753 (1999)
8. Su, D., Santare, M.H., Gazonas, G.A.: An effective medium model for elastic waves in microcrack damaged media. Eng. Fract. Mech. **75**, 4104–4116 (2008)
9. Carvalho, F.C.S., Labuz, J.F.: Experiments on effective elastic modulus of two-dimensional solids with cracks and holes. Int. J. Solids Struct. **33**, 4119–4130 (1996)
10. Portela, A., Aliabadi, M.H., Rooke, D.P.: The dual boundary element method: effective implementation for crack problems. Int. J. Num. Method Eng. **33**, 1269–1287 (1992)
11. Fedelinski, P., Aliabadi, M.H., Rooke, D.P.: The Laplace transform DBEM method for mixed-mode dynamic crack analysis. Comput. Struct. **59**, 1021–1031 (1996)

Chapter 18
Application of General Boundary Element Method for Numerical Solution of Bioheat Transfer Equation

E. Majchrzak

Abstract The heat transfer processes proceeding in domain of living tissue are discussed. The typical model of bioheat transfer bases, as a rule, on the well known Pennes equation (heat diffusion equation with additional terms corresponding to the perfusion and metabolic heat sources). Here, the other approach basing on the dual-phase-lag equation (DPLE) is considered. This equation is supplemented by the adequate boundary and initial conditions. To solve the problem the general boundary element method is adapted. The examples of computations for 2D problem are presented in the final part of the paper.

Keywords Bioheat transfer · Dual-phase lag model · Numerical methods · Boundary element method

18.1 Introduction

Knowledge of heat transfer in living tissue is very important in the planning of therapeutic treatments such as laser irradiation [1–4], cryosurgery [5–8] or electromagnetic field therapy (magnet therapy) [9–11]. The mathematical description of thermal processes proceeding in domain of biological tissue bases on the different bioheat transfer models. The most popular is the Pennes model [12] and it is very often used, among others, for numerical modelling of tissue freezing [5–8], the burns of skin tissue [13–16] and the thermal parameter estimation [17–20]. The Pennes bioheat equation bases on the classical Fourier's law with its assumption of instantaneous thermal propagation. Cattaneo and Vernotte [21, 22] proposed a thermal wave model with a relaxation time τ_q that is required for a heat flux to respond to the temperature gradient and then the thermal wave propagates through the medium with a finite speed C ($C^2 = a/\tau_q$, where a is the thermal diffusivity).

E. Majchrzak (✉)
Department for Strength of Materials and Computational Mechanics, Silesian University of Technology, Konarskiego 18a, 44-100 Gliwice, Poland
e-mail: ewa.majchrzak@polsl.pl

J. Murín et al. (eds.), *Computational Modelling and Advanced Simulations*, Computational Methods in Applied Sciences 24, DOI 10.1007/978-94-007-0317-9_18,

Because the living tissues are highly nonhomogeneous and accumulating energy to transfer to the nearest elements would take time, therefore the Cattaneo-Vernotte equation in comparison with the Pennes one describes more exactly the heat transfer proceeding in the biological tissue [23–25]. Although the thermal wave model takes into account the microscale response in time [24], the wave concept does not explain the microscale response in space. Additionally, the thermal wave model gives sometimes unusual physical solutions. To consider the effect of microstructural interactions in the fast transient process of heat transport, a phase lag τ_T for temperature gradient is introduced. In this way the dual-phase lag (DPL) heat conduction model is considered in which two time delays τ_q, τ_T (phase lags) appear [4, 24, 26].

From the mathematical point of view the DPLE is most general because for $\tau_T = 0$ it reduces to the Cattaneo-Vernotte equation, while for $\tau_q = \tau_T = 0$ it reduces to the Pennes equation.

In this work, the general boundary element method for hyperbolic heat conduction equation proposed by Liao [27–29] is adapted to solve the dual-phase-lag equation. It should be pointed out that for the DPLE the corresponding fundamental solution is either unknown or very difficult to obtain. So, the concept for solving the unsteady non-linear differential equation using a direct boundary element approach based on homotopy analysis method [27–29] is applied. For the simplicity, the heat transfer in a 2D rectangle with constant thermophysical parameters of tissue is considered. To verify the algorithm proposed, the results of computations under the assumption that $\tau_q = \tau_T = 0$ have been compared with the 1st scheme of the BEM [30, 31].

18.2 Bioheat Transfer Models

The general bioheat transfer equation is given as

$$c\frac{\partial T(x,t)}{\partial t} = -\nabla \mathbf{q}(x,t) + Q(x,t) \tag{18.1}$$

where c is the volumetric specific heat of tissue, $\mathbf{q}(x,t)$ is the heat flux, $Q(x,t)$ is the source term due to metabolism and blood perfusion, T is the temperature, x are the spatial co-ordinates and t is the time.

Heat transfer in biological systems is usually described by the equation basing on the classical Fourier law

$$\mathbf{q}(x,t) = -\lambda \nabla T(x,t) \tag{18.2}$$

where λ is the thermal conductivity of tissue and $\nabla T(x,t)$ is the temperature gradient. Introducing (18.2) into (18.1) one obtains the traditional Pennes bioheat transfer equation (for constant λ) [12]

$$c\frac{\partial T(x,t)}{\partial t} = \lambda \nabla^2 T(x,t) + Q(x,t) \tag{18.3}$$

The source term $Q(x, t)$ is of the form

$$Q(x,t) = G_B c_B [T_B - T(x,t)] + Q_m \tag{18.4}$$

where G_B is the blood perfusion rate, c_B is the volumetric specific heat of blood, T_B is the blood temperature and Q_m is the metabolic heat source.

Because the biological tissue is the material with particular nonhomogeneous inner structure therefore the others, modified heat conduction equations, should be taken into account. One of them is the Cattaneo-Vernotte model [21–24] with corresponding equation

$$\mathbf{q}(x,\ t + \tau_q) = -\lambda \nabla T(x,t) \tag{18.5}$$

where $\tau_q = a/C^2$ is defined as the relaxation time, $a = \lambda/c$ is the thermal diffusivity of tissue and C is the speed of thermal wave in the medium.

According to this equation τ_q is the phase-lag in establishing the heat flux and associated conduction through the medium.

Taking into account the first order Taylor expansion for $\mathbf{q}$

$$\mathbf{q}(x,t) + \tau_q \frac{\partial \mathbf{q}(x,t)}{\partial t} = -\lambda \nabla T(x,t) \tag{18.6}$$

one obtains the Cattaneo-Vernotte equation in the following form

$$c\left[\frac{\partial T(x,t)}{\partial t} + \tau_q \frac{\partial^2 T(x,t)}{\partial t^2}\right] = \lambda \nabla^2 T(x,t) + Q(x,t) + \tau_q \frac{\partial Q(x,t)}{\partial t} \tag{18.7}$$

According to the newest publications, e.g. [4, 26] the heat transfer in biological tissues should be described by dual-phase-lag model basing on the assumption that

$$\mathbf{q}(x, t + \tau_q) = -\lambda \nabla T(x, t + \tau_T) \tag{18.8}$$

where τ_q is the relaxation time, τ_T is the thermalization time, τ_q is the phase-lag in establishing the heat flux and associated conduction through the medium, τ_T is the phase-lag in establishing the temperature gradient across the medium during which conduction occurs through its small-scale structures [26].

Taking into account in (18.8) the first order Taylor expansions for $\mathbf{q}$ and T

$$q(x,t) + \tau_q \frac{\partial \mathbf{q}(x,t)}{\partial t} = -\lambda \nabla T(x,t) - \lambda \tau_T \frac{\partial \nabla T(x,t)}{\partial t} \tag{18.9}$$

the following bioheat transfer equation is considered

$$\begin{aligned} & c\left[\frac{\partial T(x,t)}{\partial t} + \tau_q \frac{\partial^2 T(x,t)}{\partial t^2}\right] \\ & = \lambda \nabla^2 T(x,t) + \lambda \tau_T \frac{\partial \nabla^2 T(x,t)}{\partial t} + Q(x,t) + \tau_q \frac{\partial Q(x,t)}{\partial t} \end{aligned} \tag{18.10}$$

It should be pointed out that for $\tau_T = 0$ the dual-phase-lag equation (18.10) reduces to the Cattaneo-Vernotte equation (18.7), while for $\tau_q = \tau_T = 0$ reduces to the Pennes equation (18.3).

18.3 Boundary Element Method for Pennes Equation

From the mathematical point of view the Pennes equation (18.3) is the parabolic one, the Cattaneo-Vernotte equation (18.7) is the hyperbolic one, while the DPL equation (18.10) contains a second order time derivative and higher order mixed derivative in both time and space.

To solve the Pennes equation by means of the boundary element method the several variants basing on a time marching technique have been applied. The principle of this technique is to use the solution obtained for time t^{f-1} as a starting point to get the solution for the time t^f.

In the case of the 1st scheme of the BEM [30, 31] the boundary integral equation corresponding to the Pennes equation (18.3) has the following form

$$B(\xi)T(\xi,t^f) + \frac{1}{c}\int_{t^{f-1}}^{t^f}\int_\Gamma T^*(\xi,x,t^f,t)q(x,t)\mathrm{d}\Gamma\,\mathrm{d}t$$

$$= \frac{1}{c}\int_{t^{f-1}}^{t^f}\int_\Gamma q^*(\xi,x,t^f,t)T(x,t)\mathrm{d}\Gamma\,\mathrm{d}t + \iint_\Omega T^*(\xi,x,t^f,t^{f-1})T(x,t^{f-1})\mathrm{d}\Omega$$

$$+\frac{1}{c}\int_{t^{f-1}}^{t^f}\iint_\Omega Q(x,t)T^*(\xi,x,t^f,t^{f-1})\mathrm{d}\Omega\mathrm{d}t \tag{18.11}$$

where ξ is the observation point, $B(\xi) \in (0,1]$ is the coefficient depending on the location of the point ξ, $T^*(\xi,x,t^f,t)$ is the time-dependent fundamental solution,

$q^*(\xi,x,t^f,t) = -\lambda n \cdot \nabla T^*(\xi,x,t^f,t)$ and $q(x,t) = -\lambda n \cdot \nabla T(x,t)$ where n is unit normal outward to the boundary surface.

The fundamental solution for domain oriented in Cartesian co-ordinate system has the following form

$$T^*(\xi,x,t^f,t) = \frac{1}{[4\pi a(t^f-t)]^{m/2}}\exp\left[-\frac{r^2}{4a(t^f-t)}\right] \tag{18.12}$$

where m is the problem dimension, $a = \lambda/c$ is the thermal diffusivity and r is the distance between the observation point ξ and field point x.

Function $T^*(\xi,x,t^f,t)$ fulfills the following equation

$$a\nabla^2 T^*(\xi,x,t^f,t) + \frac{\partial T^*(\xi,x,t^f,t)}{\partial t} = -\delta(\xi,x)\delta(t^f,t) \tag{18.13}$$

where $\delta(\cdot,\cdot)$ is the Dirac function.

This variant of the BEM supplemented by adequate iterative procedure of source function $Q(x, t^f)$ determination (c.f. (18.4)) has been successfully applied in the scope of bioheat transfer modelling [6–8, 13–15, 18, 19, 32].

Another approach consists in the following approximation of (18.3) [30, 31]

$$c\frac{T(x,t^f) - T(x,t^{f-1})}{\Delta t} = \lambda \nabla^2 T(x,t^f) + Q(x,t^f) \tag{18.14}$$

or

$$\nabla^2 T(x,t^f) - \frac{1}{a\Delta t} T(x,t^f) + \frac{1}{a\Delta t} T(x,t^{f-1}) + \frac{1}{\lambda} Q(x,t^f) = 0 \tag{18.15}$$

where $\Delta t = t^f - t^{f-1}$ is the time step.

In this case the following boundary integral equation can be derived

$$\begin{aligned} B(\xi)T(\xi,t^f) + \frac{1}{\lambda}\int_\Gamma T^*(\xi,x)q(x,t^f)\mathrm{d}\Gamma &= \frac{1}{\lambda}\int_\Gamma q^*(\xi,x)T(x,t^f)\mathrm{d}\Gamma \\ &+ \frac{1}{a\Delta t}\iint_\Omega T(x,t^{f-1})T^*(\xi,x)\mathrm{d}\Omega + \frac{1}{\lambda}\iint_\Omega Q(x,t^f)T^*(\xi,x)\mathrm{d}\Omega \end{aligned} \tag{18.16}$$

where $T^*(\xi,x)$ is the fundamental solution and it is independent of time, namely

$$T^*(\xi,x) = \begin{cases} \dfrac{\sqrt{a\Delta t}}{2}\exp\left(-\dfrac{r}{\sqrt{a\Delta t}}\right) & \text{for 1D problem} \\ \dfrac{1}{2\pi}\mathrm{K}_0\left(\dfrac{r}{\sqrt{a\Delta t}}\right) & \text{for 2D problem} \\ \dfrac{1}{4\pi r}\exp\left(-\dfrac{r}{\sqrt{a\Delta t}}\right) & \text{for 3D problem} \end{cases} \tag{18.17}$$

while $\mathrm{K}_0(\cdot)$ is the modified Bessel function of the second kind of order zero [30, 31].

Function $T^*(\xi,x)$ fulfills the equation

$$\nabla^2 T^*(\xi,x) - \frac{1}{a\Delta t} T^*(\xi,x) = -\delta(\xi,x) \tag{18.18}$$

This variant is called the BEM using discretization in time and also has been applied in the case of bioheat transfer problems modelling [31].

The dual reciprocity boundary element method (DRBEM) allows one to obtain the integral equation in which only the boundary integrals appear [33]. The (18.3) can be written in the form

$$\lambda \nabla^2 T(x,t^f) = S(x,t^f) \tag{18.19}$$

where

$$S(x,t^f) = c\frac{T(x,t^f) - T(x,t^{f-1})}{\Delta t} - Q(x,t^f) \tag{18.20}$$

The standard boundary element method for the Poisson equation (18.14) leads to the following integral equation [30, 31]

$$\begin{aligned} B(\xi)T(\xi,t^f) + \int_\Gamma T^*(\xi,x)q(x,t^f)\mathrm{d}\Gamma &= \int_\Gamma q^*(\xi,x)T(x,t^f)\mathrm{d}\Gamma \\ &- \iint_\Omega S(x,t^f)T^*(\xi,x)\mathrm{d}\Omega \end{aligned} \tag{18.21}$$

where $T^*(\xi,x)$ is the fundamental solution corresponding to the Laplace equation

$$\lambda\nabla^2 T^*(x,t^f) = -\delta(\xi,x) \tag{18.22}$$

this means

$$T^*(\xi,x) = \begin{cases} \dfrac{r}{2\lambda} & \text{for 1D problem} \\ \dfrac{1}{2\pi\,\lambda}\ln\dfrac{1}{r} & \text{for 2D problem} \\ \dfrac{1}{4\pi\,\lambda\,r} & \text{for 3D problem} \end{cases} \tag{18.23}$$

In the dual reciprocity method the following approximation of function $S(x,t^f)$ is proposed [33]

$$S(x,t^f) = \sum_{k=1}^{N+L} a_k(t^f)P_k(x) \tag{18.24}$$

where $a_k(t^f)$ are unknown coefficients, $P_k(x)$ are approximating functions fulfilling the equations

$$P_k(x) = \lambda\nabla^2 U_k(x) \tag{18.25}$$

In (18.24) $N+L$ corresponds to the total number of nodes, where N is the number of boundary nodes and L is the number of internal collocation points.

Using the Green second identity formula [30, 31] to the last component of (18.21) one obtains the dual-reciprocity boundary integral equation

$$\begin{aligned} B(\xi)T(\xi,t^f) + \int_\Gamma T^*(\xi,x)q(x,t^f)\mathrm{d}\Gamma &= \int_\Gamma q^*(\xi,x)\,T(x,t^f)\mathrm{d}\Gamma \\ + \sum_{k=1}^{N+L} a_k(t^f)\left[B(\xi)\,U_k(\xi) + \int_\Gamma T^*(\xi,x)\,W_k(x)\mathrm{d}\Gamma - \int_\Gamma q^*(\xi,x)\,U_k(x)\mathrm{d}\Gamma\right] \end{aligned} \tag{18.26}$$

It should be pointed out that this equation can be successfully solved under the assumption that the functions $U_k(x)$ are known [33], while $W_k(x) = -\lambda n \cdot \nabla U_k(x)$.

This variant of the BEM has been applied among others in [34, 3].

18.4 General Boundary Element Method for DPL Equation

Application of the traditional BEM for solving the Cattaneo-Vernotte equation (18.7) or DPL equation (18.10) is more complicated. The fundamental solution is difficult to obtain, thus the equivalent boundary integral equation is not available. It should be pointed out that in literature [35] one can find the solutions of Cattaneo-Vernotte equation obtained by means of the DRBEM but the DPL equation, up to the present, is not solved using the BEM.

In this work, the general boundary element method (GBEM) for hyperbolic heat conduction equation proposed by Liao [27–29] is adapted to solve the dual-phase -lag equation.

Taking into account the form (18.4) of source function the (18.10) can be expressed as follows (2D problem is considered)

$$(x, y) \in \Omega : \quad c\left(\frac{\partial T}{\partial t} + \tau_q \frac{\partial^2 T}{\partial t^2}\right) = \lambda \nabla^2 T + \lambda \tau_T \frac{\partial}{\partial t}(\nabla^2 T) + k(T_B - T) + Q_m - k\tau_q \frac{\partial T}{\partial t} \tag{18.27}$$

where $k = G_B c_B$ and $T = T(x, y, t)$.

This equation is supplemented by boundary conditions

$$\begin{aligned} (x, y) \in \Gamma_1 : &\quad T(x, y, t) = T_b(x, y, t) \\ (x, y) \in \Gamma_2 : &\quad -\lambda \frac{\partial T(x, y, t)}{\partial n} = w_b(x, y, t) \end{aligned} \tag{18.28}$$

and initial ones

$$t = 0 : \quad T(x, y, t) = T_p, \quad \left.\frac{\partial T(x, y, t)}{\partial t}\right|_{t=0} = 0 \tag{18.29}$$

where $T_b(x, y, t)$ is known boundary temperature, $w_b(x, y, t)$ is known function, T_p is the initial tissue temperature and $\partial T(x, y, t)/\partial n$ denotes the normal derivative.

Let $\beta = 1/\Delta t$ and $T^f = T(x, y, f\Delta t)$ where Δt is the time step. Then, for time $t^f = f\Delta t$ ($f \geq 2$) the following approximate form of (18.27) can be taken into account

$$\begin{aligned} &c\,[\beta\,(T^f - T^{f-1}) + \tau_q\,\beta^2 (T^f - 2T^{f-1} + T^{f-2})] = \lambda \nabla^2 T^f \\ &+\lambda \tau_T\,\beta\,(\nabla^2 T^f - \nabla^2 T^{f-1}) + k(T_B - T^f) + Q_m - k\tau_q \beta (T^f - T^{f-1}) \end{aligned} \tag{18.30}$$

or

$$\nabla^2 T^f - BT^f + C\nabla^2 T^{f-1} + DT^{f-1} + ET^{f-2} + F = 0 \tag{18.31}$$

where

$$\begin{aligned} B &= \frac{(c\beta + k)(1 + \tau_q \beta)}{\lambda(1 + \tau_T \beta)} \\ C &= -\frac{\tau_T \beta}{1 + \tau_T \beta} \\ D &= \frac{c\,\beta(1 + 2\tau_q \beta) + k\tau_q \beta}{\lambda(1 + \tau_T \beta)} \\ E &= -\frac{c\tau_q \beta^2}{\lambda(1 + \tau_T \beta)} \\ F &= kT_b + Q_m \end{aligned} \tag{18.32}$$

The boundary conditions (18.28) have the form

$$\begin{aligned} (x, y) \in \Gamma_1 : \quad & T^f = T_b(x, y, t^f) \\ (x, y) \in \Gamma_2 : \quad & -\lambda \frac{\partial T^f}{\partial n} = w_b(x, y, t^f) \end{aligned} \tag{18.33}$$

From initial conditions (18.29) results that $T(x, y, 0) = T_p$ and $T(x, y, \Delta t) = T_p$.

At first, a family of partial differential equations for $\Phi(x, y; p)$ is constructed [27–29]

$$(1 - p)\mathrm{L}[\Phi(x, y; p) - U(x, y)] = -p\mathrm{A}[\Phi(x, y; p)], \quad p \in [0, 1] \tag{18.34}$$

with boundary conditions

$$\begin{aligned} (x, y) \in \Gamma_1 : \quad & \Phi(x, y; p) = pT_b(x, y, t^f) + (1 - p)\, U(x, y) \\ (x, y) \in \Gamma_2 : -\lambda \frac{\partial \Phi(x, y; p)}{\partial n} & = pw_b(x, y, t^f) + (1 - p) \left[-\lambda \frac{\partial U(x, y)}{\partial n} \right] \end{aligned} \tag{18.35}$$

where p is a parameter, $U(x, y)$ is an initial approximation of temperature distribution T^f, L is a 2D linear operator whose fundamental solution is known and A is a non-linear operator. The linear operator is the following

$$\mathrm{L}(u) = \nabla^2 u - Bu = \frac{\partial^2 u}{\partial x^2} + \frac{\partial^2 u}{\partial y^2} - Bu \tag{18.36}$$

while the non-linear operator takes a form (c.f. (18.31))

$$\mathrm{A}[\Phi(x, y; p)] = \nabla^2 \Phi - B\,\Phi + C\nabla^2 T^{f-1} + DT^{f-1} + ET^{f-2} + F \tag{18.37}$$

For $p = 0$ one obtains (c.f. (18.34))

$$\mathrm{L}[\Phi(x, y; 0) - U(x, y)] = 0 \tag{18.38}$$

this means

$$\mathrm{L}[\Phi(x, y; 0)] = \mathrm{L}[U(x, y)] \tag{18.39}$$

and (c.f. conditions (18.35))

$$\begin{aligned} (x, y) \in \Gamma_1 : \quad & \Phi(x, y; 0) = U(x, y) \\ (x, y) \in \Gamma_2 : -\lambda \frac{\partial \Phi(x, y; 0)}{\partial n} & = -\lambda \frac{\partial U(x, y)}{\partial n} \end{aligned} \tag{18.40}$$

The solution of (18.39) and (18.40) is obvious

$$\Phi(x, y; 0) = U(x, y) \tag{18.41}$$

On the other hand, for $p = 1$ one has

$$\mathrm{A}[\Phi(x, y; 1)] = 0 \tag{18.42}$$

and

$$\begin{aligned} (x, y) \in \Gamma_1 : \quad & \Phi(x, y; 1) = T_b(x, y, t^f) \\ (x, y) \in \Gamma_2 : -\lambda \frac{\partial \Phi(x, y; 1)}{\partial n} & = w_b(x, y, t^f) \end{aligned} \tag{18.43}$$

Taking into account the form of operator A ((18.37) it is visible that

$$\Phi(x, y; 1) = T^f = T(x, y, t^f) \tag{18.44}$$

Thus, if $p = 0$ then $\Phi(x, y; p)$ corresponds to the initial approximation $U(x, y)$, while if $p = 1$ then $\Phi(x, y; p)$ corresponds to the unknown temperature $T^f = T(x, y, T^f)$. So, the (18.34) and (18.35) form a family of equations in parameter $p \in [0, 1]$ and the process of continuous change of the parameter p from 0 to 1 is the process of continuous variation of solution $\Phi(x, y; p)$ from $U(x, y)$ to $T^f = T(x, y, t^f)$.

Function $\Phi(x, y; p)$ is expanded into a Taylor series about value $p = 0$ taking into account the first derivative

$$\Phi(x, y; p) = \Phi(x, y; 0) + \left. \frac{\partial \Phi(x, y; p)}{\partial p} \right|_{p=0} (p - 0) \tag{18.45}$$

or

$$\Phi(x, y; p) = \Phi(x, y; 0) + U^{[1]}(x, y) p \tag{18.46}$$

where

$$U^{[1]}(x,y) = \left. \frac{\partial \Phi(x,y;p)}{\partial p} \right|_{p=0} \tag{18.47}$$

For $p = 1$ one obtains

$$\Phi(x,y;1) = \Phi(x,y;0) + U^{[1]}(x,y) \tag{18.48}$$

this means

$$T^f = U(x,y) + U^{[1]}(x,y) \tag{18.49}$$

Under the assumption that $U(x,y) = T^{f-1}$ one has

$$T^f = T^{f-1} + U^{[1]}(x,y) \tag{18.50}$$

Liao and Chwang [36] report that better than initial approximation $U(x,y) = T^{f-1}$ is to use the iterative formula

$$T_k^f = T_{k-1}^f + mU^{[1]}(x,y), \quad k = 1,2,3,...,K \tag{18.51}$$

where $T_0^f = T^{f-1}$, m is an iterative parameter and K is the number of iterations.

Differentiation of (18.34) and (18.35) with respect to parameter p gives

$$\begin{gathered} -\mathrm{L}[\Phi(x,y;p) - U(x,y)] + (1-p)\mathrm{L}\left[\frac{\partial \Phi(x,y;p)}{\partial p} - \frac{\partial U(x,y)}{\partial p}\right] = \\ -\mathrm{A}[\Phi(x,y;p)] - p\frac{\partial \mathrm{A}[\Phi(x,y;p)]}{\partial p} \end{gathered} \tag{18.52}$$

and

$$\begin{aligned} (x,y) \in \Gamma_1 : \quad & \frac{\partial \Phi(x,y;p)}{\partial p} = T_b(x,y,t^f) + p\frac{\partial T_b(x,y,t^f)}{\partial p} \\ & -U(x,y) + (1-p)\frac{\partial U(x,y)}{\partial p} \\ (x,y) \in \Gamma_2 : \quad & -\lambda\frac{\partial}{\partial n}\left(\frac{\partial \Phi(x,y;p)}{\partial p}\right) = w_b(x,y,t^f) + p\frac{\partial w_b(x,y,t^f)}{\partial p} \\ & +\lambda\frac{\partial U(x,y)}{\partial n} - (1-p)\lambda\frac{\partial}{\partial n}\left[\frac{\partial U(x,y)}{\partial p}\right] \end{aligned} \tag{18.53}$$

For $p = 0$ one has

$$\mathrm{L}[U^{[1]}(x,y)] = -\mathrm{A}[U(x,y)] \tag{18.54}$$

and

$$\begin{aligned} (x,y) \in \Gamma_1: &\qquad U^{[1]}(x,y) = T_b(x,y,t^f) - U(x,y) \\ (x,y) \in \Gamma_2: &-\lambda\left(\frac{\partial U^{[1]}(x,y)}{\partial n}\right) = w_b(x,y,t^f) + \lambda\frac{\partial U(x,y)}{\partial n} \end{aligned} \tag{18.55}$$

Taking into account the form of operators L and A ((18.36) and (18.37)) the (18.54) can be written as follows

$$\nabla^2 U^{[1]} - BU^{[1]} + R(U) = 0 \tag{18.56}$$

where

$$R(U) = \nabla^2 U - B\,U + C\nabla^2 T^{f-1} + DT^{f-1} + ET^{f-2} + F \tag{18.57}$$

One can notice that (18.56) with the corresponding boundary conditions (18.55) is linear with respect to the 1st order derivative $U^{[1]} = U^{[1]}(x,y)$ and it contains the 2D modified Helmholtz operator (18.36) [27] whose fundamental solution is

$$U^* = \frac{1}{2\pi}\mathrm{K}_0(r\sqrt{B}) \tag{18.58}$$

where $\mathrm{K}_0(\cdot)$ is the modified Bessel function of the second kind of order zero and r is the distance between source point (ξ,η) and field point (x, y).

So, the boundary integral equation corresponding to the (18.56) has the following form

$$\begin{aligned} &B(\xi,\eta)U^{[1]}(\xi,\eta) - \int_\Gamma U^*(\xi,\eta,x,y)\frac{\partial U^{[1]}(x,y)}{\partial n}\mathrm{d}\Gamma = \\ &-\int_\Gamma \frac{\partial U^*(\xi,\eta,x,y)}{\partial n}U^{[1]}(x,y)\mathrm{d}\Gamma + \iint_\Omega R(U)U^*(\xi,\eta,x,y)\mathrm{d}\Omega \end{aligned} \tag{18.59}$$

or

$$\begin{aligned} &B(\xi,\eta)U^{[1]}(\xi,\eta) - \frac{1}{\lambda}\int_\Gamma U^*(\xi,\eta,x,y)q^{[1]}(x,y)\mathrm{d}\Gamma \\ &= \frac{1}{\lambda}\int_\Gamma q^*(\xi,\eta,x,y)U^{[1]}(x,y)\mathrm{d}\Gamma + \iint_\Omega R(U)U^*(\xi,\eta,x,y)\mathrm{d}\Omega \end{aligned} \tag{18.60}$$

where

$$q^{[1]}(x,y) = -\lambda \frac{\partial U^{[1]}(x,y)}{\partial n}$$
$$q^*(\xi,\eta,x,y) = -\lambda \frac{\partial U^*(\xi,\eta,x,y)}{\partial n} \tag{18.61}$$

Function $q^*(\xi,\eta,x,y)$ can be calculated in analytical way and then

$$q^*(\xi,\eta,x,y) = \frac{\lambda d\sqrt{B}}{2\pi\, r} \mathrm{K}_1(r\sqrt{B}) \tag{18.62}$$

where $\mathrm{K}_1(\cdot)$ is the modified Bessel function of the second kind of order one and

$$d = (x-\xi)\cos\alpha + (y-\eta)\cos\beta \tag{18.63}$$

while $\cos\alpha$, $\cos\beta$ are directional cosines of normal vector n.

In numerical realization the traditional BEM is used to solve the (18.60). So, the boundary Γ is divided into N boundary elements, while the interior Ω is divided into L internal cells. For constant boundary elements and constant internal cells one obtains the following approximation of (18.60)

$$\sum_{j=1}^{N} G_{ij} q_j^{[1]} = \sum_{j=1}^{N} H_{ij} U_j^{[1]} + \sum_{j=1}^{L} P_{i\,l} R(U_l) \tag{18.64}$$

where

$$G_{ij} = \frac{1}{\lambda} \int_{\Gamma_j} U^*(\xi_i,\eta_i,x,y)\mathrm{d}\Gamma_j \tag{18.65}$$

and

$$H_{ij} = \begin{cases} \int_{\Gamma_j} q^*(\xi_i,\eta_i,x,y)\mathrm{d}\Gamma_j, & i \neq j \\ -0.5, & i = j \end{cases} \tag{18.66}$$

while

$$P_{il} = \iint_{\Omega} U^*(\xi_i,\eta_i,x,y)\mathrm{d}\Gamma_l \tag{18.67}$$

Introducing the boundary conditions (18.55) into the linear algebraic equations (18.64) one obtains the equations for the unknown $q^{[1]}$ on the boundary Γ_1 and $U^{[1]}$ on the boundary Γ_2. After solving the system of (18.64), the values $U^{[1]}$ at the internal points (ξ_i,η_i) are calculated using the formula

$$U_i^{[1]} = \sum_{j=1}^{N} H_{ij} U_j^{[1]} - \sum_{j=1}^{N} G_{ij} q_j^{[1]} + \sum_{j=1}^{L} P_{il} R(U_l) \tag{18.68}$$

Summing up, numerical solution of DPL equation by means of the GBEM is connected with the determination of function $U^{[1]}$ and next for transition $t^{f-1} \to t^f$ the temperature T^f is calculated using the iterative formula (18.51). It should be pointed out that for each iteration the problem described by (18.56) and boundary conditions (18.55) should be solved and then in the place of function U the value T^f_{k-1}(c.f. (18.51)) is introduced.

18.5 Results of Computations

The biological tissue domain of dimensions 0.02×0.04 m (rectangle) is considered. The following values of parameters are assumed: volumetric specific heat of tissue $c = 3\,\text{MW/(m}^3\text{K)}$, thermal conductivity of tissue $\lambda = 0.5$ W/(mK), blood perfusion rate $G_B = 0.002$ 1/s, volumetric specific heat of blood $c_B = 3.9962\,\text{MW/(m}^3\text{K)}$, blood temperature $T_B = 37°\text{C}$, metabolic heat source $Q_m = 245\,\text{W/m}^3$, relaxation time $\tau_q = 15$ s, thermalization time $\tau_T = 10$ s. Initial temperature of tissue equals $T_p = 37°\text{C}$. On the fragment of external surface of tissue ($x = 0$, $g/4 \le y \le 3g/4$, where $g = 0.4$) the Dirichlet condition in the form

$$T_b(y) = \frac{4T_p}{g^2}(2y - g)^2 - \frac{T_{\max}}{g^2}(4y - g)(4y - 3\,g) \tag{18.69}$$

is accepted, where $T_{\max} = 100°\text{C}$ corresponds to the maximum of temperature assumed at the central point of external surface and $T_p = 37°\text{C}$ corresponds to the temperature at the points $y = g/4$, $y = 3\,g/4$ as shown in Fig. 18.1. On the internal

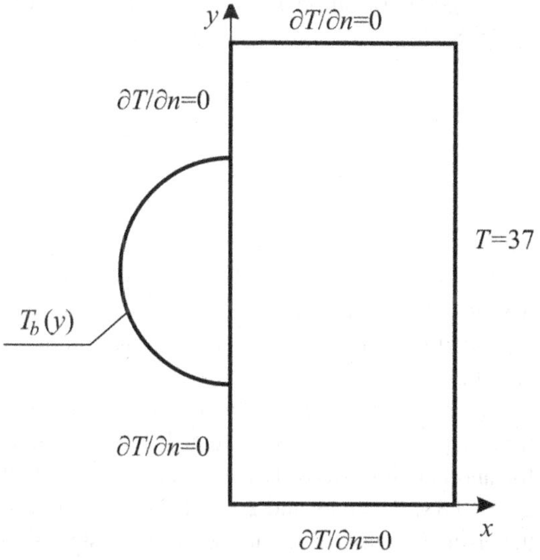

Fig. 18.1 Domain considered

Fig. 18.2 Discretization

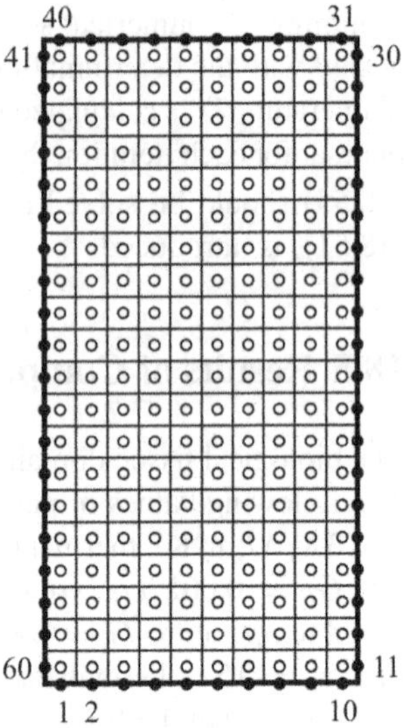

boundary ($x = 0.02$, $0 \leq y \leq 0.04$) the condition $T_b = 37°C$ is assumed, on the remaining boundaries the no-flux condition can be accepted.

The upper and lower boundaries are divided into 20 equal boundary elements, the left and right boundaries are divided into 40 equal boundary elements, so the number of constant boundary elements equals $N = 60$. The interior is divided into $L = 10 \times 20$ constant internal cells (squares) as shown in Fig. 18.2. Time step: $\Delta t = 1$ s, iterative parameter: $m = 0.9$, number of iterations: $K = 15$. In Fig. 18.3 the temperature distribution in the tissue domain for times 10, 20, 30 and 40 s is shown.

For each transition $t^{f-1} \to t^f$ the error of numerical solution is calculated (c.f. (18.56))

$$Er_k = \sqrt{\frac{1}{L^2} \sum_{l=1}^{L} \left[\nabla^2 U_l^{[1]} - BU_l^{[1]} + R(U_l) \right]_k^2} \tag{18.70}$$

where $k = 1, 2, \ldots, K$ is the number of iteration.

It should be pointed out that the number of iterations assuring the assumed exactness (e.g. 10^{-5}) is connected with the value of iterative parameter m. The testing computations show that for $m = 0.9$ the number of iterations equals 17, while for $m = 0.5$ the number of iterations equals 37.

The error Er_{15} after the last iteration under the assumption that $m = 0.9$ for each time step $t^{f-1} \to t^f$ is less than $1.2 \cdot 10^{-5}$. The computations have been also done for another time steps. For example, Fig. 18.4 illustrates the errors for time steps $\Delta t = 0.8$ s, $\Delta t = 1$ s and $\Delta t = 1.2$ s, respectively. From this figure we can see that the errors increase when the size of time step is reduced.

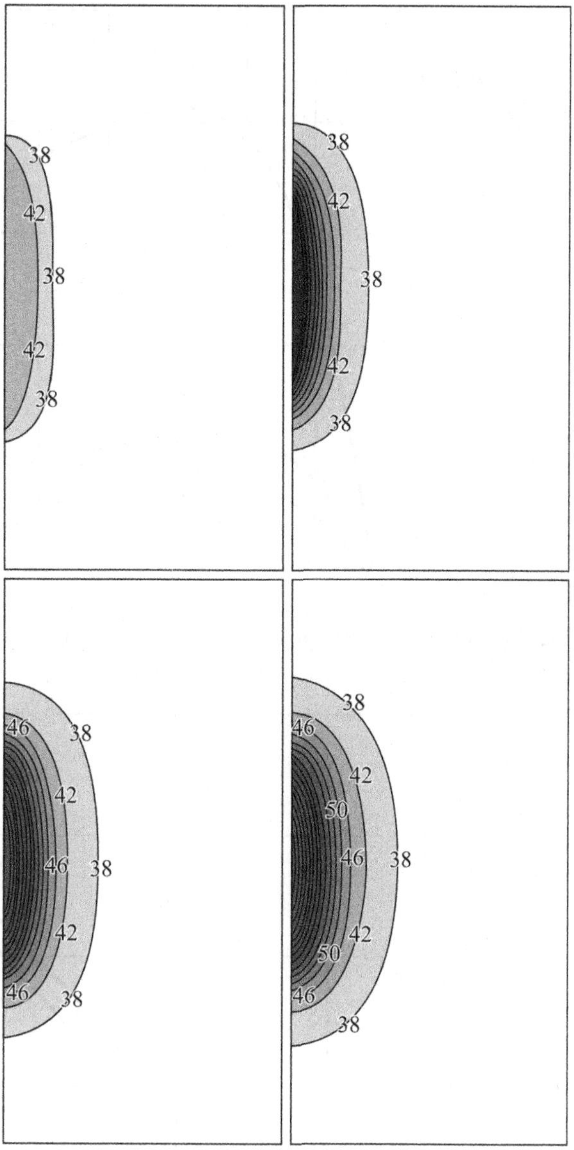

Fig. 18.3 Temperature distribution for $t = 10, 20, 30$ and 40 s

To verify the algorithm proposed, the (18.27) with boundary initial conditions (18.28) and (18.29) under the assumption that $\tau_q = \tau_T = 0$ has been also solved by means of the 1st scheme of the BEM. In Fig. 18.5 the heating curves at the points (0.005, 0.021) and (0.007, 0.021) obtained using the BEM and GBEM are shown. The difference between these solutions is less than 1°C.

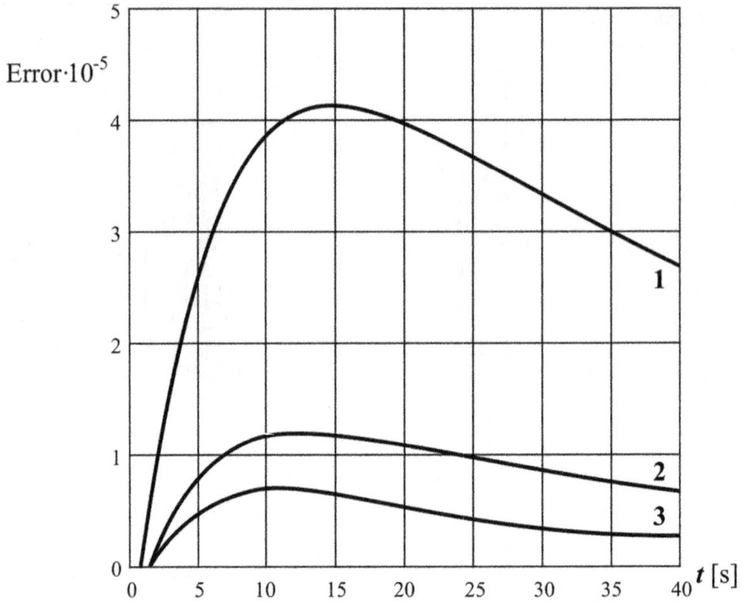

Fig. 18.4 Errors of numerical solutions with different time steps ($1 - \Delta t = 0.8\,\mathrm{s}, 2 - \Delta t = 1\,\mathrm{s}, 3 - \Delta t = 1.2\,\mathrm{s}$)

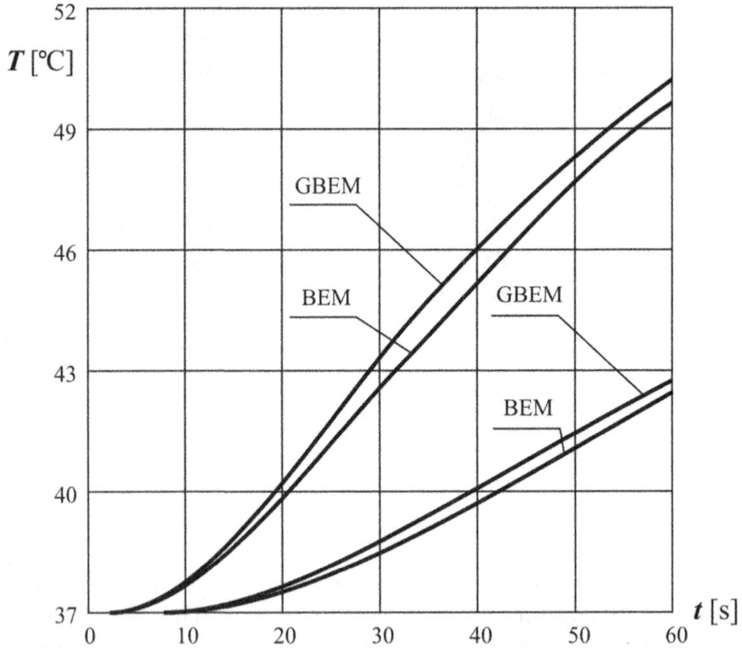

Fig. 18.5 Heating curves – comparison of solutions obtained by the BEM and GBEM

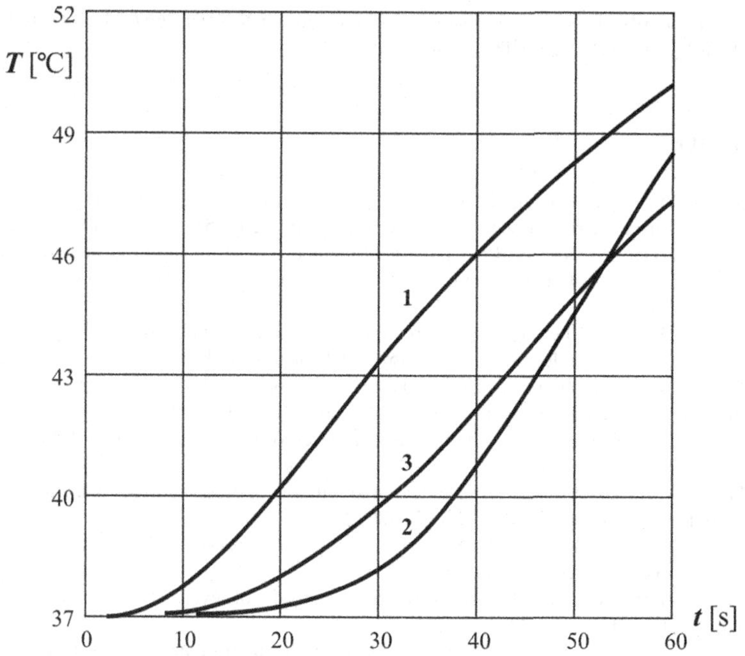

Fig. 18.6 Heating curves: 1– Pennes model, 2 – Cattaneo-Vernote model, 3 – DPL model

For comparison, Fig. 18.6 illustrates the heating curves at the point (0.006, 0.021) for Pennes model, Cattaneo-Vernotte model and DPL model, respectively. It is visible, that the choice of bioheat transfer equation has the big influence on the results of computations.

18.6 Conclusions

The general boundary element method is adapted to solve the DPL equation. In this method the solution obtained for time t^{f-1} is a starting point to get the solution for the time t^f. For each transition $t^{f-1} \rightarrow t^f$ the 1st-order deformation derivative $U^{[1]}(x, y)$ is determined from linear equation which can be easily solved by the traditional BEM and next the temperature T^f is calculated using the iterative procedure. The exactness of the method proposed is connected with the proper choice of time step Δt, discretization of the domain considered, the value of iterative parameter m and the number of iterations.

In this paper the 2D numerical solution for the rectangular domain of tissue is presented, at the same time the Dirichlet and adiabatic conditions on the boundaries are taken into account. It seems, that the next stage of investigations should concern the introduction of others boundary conditions which will be closer to the real course of thermal processes proceeding in the domain of tissue subjected to the external thermal interactions (e.g. laser irradiation, cryosurgical treatment etc.).

Acknowledgements This work was supported by Grant No N N501 3667 34 sponsored by the Polish Ministry of Science and Higher Education.

References

1. Jaunich, M., Raje, S., Kim, K., Mitra, K., Guo, Z.: Bio-heat transfer analysis during short pulse laser irradiation of tissue. Int. J. Heat Mass Transf. **51**, 5511–5521 (2008)
2. Zhang, L., Dai, W., Nassar, R.: A numerical method for optimizing laser power in the irradiation of a 3-D triple-layered cylindrical skin structure. Numer. Heat Transf A **48**, 21–41 (2005)
3. Zhou, J., Zhang, Y., Chen, J.K.: A dual reciprocity boundary element method for photothermal interactions in laser-induced thermotherapy. Int. J. Heat Mass Transf. **51**, 3869–3881 (2008)
4. Zhou, J., Chen, J.K., Zhang, Y.: Dual-phase lag effects on thermal damage to biological tissues caused by laser irradiation. Comput. Biol. Med. **39**, 286–293 (2009)
5. Comini, G., Del Giudice, L.: Thermal aspects of cryosurgery. J. Heat Transf. **98**, 543–549 (1976)
6. Majchrzak, E., Dziewonski, M.: Numerical simulation of freezing process using the BEM. Comput. Assist. Mech. Eng. Sci. **7**, 667–676 (2000)
7. Majchrzak, E., Dziewonski, M.: Numerical Analysis of Biological Tissue Freezing Process, pp. 189–200. Kluwer, Dordrecht (2001)
8. Mochnacki, B., Dziewonski, M.: Numerical analysis of cyclic freezing. Acta Bioeng. Biomech. **6**(1), 476–479 (2004)
9. Lv, Y.G., Deng, Z.S., Liu, J.: 3D numerical study on the induced heating effects of embedded micro/nanoparticles on human body subject to external medical electromagnetic field. IEEE Trans. Nanobiosci. **4**(4), 284–294 (2005).
10. Majchrzak, E., Dziatkiewicz, G., Paruch, M.: The modelling of heating a tissue subjected to external electromagnetic field. Acta of Bioeng. Biomech. **10**(2), 29–37 (2008)
11. Wang, H., Dai, W., Bejan, A.: Optimal temperature distribution in 3D triple-layered skin structure embedded with artery and vein vasculature and induced by electromagnetic radiation. Int. J. Heat Mass Transf. **50**, 1843–1854 (2007)
12. Pennes, H.H.: Analysis of tissue and arterial blood temperatures in the resting human forearm. J. Appl. Physiol. **1**, 93–122 (1948)
13. Majchrzak, E., Jasinski, M.: Numerical estimation of burn degree of skin tissue using the sensitivity analysis methods. Acta Bioeng. Biomech. **5**(1), 93–108 (2003)
14. Majchrzak, E., Jasinski, M.: Sensitivity analysis of burns integrals. Comput. Assist. Mech. Eng. Sci. **11**(2/3) 125–136 (2004)
15. Mochnacki, B., Majchrzak, E.L.: Sensitivity of the skin tissue on the activity of external heat sources. Comput. Model. Eng. Sci. 4(3/4), 431–438 (2003)
16. Torvi, D.A., Dale, J.D.: A finite element model of skin subjected to a flash fire. J. Biomech. Eng. **116**, 250–255 (1994)
17. Erhart, K., Divo, E., Kassab, A.: An evolutionary-based inverse approach for the identification of non-linear heat generation rates in living tissues using localized meshless method. Int. J. Numer. Methods Heat Fluid Flow **18**(3), 401–414 (2008)
18. Majchrzak, E., Jasinski, M., Janisz, D.: Identification of thermal parameters of biological tissue. Acta Bioeng. Biomech. **6**(1), 467–470 (2004)
19. Majchrzak, E., Mochnacki, B.: Sensitivity analysis and inverse problems in bio-heat transfer modelling. Comput. Assist. Mech. Eng Sci. **13**, 85–108 (2006)
20. Paruch, M., Majchrzak, E.: Identification of tumor region parameters using evolutionary algorithm and multiple reciprocity boundary element method. Eng. Appl. Artif. Intell. **20**: 647–655 (2007)
21. Cattaneo, C.: A form of heat conduction equation which eliminates the paradox of instantaneous propagation. Comp. Rend. **247**, 431–433 (1958)

22. Vernotte, P.: Les paradoxes de la theorie continue de l'equation de la chaleur. Comp. Rend. **246**, 3154–3155 (1958)
23. Kaminski, W.: Hyperbolic heat conduction equation for materials with a nonhomogeneous inner structure. J. Heat Transf. **112**, 555–560 (1990)
24. Özişik, M.N., Tzou, D.Y.: On the wave theory in heat conduction. J. Heat Transf. **116**, 526–535 (1994)
25. Tamma, K.K., Zhou, X.: Macroscale and microscale thermal transport and thermo-mechanical interactions: some noteworthy perspectives. J. Therm. Stresses **21**, 405–449 (1998)
26. Xu, F., Seffen, K.A., Lu, T.J.: Non-Fourier analysis of skin biothermomechanics. Int. J. Heat Mass Transf. **51**, 2237–2259 (2008)
27. Liao, S.: General boundary element method for non-linear heat transfer problems governed by hyperbolic hear conduction equation. Comput. Mech. **20**, 397–406 (1997)
28. Liao, S.: Boundary element method for general nonlinear differential operators. Eng. Anal. Bound. Elem. **20**(2), 91–99 (1997)
29. Liao, S.: A direct boundary element approach for unsteady non-linear heat transfer problems. Eng. Anal. Bound.ry Elem. **26**, 55–59 (2002)
30. Brebbia, C.A., Telles, J.C.F., Wrobel, L.C.: Boundary Element Techniques. Springer, Berlin, New York (1984)
31. Majchrzak, E.: Boundary Element Method in Heat Transfer. Publication of Czestochowa University of Technology, Czestochowa (in Polish) (2001)
32. Majchrzak, E.: Numerical modelling of bio-heat transfer using the boundary element method. J. Theor. Appl. Mech. **2**(36), 437–455 (1998)
33. Partridge, P.W., Brebbia, C.A., Wrobel, L.C.: The Dual Reciprocity Boundary Element Method. Computational Mechanics Publications, London, New York (1992)
34. Liu, J., Xu, L.X.: Boundary information based diagnostics on the thermal states of biological bodies. J. Heat Mass Transf. **43**, 2827–2839 (2000)
35. Lu, W.Q., Liu, J., Zeng, Y.: Simulation of thermal wave propagation in biological tissues by the dual reciprocity boundary element method. Eng. Anal. Bound. Elem. **22**, 167–174 (1998)
36. Liao, S., Chwang, A.: General boundary element method for unsteady nonlinear heat transfer problems. Numer. Heat Transf. B **35**, 225–242 (1999)

Index

J. Murín et al. (eds.), *Computational Modelling and Advanced Simulations*, Computational Methods in Applied Sciences 24, DOI 10.1007/978-94-007-0317-9,